Computing Supplementum 8

G. Farin, H. Hagen, H. Noltemeier (eds.)
in cooperation with W. Knödel

Geometric Modelling

Springer-Verlag Wien New York

Prof. Dr. G. Farin
Department of Computer Science
Arizona State University
Tempe, U.S.A.

Prof. Dr. H. Hagen
Fachbereich Informatik
Universität Kaiserslautern
Federal Republic of Germany

Prof. Dr. H. Noltemeier
Lehrstuhl für Informatik I
Universität Würzburg
Federal Republic of Germany

Prof. Dr. W. Knödel
Institut für Informatik
Universität Leipzig
Federal Republic of Germany

© 1993 Springer-Verlag/Wien

Typesetting: Asco Trade Typesetting Limited, Hong Kong

Printed on acid-free paper

With 175 Figures and 6 Plates

ISSN 0344-8029
ISBN-13:978-3-211-82399-6   e-ISBN-13:978-3-7091-6916-2
DOI:10.1007/978-3-7091-6916-2

Preface

This book is based on lectures presented at the first Dagstuhl-seminar on Geometric Modelling organized by Gerald Farin (Arizona State University), Hans Hagen (Universität Kaiserslautern) and Hartmut Noltemeier (Universität Würzburg). International experts from academia and industry were selected to speak on the most interesting topics in geometric modelling. The resulting papers, published in this volume, give a state-of-the art survey of the relevant problems and issues. The following topics are discussed:

- solid modelling
- geometry processing
- feature modelling
- product modelling
- surfaces over arbitrary topologies
- blending methods
- scattered data algorithms
- smoothing and fairing algorithms
- NURBS

The discussion between industry and university has proven to be very fruitful. The scientists from industry were able to give many important and practicable impulses for new research; in the opposite direction the university researchers have developed many new technologies, solving industrial problems, which may be transfered back to industry. Everybody was impressed by the quality of the presentations; they acknowledged the importance of such research exchange between the various partners.

We would like to thank all participating speakers, and the audience, for what appears to have been a very successful workshop. Thanks also to Stefanie Hahmann for her help in editing this book.

Tempe/Kaiserslautern/Würzburg,
July 1992

Gerald Farin
Hans Hagen
Hartmut Noltemeier

Contents

Listed in Current Contents

Computing Suppl. 8, 1–20 (1993)

Constant-Radius Blending of Parametric Surfaces

Robert E. Barnhill, Gerald E. Farin, and **Qian Chen,** Tempe

Abstract. A method for blending two parametric surfaces is presented. It is based an an algorithm which calculates the intersection of two offset surfaces using only the first-order derivatives of the progenitors. The method converges quadratically in non-singular cases.

Key words: Offset surfaces, NURBS, fillets, intersections.

1. Introduction

In CAD/CAM industry, blending surfaces refer to a kind of secondary surfaces whose purpose is to smoothly connect the functional, primary surfaces and whose shapes are, on the other hand, rarely defined. There are two types of blends: those forming smooth transitions between different, but intersecting surfaces, and those which smoothly join two or more disconnected surfaces. Both types are considered in this paper.

An early paper addressing this topic was [Rossignac & Requicha '84]. It discussed various aspects of the problem in view of geometric modeling. Many of its ideas were adopted later by other researchers. For example, it mentioned Ricci's work [Ricci '73] which could be thought as the beginning of expressing the blend in an implicit form using the functions (implicit forms too) of the surfaces to be blended. It also proposed the so called "rolling-ball" concept and pointed out the potentiality of generating fillets by sweeping and varying cross sections. In addition, the authors foresaw that the surface/surface intersection(SSI) might be a major problem to be solved.

In [Hoffman & Hopcroft '86, '87] and [Rockwood & Owen '87], blends for implicitly defined surfaces were obtained by substituting the surface functions for the variables of a blending function to lines on a two dimensional plane. Hoffman et al. used a conic function, while Rockwood et al. used a super-elliptic function as the basic blending function. Hoffman et al. only talked about blending quadrics and the proposed method met problems if the surface was composed of different patches. On the other hand, although Rockwood et al.'s construction of the blend was less intuitive, its characteristics were more understood [Rockwood '89]. Particularly, the modified blending function had the property of "blending on blends" so that corners could be blended recursively. Other popular literatures about blending implicit surfaces include [Middleditch & Sears '85] and [Zhang & Bowyer '86].

To blend parametric surfaces, the possibility of implicitizing them first and then applying the above techniques is ruled out by Sederberg et al.'s paper [Sederberg et al. '84] because the algebraic degrees of the resulting patches would be too high to form a stable algorithm. Therefore, other methods have to be sought.

In [Filip '89], the blending surface between two rail curves was calculated using Hermite functions. Shape control was also available. The rail curves were obtained by pulling back points on the intersection of offset surfaces. But apparently, the paper did not include any information as to how the offset intersection curve was found at the first place. In [Varady et al. '89], the trimlines (equivalent to rail curves mentioned above) were defined as intersections of the envelop of a family of ellipsis whose centers move along the intersection of the surfaces with the surfaces themselves. The drawbacks are, first of all, it needs to calculate three intersection curves; secondly, it pre-assumes the surfaces or their extensions intersect; and finally, it cannot produce the rolling-ball blend. Choi and Ju [Choi & Ju '89] devised a method to calculate the offset surface intersection curve(OSIC) by applying the SSI algorithm [Barnhill et al. '87] directly to offset surfaces which, as a pre-requisite, needed to calculate the second order derivatives as well as the twists of the original surfaces. A recent paper [Sanglikar et al. '90] used the rolling ball concept again to blend parametric quadrics. The problem of finding OSIC however was still unsolved in a generic sense. In [Frost '91], a piecewise quartic triangular Bezier approximation is found first for the offset surfaces. The SSI algorithm is applied then to find the OSIC. Since only the intersection is needed for the sake of constructing a blending surface, this appears to be too waste an operation, esp. when the offset surface contains degenerated points because most of the time will be spent on the subdivision near the C^0 areas which may far from the intersection areas and therefore is unnecessary. A relatively new and comprehensive summary of blending techniques is [Woodwark '87]. But it contains no more information than Varady's in terms of blending parametric surfaces.

Bloor and Wilson [Bloor & Wilson '89] considered this problem from a different angle.The blend was deemed as the solution to the fourth-order elliptic differential equation with boundary conditions of point and derivative continuity between the blend and the surfaces. The boundaries were just the contract lines and were given in the paper.

The previous overview leads to an assertion that the settlement of the blending problem largely depends on an algorithm which is capable of finding the intersection of two offset surfaces. Because once this is done, the rolling ball blend can be very easily constructed. In addition, it generates reasonable contact lines so that other forms of blends can also be constructed. In this paper, we are going to develop and implement such an algorithm which only requires the first order derivatives of the original surfaces. It is developed from the enhanced SSI algorithm [Barnhill et al. '90] and in return considers the latter as a special case where the offset distance is zero. It is mathematically equivalent to the former. Consequently, as proved in [Mullenheim '90], the convergence speed is quadratic in mon-singular situations.

Some symbol conventions are adopted through out the paper: capitalized and lowercase bold face characters denote points, lowercase bold face characters with arrow bars on top denote vectors, lowercase plain face characters denote real numbers.

2. The Point/Surface Problem

Definition 1. Let $S = F(u,v)$ be a parametric surface, (u',v') be a 2D point in S's domain. Then (u',v') is called the preimage of $F(u',v')$.

The following describes the point/surface problem:

Given: A 3D point P and a parametric surface $S = X(u,v)$.

Find: $(\bar{u}, \bar{v})$ so that $X(\bar{u}, \bar{v})$ has the shortest distance towards P. $X(\bar{u}, \bar{v})$ is called P's projection on the surface S.

Many references [Faux '79, Mortenson '87, Barnhill et al. '87, '90 and Frost '91] talked about solving this type of problem using Newton-Raphson's method: denote $R(u,v) = P - X(u,v)$ and approximate its derivative by

$$R(u_{i+1}, v_{i+1}) - R(u_i, v_i) = -X'_u(u_i, v_i) \cdot \Delta u - X'_v(u_i, v_i) \cdot \Delta v. \tag{1}$$

Forcing $R(u_{i+1}, v_{i+1}) = 0$ leads to

$$X'_u(u_i, v_i) \cdot \Delta u + X'_v(u_i, v_i) \cdot \Delta v = P - X(u_i, v_i). \tag{2}$$

In the above i denotes the iteration level, Δu and Δv denote the correction factors. (2) is over-determined and we take its least squares approximation:

$$\begin{bmatrix} X'_u(u_i, v_i) \cdot X'_u(u_i, v_i) & X'_u(u_i, v_i) \cdot X'_v(u_i, v_i) \\ X'_v(u_i, v_i) \cdot X'_u(u_i, v_i) & X'_v(u_i, v_i) \cdot X'_v(u_i, v_i) \end{bmatrix} \begin{bmatrix} \Delta u \\ \Delta v \end{bmatrix} = \begin{bmatrix} (P - X(u_i, v_i)) \cdot X'_u(u_i, v_i) \\ (P - X(u_i, v_i)) \cdot X'_v(u_i, v_i) \end{bmatrix}. \tag{3}$$

(3) is singular if and only if the determinant of the coefficient matrix is zero, that is

$$\|X'_u\|^2 \|X'_v\|^2 - (\|X'_u\| \|X'_v\| \cos \alpha)^2 = 0 \tag{4}$$

where α is the angle between X'_u and X'_v. (4) corresponds to $\|X'_u\| = 0$, $\|X'_v\| = 0$ or $\alpha = n\pi$ which is identical to the degenerated condition of the surface S. If that is avoided, then a solution for (3) always exists. Notice that (3) also means $(P - X(u_i, v_i)) - (X'_u(u_i, v_i)\Delta u + X'_v(u_i, v_i)\Delta v)$ is perpendicular to X'_u and X'_v, resulting the minimized error of (2) in a least squares sense. Now assuming $\lim_{i \to \infty}(u_i, v_i) = (\bar{u}, \bar{v})$ leads to:

$$(P - X(\bar{u}, \bar{v})) \cdot X'_u(\bar{u}, \bar{v}) = (P - X(\bar{u}, \bar{v})) \cdot X'_v(\bar{u}, \bar{v}) = 0. \tag{5}$$

The geometric explanation of this solution is: project $P - X(u_i, v_i)$ onto the tangent

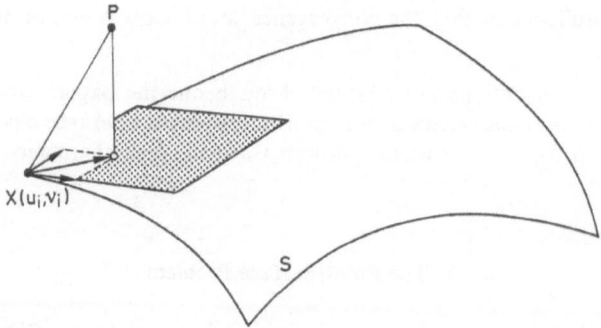

Figure 1. Obtaining the correction factors for the point/surface problem by projecting $P - X(u_i, v_i)$ onto the tangent plane

plane passing through $X(u_i, v_i)$ and write the projection in terms of $X'_u(u_i, v_i)$ and $X'_v(u_i, v_i)$, then the signed coefficients are the corresponding correction factors (see Fig. 1).

3. Point Refinement for SSI

To relax a guess point to the true intersection position is often called "point refinement":

Given: Two parametric surfaces $S1 = F(u, v)$ and $S2 = G(s, t)$, an initial point I_0 and its guess preimages (u_0, v_0), (s_0, t_0).

Find: The true intersection I together with its preimages $(\bar{u}, \bar{v})$ and $(\bar{s}, \bar{t})$ in a vicinity of I_0. In the following, we are going to briefly describe a Newton-like method for solving this problem. The readers are referred to [Barnhill & Kersey '90] for details. Currently, we concentrate on the deduction formula, that is to calculate I_{i+1} form I_i where the subscript denotes the iteration level.

Using the technique introduced in the previous section, let $R(u, v, s, t) = F(u, v) - G(s, t)$, and write the derivative of it as

$$R_{i+1} - R_i = F'_u(u_i, v_i)\Delta u + F'_v(u_i, v_i)\Delta v - G'_s(s_i, t_i)\tau - H_t(s_i, t_i)\Delta t. \quad (6)$$

Having $R_{i+1} = 0$ yields

$$F(u_i, v_i) + F'_u(u_i, v_i)\Delta u + F'_v(u_i, v_i)\Delta v = G(s_i, t_i) + G'_s(s_i, t_i)\Delta s + G'_t(s_i, t_i)\Delta t. \quad (7)$$

The two sides of this equation represent the tangent plane of each surface passing through $F(u_i, v_i)$ and $G(s_i, t_i)$ respectively. Any point on the intersection line of the tangent planes corresponds to a group of correction factors. Among them we are pursuing the solution near the previous one. Let us assume that it is on a third plane with normal $\vec{n}1 \times \vec{n}2$ and passing through the point:

$$\alpha F(u_i, v_i) + \beta G(s_i, t_i) + \gamma I_i.$$

Here $\vec{n}1$ and $\vec{n}2$ are the surfaces' normals:

$$\vec{n}1 = F'_u(u_i, v_i) \times F'_v(u_i, v_i)$$

$$\vec{n}2 = G'_s(s_i, t_i) \times G'_t(s_i, t_i)$$

α, β, γ can be any numbers satisfying:

$$0 \leqslant \alpha, \beta, \gamma \leqslant 1 \quad \text{and} \quad \alpha + \beta + \gamma = 1.$$

Thus I_{i+1} is the solution to

$$(I_{i+1} - F(u_i, v_i)) \cdot \vec{n}1 = 0 \tag{8a}$$

$$(I_{i+1} - G(s_i, t_i)) \cdot \vec{n}2 = 0 \tag{8b}$$

$$(I_{i+1} - (\alpha F(u_i, v_i) + \beta G(s_i, t_i) + \gamma I_i)) \cdot \vec{n}1 \times \vec{n}2 = 0. \tag{8c}$$

Because of this, the method is called Three-Plane-Method(TPM). After I_{i+1} is found, the correction factors can be solved from:

$$F'_u(u_i, v_i) \Delta u + F'_v(u_i, v_i) \Delta v = I_{i+1} - F(u_i, v_i)$$

$$G'_s(s_i, t_i) \Delta s + G'_t(s_i, t_i) \Delta t = I_{i+1} - G(s_i, t_i). \tag{9}$$

For stability considerations, we project I_i onto both surfaces which refines the guess preimages and insert into (8) with the refined values. This amounts to solving two point/surface problems through:

$$F'_u(u_i, v_i) \Delta u + F'_v(u_i, v_i) \Delta v = I_i - F(u_i, v_i)$$

$$G'_s(s_i, t_i) \Delta s + G'_t(s_i, t_i) \Delta t = I_i - G(s_i, t_i). \tag{10}$$

If this is done, correction factors are already found so there is no need to solve (9). Now the plane defined by (8c) is identical to that passing through I_i, $F(u_{i+1}, v_{i+1})$ and $G(s_{i+1}, t_{i+1})$. See Fig. 2 for an illustration. In Barnhill et al.'s paper, it was claimed that projecting once would be enough for the sake of relaxing the guess point. However when using TPM for offset surfaces' intersection (to be discussed in next section), the quality of this process has large influence on the stability of the marching process. The converge speed of (8) is proved to be quadratic in [Mullenheim '90] except under the singular condition when $\vec{n}1 \times \vec{n}2 \cdot \vec{n}1 \times \vec{n}2 = 0$ which means either the degeneracy of one of the surfaces or the coincidence of the normals at the touching points.

4. Offset Surface Intersection

4.1 Some Mathematical Descriptions

Definition 2. Let a parametric surface, called progenitor, be denoted as $S = F(u, v)$. An offset surface O with offset distance d is defined by:

$$O(u, v) = F(u, v) + d\vec{n}(u, v). \tag{11}$$

where $\vec{n}(u, v)$ is the unit surface normal of S:

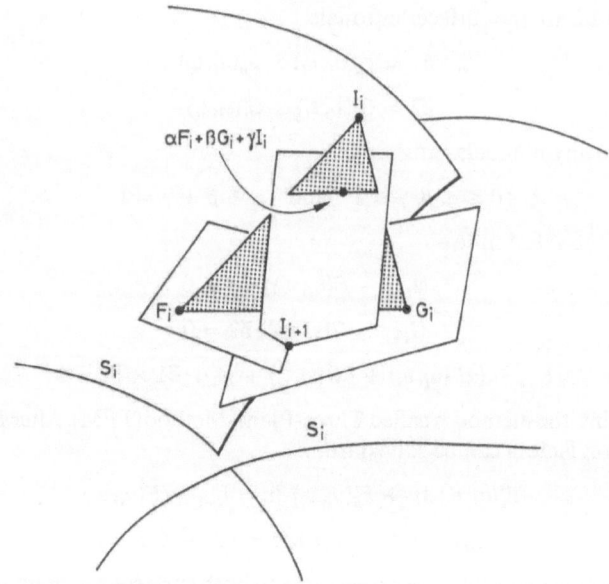

Figure 2. The Three-Plane-Method—I_{i+1} is the intersection of three planes: two of them are the tangent planes; the third one reflects the "local" property

$$\vec{n}(u, v) = \frac{F'_u(u, v) \times F'_v(u, v)}{\|F'_u(u, v) \times F'_v(u, v)\|}.$$

Given a pair of parametric value $(\bar{u}, \bar{v})$ in the domain, $F(\bar{u}, \bar{v})$ and $O(\bar{u}, \bar{v})$ are called the corresponding points, $(\bar{u}, \bar{v})$ is called the preimage of $O(\bar{u}, \bar{v})$.

In the following of this paper, sometimes $\vec{n}(u, v)$ is simply written as $\vec{n}$. Other conventions are: F'_u, F'_v, O'_u and O'_v which stand for the partial derivatives of the surfaces with respect to the parameters.

Theorem 1. *If both the offset surface and its progenitor are non-degenerated at the corresponding points, their normals have the same direction there:*

$$O'_u(u, v) \times O'_v(u, v) = \lambda(F'_u(u, v) \times F'_v(u, v)), \quad \text{for} \quad \lambda \in \mathbf{R} \quad \text{and} \quad \lambda \neq 0. \quad (12)$$

Proof. Take the cross product of the partial derivatives of (11),

$$O'_u(u, v) \times O'_v(u, v) = F'_u \times F'_v + d(F'_u \times \vec{n}'_v) + d(\vec{n}'_u \times F'_v) + d^2(\vec{n}'_u \times \vec{n}'_v).$$

Using $\vec{n}'_u \cdot \vec{n} = \vec{n}'_v \cdot \vec{n} = 0$, we obtain at once:

$$\vec{n}'_u \times \vec{n}'_v = \sigma_1 \vec{n}.$$

Since $(\vec{n}'_u \times F'_v) \cdot F'_u = \vec{n}'_u \cdot (-F'_u \times F'_v) = -\vec{n}'_u \cdot \vec{n} \|F'_u \times F'_v\| = 0$, and $(\vec{n}'_u \times F'_v) \cdot F'_v = \vec{n}'_u \cdot (F'_v \times F'_v) = 0$, it follows that: $(\vec{n}'_u \times F'_v) = \sigma_2(F'_u \times F'_v)$. By the same token: $(F'_u \times \vec{n}'_n) = \sigma_3(F'_u \times F'_v)$. Therefore

$$O'_u(u, v) \times O'_v(u, v) = \left(1 + d \cdot \sigma_2 + d \cdot \sigma_3 + \frac{d^2 \cdot \sigma_1}{\|F'_u \times F'_v\|}\right) F'_u \times F'_v$$

$$= \lambda(F'_u \times F'_v),$$

if we let $\lambda = \left(1 + d \cdot \sigma_2 + d \cdot \sigma_3 + \dfrac{d^2 \cdot \sigma_1}{\|F'_u \times F'_v\|}\right).$

As for the sign of λ, notice that it is independent of d provided the latter is small. In addition, $\lim_{d \to 0} \lambda = 1 > 0$, therefore $\lambda > 0$. If d exceeds one of the principal radii of curvature, then $\lambda < 0$, which means the normals have opposite sense [Willmore '59].

Before going any farther, some notations need to be specified: let $S1 = F(u, v)$ and $S2 = G(s, t)$ be two progenitors, $O1 = O_1(u, v)$ and $O2 = O_2(s, t)$ be offset from them by d_1 and d_2 respectively. $I(\tau)$ is the intersection of $O1$ and $O2$.

Definition 3. Given τ, there exists $(u(\tau), v(\tau))$ in $S1$'s domain so that

$$F(u(\tau), v(\tau)) + d_1 \bar{n}(u(\tau), v(\tau)) = I(\tau).$$

$F(u(\tau), v(\tau))$ is called $I(\tau)$'s projection on $S1$ and $(u(\tau), v(\tau))$ is its preimage.

Definition 4. The point set

$$\{F(u(\tau), v(\tau)) | F(u(\tau), v(\tau)) \text{ is } I(\tau)\text{'s projection on } S1\}$$

is called the projection of the whole curve $I(\tau)$ on $S1$.

Needless to say, $I(\tau)$ has another projection on $S2$ too. In the following, we will use $P_1(\tau)$ and $P_2(\tau)$ to denote $I(\tau)$'s projections on $S1$ and $S2$ respectively.

Theorem 2. *Given a parametric surface $S = F(u, v)$, offset distance d and offset surface $O = O(u, v)$ satisfying (11); I is a 3D point; P_F is its projection on S; P_O is its projection on O. Then $P_O = P_F + d\bar{n}$ where $\bar{n}$ is S's unit normal at P_F.*

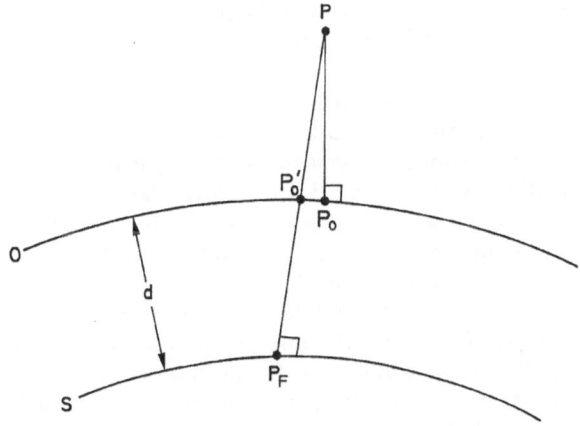

Figure 3. Project a point onto an offset surface: P_O and P'_O are actually coincident

Proof. Refer to Fig. 3. Let $P_O' = P_F + d\vec{n}$. Since P_O' is the corresponding point of P_F, by Theorem 1, $I - P_F$ is normal to O at P_O'. However, since I has unique projection locally, $P_O = P_O'$, which proves the conclusion.

Remarks.

i). In order to project a point onto an offset surface, we first project it on the progenitor, then offset that result by the given distance.

ii). The progenitor itself can be thought as a special offset surface whose offset distance is zero. Based on this view, the SSI problem is a special case of OSI instead of the reverse as usually thought.

Theorem 3. *The derivative of the OSIC at a point is in the direction of the cross-product of its corresponding points' normals:*

$$\dot{I}(\tau) = \lambda(\vec{n}1(u(\tau), v(\tau)) \times \vec{n}2(s(\tau), t(\tau))), \quad \text{for some } \lambda \in \mathbf{R}. \tag{13}$$

Proof. Let $I(\tau) = P_1(\tau) + d_1\vec{n}1(u(\tau), v(\tau)) = P_2(\tau) + d_2\vec{n}2(s(\tau), t(\tau))$, then

$$\dot{I}(\tau) = F_u'\dot{u} + F_v'\dot{v} + d_1(\vec{n}1_u'\dot{u} + \vec{n}1_v'\dot{v})$$

$$= G_s'\dot{s} + G_t'\dot{t} + d_2(\vec{n}2_s'\dot{s} + \vec{n}2_t'\dot{t})$$

where the top dot represents the derivative with respect to τ. Multiply $\vec{n}1$ with it,

$$\dot{I}(\tau) \cdot \vec{n}1 = (F_u' \cdot \vec{n}1)\dot{u} + (F_v' \cdot \vec{n}1)\dot{v} + d_1(\vec{n}1_u' \cdot \vec{n}1)\dot{u} + d_1(\vec{n}1_v' \cdot \vec{n}1)\dot{v} \overset{=}{=} 0.$$

Similar inference leads to $\dot{I}(\tau) \cdot \vec{n}2 = 0$. Consequently, $\dot{I} = \lambda(\vec{n}1 \times \vec{n}2)$.

4.2 Determination of the Step Length

It is already know that the OSIC rarely has a closed algebraic form, therefore it can only be approximated. In our case, a piecewise linear approximation is sought whose vertices are points on the curve. The algorithm devised to calculate the approximation composes with two steps: firstly, find a starting point; secondly, march along the curve and trace out the vertex points. The second step can be further divided into two sub-steps: first of all, find a guess position; then, refine it to the true intersection. The next section will discuss how to do point refinement for offset surfaces. Currently, let us face such a problem: suppose I_i has been located, how to compute the next guess point. Here I_i denotes the ith vertex of the linear approximation. Since Theorem 3 already gives the tangent direction at I_i, the problem becomes simple: how to decide the step length. Given an error tolerance, the step length can be estimated through some formula involving the local curvature which has to be estimated beforehand [Faux et al. '79 and Barnhill et al. '90]. The latter paper is of interest at this point because it uses the angle between the unit tangent vectors at two successive vertex points as the error measure, instead of some Euclidean norm. This angle will be called "transition angle" from now on in this paper. In practice, however, the tolerance cannot be exactly satisfied due to: 1, the curvature is estimated; 2, the curve is approximated locally by an arc on the osculating circle which results in error itself; 3, after the guess point is pulled back to the intersection, the step length is changed. Thus the step length can be determined in a simpler way: Let ε be the given tolerance, T_i the unit tangent vector associated with I_i, and $l_i = \|I_i - I_{i-1}\|$, the ith chord length. Then $l_{i+1} = l_i * \varepsilon / arccos(T_i \cdot T_{i-1})$ is the wanted one. The first step length I_0 has to be decided in a different manner such as curvature-estimation. If the OSIC is a line segment, $T_i = T_{i-1}$ which makes l_{i+1} infinite. This suggests a upper-bound be set for the adjustment factor. Some valuable references are [Crampin et al. '85] and [Hoschek '88].

4.3 Point Refinement on an Offset Surface

Given: $S1, S2, O1, O2$ defined as before. A guess offset intersection point I_0 and its rough preimages $(u_0, v_0), (s_0, t_0)$.

Find: The true intersection point I near the vicinity of I_0 and its preimages.

Theorem 2 essentially establishes the base for point projection on an offset surface. As a result, TPM is applicable here. The equations are:

$$(I_{i+1} - O_1(u_i, v_i)) \cdot \vec{n}1(u_i, v_i) = 0 \tag{14a}$$

$$(I_{i+1} - O_2(s_i, t_i)) \cdot \vec{n}2(s_i, t_i) = 0 \tag{14b}$$

$$(I_{i+1} - (\alpha O_1(u_i, v_i) + \beta O_2(s_i, t_i) + \gamma I_i)) \cdot \vec{n}1 \times \vec{n}2 = 0. \tag{14c}$$

The singular condition of (14) is identical to that of (8). Providing the progenitors are nondegenerated, it simply implies the coincidence of their normals at the corresponding points which, as will soon be shown in section 6, does not give too much numerical problems. Similar to the implication of (10), $O_1(u_i, v_i)$ and $O_2(s_i, t_i)$ can be refined to I_{i+1}'s projections on the offset surfaces before being plugged into (14). This is more numerically stable but paid for by more computing time. To see the reason, we understand that the deviation from the true closest point in the progenitor is scaled by a factor proportional to the offset distance if offsetting in the convex side. The longer the distance, the more times of the refinement is needed.

4.4 A Marching Algorithm for the OSIC

We are now ready for a marching algorithm to find a linear approximation of the intersection between two offset surfaces: Theorem 3 tells us how to calculate the marching direction and 4.2, the step length; Theorem 2 tells us how to project a point onto an offset surface; and finally, (14) tells us how to do point refinement.

Given: $S1, S2, O1, O2$ defined as before; a three-dimensional point I on $I(\tau)$ together with its preimages (u, v) and (s, t); a tolerance ε for the transition angle.
Find: A linear approximation to $I(\tau)$.

The marching algorithm (a draft).

0. *Set* $i = 0, I[0] = I$, *Reach_Boundary* = *Closed* = *FALSE;*
1. *Calculate the initial step length* $\Delta l;$
2. *While (not Reach_Boundary and not Closed)*
       begin {while}
           calculate the unit normals: n1 *and* n2;
           calculate the tangent vector: $T = n1 \times n2;$
           normalize $T;$
           $i = i + 1;$
           $I[i] = I[i - 1] + T^*\Delta l;$
           Refine $I[i]$ *to the true intersection position;*
           Calculate the current transition angle $\Delta;$
           $\Delta l = \Delta l^* min(\varepsilon/\Delta, Up\_Bound);$
           if $(I = I[i])$
             then Closed = *TRUE;*
             else Closed = *FALSE;*
           if (meet a boundary of O1 or O2)
             then Reach_Boundary = *TRUE;*
             else Reach_Boundary = *FALSE;*
           end; {while}

We make following comments about this draft:

i). During the point refinement, $I[i]$'s preimages are discovered too;

ii). To calculate Δ, the current transition angle, the previous tangent vector needs to be saved;

iii). If the algorithm terminates without returning *Closed* = *TRUE* which means part of an open curve has been approximated, it must be called again but in an opposite direction to find the part left.

The appropriate corrections are added into the modified version below:

The modified marching algorithm.

0. *Set* $i = 0$, *Reach_Boundary* = *Closed* = *FALSE*,
   $I[0] = I$, $Pre1[0] = (u, v)$, $Pre2[0] = (s, t)$,
   $n1 = \vec{n}1(u, v)$, $n2 = \vec{n}2(s, t)$, $T[0] = \dfrac{n1 \times n2}{\|n1 \times n2\|};$
1. *Calculate the initial step length* Δl, *set Direction* = $1;$
2. *While (not Reach_Boundary and not Closed)*
       begin {while}
           $i = i + 1;$
           $I[i] = I[i - 1] + Direction^*T[i - 1]^*\Delta l;$
           Refine $I[i]$ *and obtain its preimages* (u, v), $(s, t);$
           $Pre1[i] = (u, v)$, $Pre2[i] = (s, t);$
           $n1 = \vec{n}1(u, v)$, $n2 = \vec{n}2(s, t)$, $T[i] = \dfrac{n1 \times n2}{\|n1 \times n2\|};$
           $\Delta l = \Delta l^*\varepsilon/arccos(T[i] \cdot T[i - 1], Up\_Bound);$
           if $(I = I[i])$

```
    then Closed = TRUE;
    else Closed = FALSE;
  if (meet a boundary of O1 or O2)
    then Reach_Boundary = TRUE;
    else Reach_Boundary = FALSE;
end; {while}
```

The algorithm will be called again with *Direction* set to be −1 if *Closed* is return as *FALSE*. In this case, the two pieces ought to be combined when both have been found.

5. Formulas of the Fillet

Informally, a blending surface is defined as part of a curved cylinder whose central line is the intersection of the offset surfaces (whose progenitors are being blended) and bounded by the projections of that central line upon the two progenitors.

Definition 5. (Formally) Let *O1* and *O2* be offset from *S1* and *S2* respectively by a common distance *d*. $O(\tau)$ is the intersection of *O1* and *O2*, $P1(\tau)$, $P2(\tau)$ are $O(\tau)$'s projections on the progenitors. The blending surface for *S1* and *S2* is the small part of a curved cylinder with central line $O(\tau)$, radius *d* and bounded by $P1(\tau)$ and $P2(\tau)$.

With this definition and referring to Fig. 4, a parametric equation for the fillet is:

$$F(\tau, \eta) = O(\tau) + \vec{d}(\tau, \theta(\tau, \eta)) \tag{15}$$

where $\tau \in [0$, the length of $O(\tau)]$,

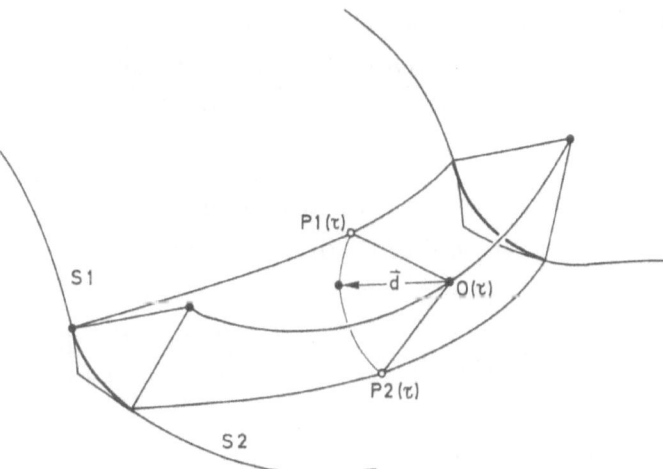

Figure 4. The blending surface is the small part of a curved cylinder whose central line corresponds to the intersection of the offset surfaces and whose offset distance equals the radius

$$\eta \in [0,1],$$

$\theta(\tau,\eta) = \eta \cdot \varphi(\tau)$, and

$$\varphi(\tau) = \arccos\left(\frac{(P1(\tau) + O(\tau)) \cdot (P2(\tau) - O(\tau))}{d^2}\right).$$

$\vec{d}(\tau,\theta)$ is the vector between $P1(\tau) - O(\tau)$ and $P2(\tau) - O(\tau)$ whose length is d and forms angle θ with $P1(\tau) - O(\tau)$. For convenience, $\vec{d}$, θ, φ wil be used in the following deduction which refer to the functions specified in (15).

Since $\vec{d}$, $(P1 - O)$ and $(P2 - O)$ are co-planar, there exist α and β so that

$$\vec{d} = \alpha(P1 - O) + \beta(P2 - O). \tag{16}$$

Therefore

$$\vec{d} \cdot (P1 - O) = d^2 \cos\theta = \alpha d^2 + \beta d^2 \cos\varphi$$
$$\vec{d} \cdot (P2 - O) = d^2 \cos(\varphi - \theta) = \alpha d^2 \cos\varphi + \beta d^2 \tag{17}$$

α and β are then solved from it and thus we obtain the formula for $\vec{d}$:

$$\vec{d} = \frac{(P1 - O)\sin(\varphi - \theta) + (P2 - O)\sin\theta}{\sin\varphi}. \tag{18}$$

In case φ (and thus θ) is very small, $\sin(\varphi - \theta) \approx \varphi - \theta$, $\sin\theta \approx \theta$ and $\sin\varphi \approx \varphi$, consequently

$$\vec{d} \approx (P1 - O)(1 - \eta) + (P2 - O)\eta. \tag{19}$$

Formula 2.

Since an arc can be expressed as a rational quadratic Bezier curve [Farin '89 and '90], we are able to write the fillet directly in a rational form:

$$F(u) = \frac{P1 B_{0,2}(u) + wX B_{1,2}(u) + P2 B_{2,2}(u)}{B_{0,2}(u) + w B_{1,2}(u) + B_{2,2}(u)} \tag{20}$$

where $B_{i,2}(u)$, $i = 0, 1, 2$, stand for the Bernstein base functions of order 2. X is the intersection of the tangent lines touching $P1$ and $P2$ respectively. Since $X - O$ is coplanar with $P1 - O$ and $P2 - O$ and is symmetrical,

$$X - 0 = \alpha(P1 - O) + \alpha(P2 - O). \tag{21}$$

Multiply it by either $P1 - O$ or $P2 - O$ to solve α:

$$X = O + \frac{1}{1 + \cos\varphi}(P1 - O) + \frac{1}{1 + \cos\varphi}(P2 - O) = \frac{P1 + P2 + (\cos\varphi - 1)O}{1 + \cos\varphi}. \tag{22}$$

The weight w in (20) is determined by:

$$w = \frac{1}{2}\sqrt{\frac{area(\mathbf{p}, P1, P2) \cdot area(\mathbf{p}, P1, P2)}{area(\mathbf{p}, P1, X) \cdot area(\mathbf{p}, P2, X)}} \tag{23}$$

where **p** can be any point on the arc. Without losing any generality, let us take **p** as the arc's middle point and thus obtain

$$w = \cos\frac{\varphi}{2}. \tag{24}$$

Neither of the above formulas applies if $\varphi = \pi$, which happens at one of the singular cases of (14). It can be handled by either giving a third vector to direct the fillet (the picture of Fig. 13 is generated in this way), or choosing a larger radius which avoids the problem at all.

6. Implementations and Results

6.1 Data Structure

Since we only need a linear approximation to the intersection curve, we want a data structure to store a sequence of points on it. A link list will be a good choice. Algorithm 2 tells us that each node of the list should have the fields to store the vertex point's position, tangent vector and preimages. This gives rise to:

```
typedef double Point_3d[3], Point_2d[2];
struct node {
    Point_3d position, tangent;
    Point_2d preimage1[2], preimage2[2];
    struct node *next;
}   Node;
Node *intersection;
```

6.2 Finding the Start Points

Using the bounding box concept [Houghton et al. '84, Lasser '86 and Barnhill et al. '90], we recursively subdivide the surfaces' domains which in return corresponds to the subdivision of the surfaces themselves. The area of a surface is culled out and ignored in next recursion if its bounding box is not overlapped by any from its counterpart. The bounding box is simply defined as a rectangular parallelepiped with minimum volume containing the current area. In practice, the bounding box is roughly taken as the minimum volume containing the points chosen from discrete positions on that area. In our case, five points are used which are composed of four corners plus the center. Experiments have shown that this choice already generates enough number of starting points for the marching stage. The subdivision process stops when all boxes have been exhausted or a prescribed subdivision level is reached. A guess point is the average of the centers from an overlapped pair. Its preimages are also selected as that of the two centers. These results are then fed into the point refinement routine to obtain the values of the true intersection. All so found points are stored in another link list with the same node structure called *start*.

6.3 Marching

This is an iteration process. Each time a point is removed from *start*. Starting from there, the curve is traced out using Algorithm 2. When a new intersection point is discovered, memory is allocated for it. The new node is then appended to the list named *intersection*. During the process, if any point is found also belonging to the *start* list, it is removed from there. When it stops, either a closed curve has been generated if the current piece is connected to itself, or one boundary is reached. If the latter is the case, the marching routine is called again with the same initial point but in an opposite direction. These two pieces are concatenated after another boundary is met. This completes the discovery of one curve. The whole process is carried out a second time for another curve until *start* becomes empty. All intersection curves can be found in this way provided each of them contributes a point in the *start* list.

6.4 Results

In the experiments, all floating point variables are defined with double precision. The *Same Point Tolerance(SPT)*[1] is set to 10^{-5}, the *Transition Angle Tolerance.(TAT)* is $3°$ and the *Up_Bound* is 1.5. For the singular case, it follows from the end of 4.4 that the singular condition is equivalent to the coincidence of the surfaces' normals. This means either the two offset surfaces are parallel to but apart from, or touch with each other (at the singular pair of points). During the tracing process, the first situation never happens along the marching path. As a result, the degeneration of the denominator implies approaching a solution too; the Newton's iteration process is stopped before it gets into oscillation. For example, the picture of Fig. 13 is obtained by setting the threshold to 10^{-8}. This criterion was used in [Barnhill et al. '90] also. For materials dealing with the singular cases, the readers are refereed to [Powell '72, Deuflhard '74 and Gill & Murray '76]. However they must remember that those methods use the surfaces' second-order derivatives whose computation may not be feasible for NURBS surfaces unfortunately.

We have tested different types of data, all expressed in the form of Non-Uniform Rational B-Splines(NURBS):

Data1—Two intersecting cyclinder-shape surfaces;
Data2—Two disjoint cylinder-shape surfaces;
Data3—A self-intersecting surface;
Data4—A surface with high curvature, the offset surface has a C^0 edge and is self-intersected;
Data5—A torus-shape surface intersecting with an open surface;
Data6—Two intersecting saddle surfaces;
Data7—A ball-shape and a revolution-shape surfaces;
Data8—A torus-shape surface being parallel with a plane.

---

[1] The distance between two points within which the two points are considered as one.

Time record for OSI algorithms

OSI	data1	data2	data3	data4	data5	data6	data7	data8
TPM1	dead	20	4	2	22, dead	6	16	dead
TPM2	106	85	9	6	95, 201	15	47	41

The previous table summarizes the time (recorded in seconds) exhausted for each combination where TPM1 stands for using formula (14) directly, TPM2 for refining the guess preimages beforehand. The program was run on a Personal IRIS workstation.

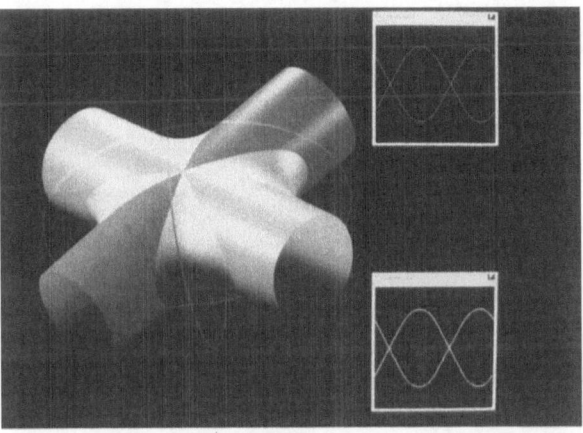

Figure 5. Blending two intersecting cyclinder-shape NURBS surfaces

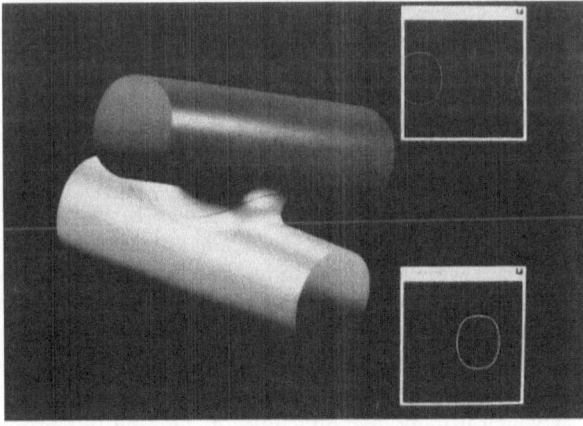

Figure 6. Blending two disjoint cylinder-shape NURBS surfaces

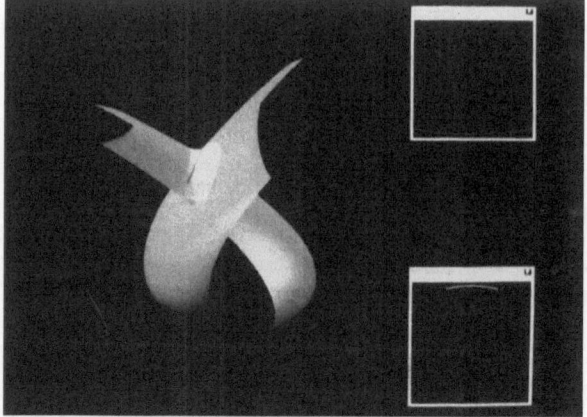

Figure 7. Self-blending I

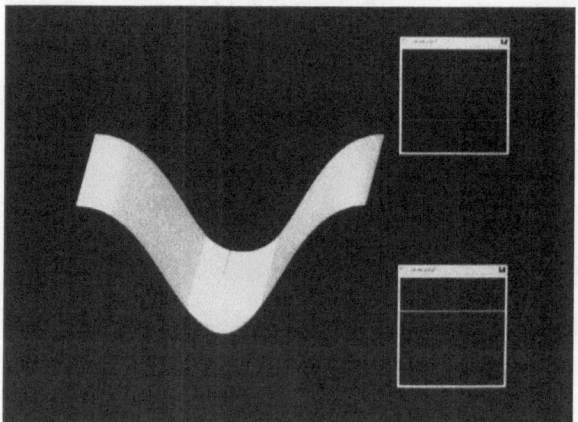

Figure 8. Self-blending II—the offset surface has a C^0 edge

The color pictures display the visual results where the surfaces are shown in the main screen, the blue curves are the intersection curves whose preimages are shown in the small boxes on the right and are color coded. In Fig. 5, the surfaces have common normals at the crossing points which, as we understood in section 4, are singular points. TPM1 will work if the radius is reduced by half. The self-intersections in Figs. 7 and 8 are obtained by feeding two identical surfaces into the routines. Particularly, the offset distance in Fig. 8 is set to so large a value that the offset surface is degenerated, containing a C^0 edge. The offset distance (or the radius of the fillet) in Fig. 9 is one-half of that in Fig. 10. Figure 11 shows the blending of two saddle surfaces which usually give difficulties if thinking this problem as solving non-linear equations [Gill & Murray '76 and Faux '79]. In Fig. 12, a ball is

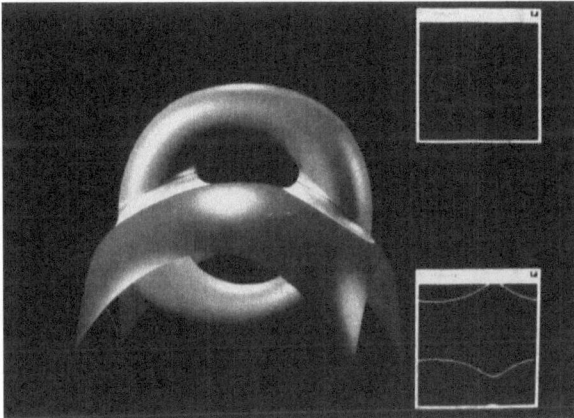

Figure 9. Blending a torus-shape and an open NURBS surfaces

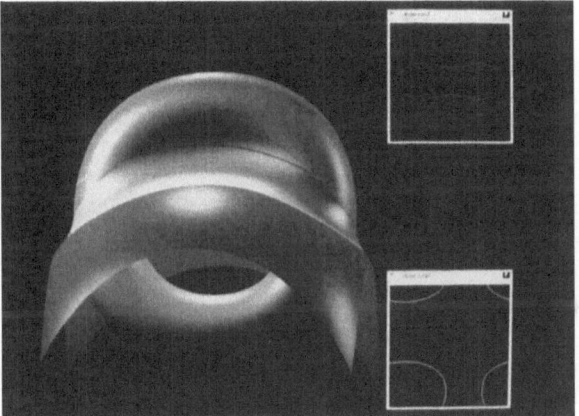

Figure 10. Blending the same data but the radius is doubled

simulated which cannot be done by an integral B-spline surface. In Fig. 13, the blend radius is half of the shortest distance between the torus and the plane. The corresponding offset surfaces are thus touching each other. The concave fillet is computed and displayed giving inward directions. In real situations, however, seldom will the same shape be designed as a blending surface.

7. Conclusions

We have developed and implemented a method to compute the constant-radius blend between two parametric surfaces, provided they are non-degenerated and

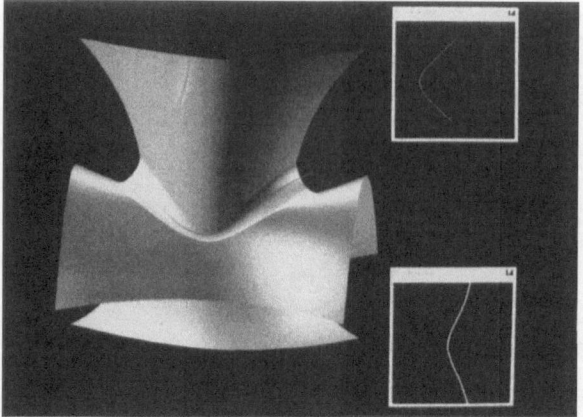

Figure 11. Blending two saddle NURBS surfaces

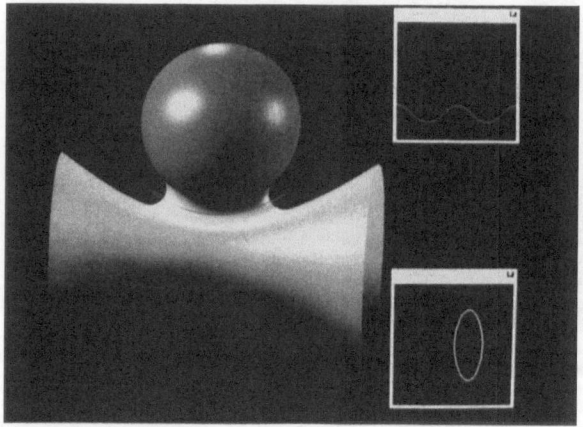

Figure 12. Blending a ball- and a revolution shape NURBS surfaces

their first order derivatives are computable. We have also devised an algorithm to calculate the offset surface intersection curve which only involves the computation of first order derivatives of the original surfaces, the offset surfaces themselves can even contain C^0 points.

We want to point out that, although only a linear approximation to the central line is pursued in our research, the algorithm has produced enough information to fit a spline curve to interpolate those points. The fillet can be easily expressed as a NURBS surface then.

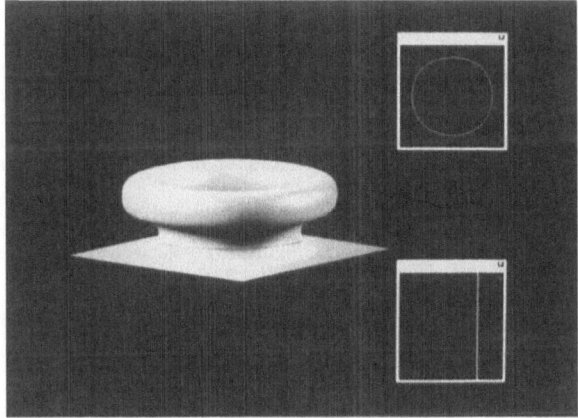

Figure 13. Blending a torus-shape NURBS surface and a plane

Some future work could be: corner-blending, shape-controlling and handling the singular situations.

Acknowledgements

This work was supported in part by NSF Grant DMC-8807747 and DOE contract DEFG0287ER25041, both awarded to Arizona State University.

References

[1] Barnhill, R. E., Farin, G., Jordan, M., Piper, B. R.: Surface/surface intersection. Computer Aided Geometric Design 4, 3–16 (1987).
[2] Barnhill, R. E., Kersey, S. N.: A marching method for parametric surface/surface intersection. Computer Aided Geometric Design 7, 257–280 (1990).
[3] Bloor, M. I. G., Wilson, M. J.: Generating blend surfaces using partial differential equations. Computer Aided Design 21(3), 165–171 (1989).
[4] Choi, B. K., Ju, S. Y.: Constant-radius blending in surface modeling. Computer Aided Design 21(4), 213–220 (1989).
[5] Crampin, M., Guifo, R., Read, G. A.: Linear approximation of curves with bounded curvature and a data reduction algorithm. Computer Aided Design 17, 257–261 (1985).
[6] Deuflhard, P.: A modified Newton method for the solution of ill-conditioned systems of non-linear equations with applications to multiple shooting. Numer. Math. 22, 289–315 (1974).
[7] Farin, G. E.: Rational curves and surfaces. In: Lyche, T., Schumaker, L. L., (eds.) Mathematical methods in computer aided geometric design, pp. 215–238. Boston: Academic Press 1989.
[8] Farin, G. E.: Curves and surfaces for computer aided geometric design, 2nd ed. Boston: Academic Press 1990.
[9] Faux, I. D., Pratt, M. J.: Computational geometry for design and manufacturing. Chichester, UK: Ellis Horwood 1979.
[10] Filip, D. J.: Blending parametric surfaces. ACM Transactions on Graphics 8(3), 165–173 (1989).
[11] Frost, T. M.: Parametric offset surfaces. Master Thesis, Arizona State University, 1991.
[12] Gill, P. E., Murray, W.: Algorithm for the solution of non-linear least squares problem. Report NAC71, National Physical Laboratory, Teddington, Middlesex (1976).
[13] Hoffman, C., Hopcroft, J.: Automatic surface generation in computer aided design. The Visual Computer 1(2), 92–100 (1985).

[14] Hoffman, C., Hopcroft, J.: Quadric blending surfaces. Computer Aided Design *18*, 301–306 (1986).

[15] Hoffman, C., Hopcroft, J.: The potential method for blending surface and curves. In: Farin G. E. (ed.) Geometric modeling—algorithms and new trends, pp. 347–366. SIAM 1987.

[16] Hoschek, J.: Intrinsic parametrization for approximation. Computer Aided Geometric Design *5*, 27–31 (1988).

[17] Houghton, E. G., Emnett, R. F., Factor, J. D., Sabharwal, C. L.: Implementation of a divided-and-conquer method for intersection of parametric surfaces. In: Barnhill, R. E., Boehm, W. (eds.) pp. 173–183. Amsterdam: North-Holland 1984.

[18] Lasser, D.: Intersection of parametric surfaces in the Bernstein-Bezier representation. Computer Aided Design *18*(4), 186–192 (1986).

[19] Middleditch, A., Sears, K.: Blend surfaces for set theoretic volume modelling systems. SIGRAPH Computer Graphics *19*, 161–170 (1985).

[20] Mortenson, M. E.: Geometric modeling. New York: John Wiley & Sons 1985.

[21] Mullenheim, G.: Convergence of a surface/surface intersection algorithm. Computer Aided Geometric Design *7*, 399–414 (1990).

[22] Peng, Q. S.: An algorithm for finding the intersection lines between two B-spline surfaces. Computer Aided Design *16*(4), 191–196 (1984).

[23] Powell, M. J. D.: Problems related to unconstrained optimization. In: Murray, W. (ed.) Numerical methods for unconstrained optimization. Boston: Academic Press 1992.

[24] Pratt, M. J., Geisow, A. D.: Surface/surface intersection problems. In: Gregory, J. A. (ed.) The mathematics of surfaces. Oxford: Oxford University Press 1986.

[25] Ricci, A.: A constructive geometry for computer graphics. Computer Journal *6*(2), 157–160 (1973).

[26] Rockwood, A. P., Owen, J. C.: Blending surfaces in solid modeling. In: Farin, G. E. (ed.) Geometric modeling—algorithms and new trends, pp. 367–384. SIAM 1987.

[27] Rockwood, A. P.: The displacement method for implicit blending surfaces. ACM Transactions on Graphics *8*(4), 279–297 (1989).

[28] Rossignac, J., Requicha, A.: Constant-radius blending in solid modelling. Computers in Mech. Engr. *3*, 65–73 (1984).

[29] Sanglikar, M. A., Koparkar, P., Joshi, V. N.: Modelling rolling ball blends for computer aided geometric design. Computer Aided Geometric Design *7*, 399–414 (1990).

[30] Sederberg, T. W., Anderson, D. C., Goldman, R. N.: Implicit representation of parametric curves and surfaces. Computer Vision, Graphics and Image Processing *28*, 77–84 (1984).

[31] Varady, T., Vida, J., Martin, R. R.: Parametric blending in a boundary representation solid modeller. Mathematics of surfaces III. Oxford: Oxford University Press, 1989.

[32] Willmore, T. J.: Differential geometry. Oxford: Oxford University Press 1959.

[33] Woodwark, J. R.: Blends in geometric modeling. Mathematics of surfaces II. Oxford: Oxford University Press, 1987.

[34] Yamaguchi, F.: Curves and surfaces in computer aided geometric design. Berlin, Heidelberg: Springer 1988.

[35] Zhang, D., Bowyer, A.: CSG set theoretic solid modelling and NC machining of blend surfaces. ACM symposium on computational geometry, New York, 1986.

Robert E. Barnhill
Computer Science Department
Arizona State University
Tempe, AZ 85287, U.S.A.

Computing Suppl. 8, 21–42 (1993)

Computing
© Springer-Verlag 1993

Functionality in Solids Obtained from Partial Differential Equations

M. I. G. Bloor and **M. J. Wilson**, Leeds

Abstract. A method is presented in this paper which allows solids, defined in terms of parametric bounding surfaces, to be mapped using a partial differential equation on to some simple object in parametric space, typically a cuboid. The isoparametric surfaces within the solid define a mesh system and it is shown how various features of the chosen PDE may be used to modify the mapping, and in particular the generated mesh, and thus facilitate analysis of mass properties or physical processes required of the object. The physics can be conveniently solved in parameter space, after a suitable transformation of the governing equations, and an example of this involving heat transfer is presented. Alternatively, physical properties of the object could be calculated via a finite element analysis, the mesh generated by the mapping forming the basis for a suitably defined finite element mesh. Thus the geometry used in the design of the object is used directly in the calculation of the physical properties or functional performance of the object.

Key words: Solids, functionality, PDE, mesh.

1. Introduction

An object's mechanical interactions with its surroundings are described in terms of forces and, apart from the exception of body forces, e.g. gravity, an object's surface is the 'interface' through which it is influenced by its surroundings. Thus, it is surface properties such as shape which are of paramount importance in determining the nature of those interactions, and so the surfaces of manufactured items inevitably reflect their functionality. Even mechanical objects, although their precise shape is often not uniquely specified by their intended purpose, have a morphology suited in some way to their alloted task. However, an object's surface shape is not the entire story. To determine its behaviour under the influence of forces or physical conditions, its properties as a solid object must be known. For instance, from the point of view of manufacture, an object's mass must be known in order to ascertain the cost of its production in terms of raw material, and also its behaviour under the action of applied stresses must be calculated if one is to be sure that it will survive the performance of its intended function.

The search for mathematical tools with sufficient flexibility to represent the surfaces of real objects has led to the techniques of free-form surface design [1], while the need to represent the properties of an object as a solid has led to the methods of solid modelling. There are currently two main varieties of solid modelling: Constructive Solid Geometry (CSG), which entails building a valid solid model by using

Boolean operations to combine simple primitive solids; and Boundary Representation, in which a solid object is described in terms of its surface as characterized by sets of bounding faces [2, 3]. Given the importance attached to both the shape of an object and its solid properties, there is a need for a technique that will integrate surface design with solid modelling, in order to meet certain functional criteria.

Much of the reason for having a solid model of an object is so that its physical properties can be calculated, which, in engineering practise, is achieved by means of solving field equations over the interior of the volume occupied by the solid subject to boundary conditions on the surface of the object. Of course, solutions for both interior and exterior domains are sometimes required but this is a straight-forward extension of the methods presented here. This paper describes how a boundary-value approach can be used to parametrize the space enclosed by a given set of surfaces. The method is a generalization to three dimensions of an idea originally introduced in a two-dimensional form as technique for surface design and generation [4, 5]. Examples are given in this paper of how a boundary-value technique can be used to develop an integrated approach to surface and solid modelling. In particular, we shall show how, given a set of surfaces bounding a closed volume of space, one can use these surfaces as boundary data for a three-dimensional partial differential equation (PDE) from whose solution one can generate a parametrization of the space enclosed by these surfaces. This is exactly analogous to the case of surface generation where a two-dimensional partial differential equation is solved, subject to boundary-data specified along curves, to create a surface [6]. This parametrization of the solid's volume can then be used as the basis for any investigation of its physical properties by transformation of the appropriate field equations from physical to parametric space.

The next section introduces the basic ideas behind the method by showing how a unit cube in (u, v, w) parameter space may be mapped to a hexahedral solid in physical space. These ideas are then extended in the next two sections to cover the case of non-hexahedral solids and also to allow control over the resulting reparametrization. Finally, the way in which the techniques may be used in practical applications is explained and examples given of calculating mass properties and heat transfer for particular solids.

2. From Surfaces to Solids Using PDES

We shall suppose that we are given a surface or set of surfaces S that enclose some volume V which is the solid we wish to consider, at this stage assumed to be simply connected, i.e. without holes. Furthermore, the surface is given in parametric form so that, for example, a point $\underline{X}_s$ on the surface is given in terms of two parametric coordinates which lie in that surface. Note that this will certainly be the case if the surfaces S have all been produced by the PDE method. To define the solid which is formed, we need to find a point $\underline{X}$ within the solid which will be given parametrically, now in terms of three parameters of course. Let us take u, v and w to be the three parameters used to define the solid so that $\underline{X} = \underline{X}(u, v, w)$ effectively gives a mapping to the interior of the solid in the physical (x, y, z) space from the interior of some region, preferably simple, in the (u, v, w) parameter space. It is desirable that any part of the bounding surface of the volume V is expressible directly in terms of two of the parameters u, v and w.

Let us try to illustrate the problem and its solution by reference to a rather general type of geometry which nevertheless allows us to be more specific about the bounding surfaces and the region to which V is mapped in the parameter space. Later we shall discuss other cases where there may be complications and also cases where there are considerable simplifications.

We choose a bounding surface S which can be thought of as being made up of six distinct surfaces S_i where $i = 1$ to 6. Bear in mind that this does not necessarily mean that the physical surface has six identifiably distinct faces, for example we could be considering a sphere the whole surface of which we have defined by six patches rather like panels drawn on a (European) football. Now denote the coordinate of a point on the surface S_i by $\underline{X}_{si}$. If we choose the labelling so that S_1 and S_2 can be thought of as 'opposite' faces, we take $\underline{X}_{s1}$ and $\underline{X}_{s2}$ to be functions of u and v only. Similarly, we take $\underline{X}_{s3}$ and $\underline{X}_{s4}$ to be functions of u and w only, and $\underline{X}_{s5}$ and $\underline{X}_{s6}$ are functions of v and w. We now proceed to map the solid defined by these bounding surfaces into the interior of the unit cube in (u, v, w) space, namely

$$
\begin{aligned}
\underline{X}_{s1} &= \underline{X}_{s1}(u, v), & 0 \le u \le 1; & \quad 0 \le v \le 1; & \quad w = 0. \\
\underline{X}_{s2} &= \underline{X}_{s2}(u, v), & 0 \le u \le 1; & \quad 0 \le v \le 1; & \quad w = 1. \\
\underline{X}_{s3} &= \underline{X}_{s3}(u, w), & 0 \le u \le 1; & \quad 0 \le w \le 1; & \quad v = 0. \\
\underline{X}_{s4} &= \underline{X}_{s4}(u, w), & 0 \le u \le 1; & \quad 0 \le w \le 1; & \quad v = 1. \\
\underline{X}_{s5} &= \underline{X}_{s5}(v, w), & 0 \le v \le 1; & \quad 0 \le w \le 1; & \quad u = 0. \\
\underline{X}_{s6} &= \underline{X}_{s6}(v, w), & 0 \le v \le 1; & \quad 0 \le w \le 1; & \quad u = 1.
\end{aligned}
\tag{1}
$$

This parametrisation is illustrated schematically in Fig. 1. We can now extend the idea we have used previously [13, 14, 15] for generating surfaces from boundary data using an elliptic partial differential equation of the appropriate order. In particular we shall treat each Cartesian coordinate independently as before. In this case, Eq. (1) simply give prescribed boundary data, the X_{si} on each face of the unit cube in the (u, v, w) space.

Thus we can obtain a suitable mapping of the solid in the physical space into the unit cube in parameter space by considering $\underline{X}(u, v, w)$ to be the solution of an elliptic PDE with u, v and w as the independent variables. If Eq. (1) represents all the conditions that we wish to impose on the mapping, the problem reduces to solving

$$
D^2 \underline{X} = F(u, v, w)
\tag{2}
$$

where D^2 is a second order elliptic differential operator, $F(u, v, w)$ is a vector-valued function defined over the interior of the unit cube in parameter space, and

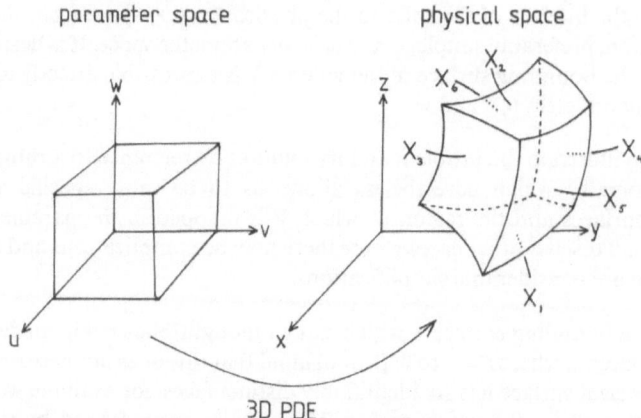

Figure 1. Schematic correspondence between the faces of a a physical object and the (u, v, w) cube

$$\underline{X}(u, v, 0) = \underline{X}_{s1}(u, v)$$

$$\underline{X}(u, v, 1) = \underline{X}_{s2}(u, v)$$

$$\underline{X}(u, 0, w) = \underline{X}_{s3}(u, w)$$

$$\underline{X}(u, 1, w) = \underline{X}_{s4}(u, w) \tag{3}$$

$$\underline{X}(0, v, w) = \underline{X}_{s5}(v, w)$$

$$\underline{X}(1, v, w) = \underline{X}_{s6}(v, w).$$

It is possible to impose further conditions to be satisfied by $\underline{X}$ on the boundaries if more control needs to be exercised on the location and spacing of the isoparametric surfaces in the solid. This will arise when calculations required for the physical problem then impose constraints on the parametrization so that the required degree of accuracy can be achieved. When such extra conditions are required, it might be possible to satisfy them by increasing the order of the PDE which determines $\underline{X}$. Another means of exercising control is to modify the operator D and the 'source' term $\underline{F}(u, v, w)$. In certain circumstances, it is attractive to increase the order of the operator so that particularly simple solutions for $\underline{X}$ can be found. This arises in a limited number of cases when the bounding surfaces have themselves been formed from solutions of PDEs using the methods previously developed by the authors. We shall return to these points later. Firstly let us construct a simple and somewhat contrived example which will illustrate the process just described.

Consider the solid bounded by the surfaces given by

$$\underline{X}_{s1} = (u, v, 0)$$

$$\underline{X}_{s2} = (u, v, 2 + u^2 - v^2)$$

$$\underline{X}_{s3} = (u, 0, w[2 + u^2])$$

$$\underline{X}_{s4} = (u, 1, w[1 + u^2]) \tag{4}$$

Plate 1. Shaded image of the exterior of a simple analytic solid (see Eq. (6) generated as a solution to Eq. (5) subject to the boundary conditions (4)

Plate 2. Shaded image of the exterior of a simple analytic solid (see Eq. (11))

$$\underline{X}_{s5} = (0, v, w[2 - v^2])$$
$$\underline{X}_{s6} = (1, v, w[3 - v^2]).$$

These give the eighteen boundary conditions on x, y and z on the six faces of the unit cube in parameter space, and no further conditions are to be imposed. The volume which is bounded externally by these surfaces is shown as a shaded solid in Plate 1. We shall take a simple form for the equation which determines $\underline{X}(u, v, w)$, namely Laplace's equation. Thus

$$\left(\frac{\partial^2}{\partial u^2} + \frac{\partial^2}{\partial v^2} + \frac{\partial^2}{\partial w^2}\right)\underline{X} = 0. \tag{5}$$

The solution is found to be

$$\underline{X} = (u, v, w[2 + u^2 - v^2]), \tag{6}$$

which maps the unit cube shown in Plate 1 into the solid object. Figure 2 shows the intersection curves of the parametric surfaces $u = $ constant, $v = $ constant on the

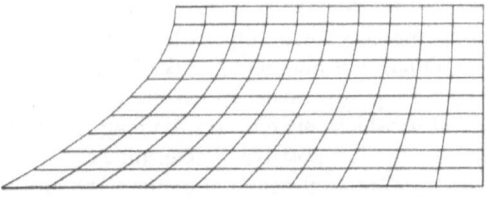

Figure 2. The intersection curves of the surfaces $u = $ constant (running left to right), $w = $ constant on the $v = 1/2$ plane for the solid shown in Plate 1

$v = 1/2$ plane. Note that the u = constant and the v = constant planes coincide with the planes x = constant, y = constant. The parametric mesh which has now been generated in the solid can be used to facilitate the calculation of physical properties of the object, for example to give mass properties, temperature distribution or a stress analysis.

There is a similarity between this boundary-value approach and a certain class of methods for mesh generation which are popular in Computational Fluid Dynamics (CFD); a typical problem in CFD involves calculating the flow about a body, and this needs some means of producing a mesh within the flow field. These ideas are typified in the work of Thompson [7, 8, 9] who seeks to generate a curvilinear coordinate system ξ^i, $i = 1, 2, 3$ which satifies the Laplacian system $\nabla^2 \xi^i = 0$, $i = 1$, 2, 3 subject to the boundary condition that there is a coincidence between each segment of the physical boundary and a coordinate surface, rather as if each physical boundary was a surface of constant ξ 'potential'. Rather than solve for the curvilinear coordinates ξ^i in terms of the Cartesian coordinates x_i, Thompson treats the Cartesian coordinates as dependent variables and solves for them in terms of the curvilinear coordinates. This approach results in a system of three quasi-linear partial differential equations which must be solved simultaneously in order to obtain the correct coordinate mesh, unlike the present method which gives rise to three linear partial differential equations which can be solved independently. Note that Thompson [9] can exercise control over his mesh by the addition of control functions P^i to the right hand side of the PDE, a technique whose relevance to the present work will be considered later on when 'forcing-functions' are discussed.

3. Non-Hexahedral Solids—more than six Faces

Let us now consider other situations of given bounding surfaces which do not so conveniently fit the recipe we have just outlined. If there are more than six faces specified then a number of adjacent faces can be amalgamated and appropriately reparametrised so that the problem can be reduced to the form of the one just dealt with. This manipulation of the system is by no means unique and the actual form that it takes will be governed to a large extent by the requirements of the resulting parametric mesh in the solid. In addition to the singularities associated with those sharp edges which may form part of the solid and are mapped to the edges of the cube in parameter space, other singular points or lines in the mapping may arise. If the PDE is of higher order than two, when supplementary unconstrained boundary conditions can be imposed, the nature of these singularities can be controlled to some extent. This situation is analogous to the filling of a six-sided hole with a surface generated by a PDE, a problem which the authors have dealt with in an earlier paper [10] and to which reference can be made for further clarification of this point. Suffice to say that it is possible to determine conditions which will avoid the difficulty and furthermore, in the present context, the problem can be avoided by using a second order equation when the boundary conditions for a well defined solid will have the appropriate continuity.

4. Non-Hexahedral Solids—less than six Faces

The other case we must consider is when there are fewer than six faces bounding the volume V. In this case we must allow for degenerate surfaces when a face may be regarded as having been shrunk to a curve or point. We shall illustrate this by means of a simple example as this is likely to be a fairly common occurrence when the bounding surfaces have been generated using the PDE method. Again we shall take Eq. (5) to be the equation that $\underline{X}$ must satisfy.

Plate 2 gives a shaded image of the solid object for which the bounding surfaces are specified. The z axis is coincident with the axis of symmetry of the object. The smaller circular face of unit radius lies in the plane $z = 0$ and is mapped to the $u = 0$ face of the unit cube while the face of radius R (>1) at $z = 1$ and is mapped to the $u = 1$ face. We shall take the curved surface which completes the definition of the bounding surface to be mapped to the $w = 1$ face of the unit cube. For convenience we take the surfaces to be parametrised as follows:-

$$\underline{X}_{s5} = (w \cos 2\pi v, w \sin 2\pi v, 0),$$

$$\underline{X}_{s6} = (Rw \cos 2\pi v, Rw \sin 2\pi v, 1), \tag{7}$$

$$\underline{X}_{s2} = (f(u) \cos 2\pi v, f(u) \sin 2\pi v, u),$$

where

$$f(u) = R \cosh u/\cosh 1. \tag{8}$$

It is clear that we must establish conditions to be applied on the remaining three surfaces of the unit cube before a solution to Eq. (5) can be found. We take the $w = 0$ face of the unit cube to be the degenerate surface consisting of the axis of symmetry from $z = 0$ to $z = 1$. Hence we have,

$$\underline{X}_{s1} = (0, 0, u), \tag{9}$$

It is then evident that the conditions required on $v = 0$ and $v = 1$ are simply those of periodicity. Thus

$$\underline{X}_{s3}(u, w) = \underline{X}_{s4}(u, w). \tag{10}$$

The correspondence between the faces of the (u, v, w) cube and the surfaces of the object in E^3 is shown in Fig. 3.

The solution of Eq. (5) which satisfies these boundary conditions is

$$\underline{X}(u, v, w) = (wf(u) \cos 2\pi v, wf(u) \sin 2\pi v, u). \tag{11}$$

Figure 4 shows the intersection curves of the $w = $ constant and the $u = $ constant planes on the v-planes $v = 0$ and $v = 1/2$ (which form a cross-section containing the symmetry axis).

It is worth considering a restricted class of solids at this stage because of the simplicity of the mapping that can be obtained. These are solids where it is possible to identify or define two parametric surfaces, say $\underline{X}_{s1}$ and $\underline{X}_{s2}$, bridged by ruled

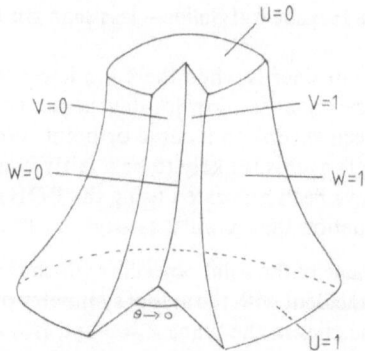

Figure 3. The relationship between the (u, v, w)
cube and the object shown in Plate 2

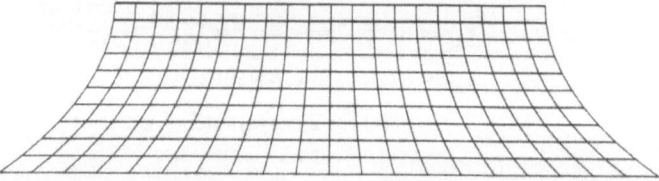

Figure 4. The intersection curves of the surfaces u = constant (running left to right), w = constant on
the $v = 0$ and $v = 1/2$ planes for the solid shown in Plate 2

surfaces. Under these circumstances the mapping can be taken to be

$$\underline{X}(u, v, w) = g(w)\underline{X}_{s2}(u, v) + [1 - g(w)]\underline{X}_{s1}(u, v), \tag{12}$$

where $g(0) = 0$ and $g(1) = 1$, and the form of g is chosen to provide the local scaling
in w that is deemed desirable from any mesh considerations associated with the
geometry or physics of the problem. The simple examples already given in this paper
fall into this category with $g(w) = w$, although the approach used did not rely on
such restricted solids. Clearly if the surfaces $\underline{X}_{s1}$ and $\underline{X}_{s2}$ have been produced by
the PDE method then $\underline{X}$ can be thought of as a solution to an appropriately chosen
PDE with rather tightly controlled boundary conditions that are easily deduced
from Eqs. (5) and (12). This interpretation is only of value in that properties of the
PDE supply information about the properties of the mapping.

5. Non-Simply Connected Solids

We can now broach the problem of solids which are not simply connected. Let us
illustrate how this difficulty may be overcome by reference to a simple example from
which we can deduce a more general answer. Consider the case of a toroid the
surface of which can be represented parametrically by

$$\underline{X}_s = [(a + b \cos v)\cos u, (a + b \cos v)\sin u, b \sin v]. \tag{13}$$

This can be made simply connected by inserting a 'cut' C_s on the plane $u = $ constant, and then parametrizing both faces of C_s in terms of v and w and demanding that, across C_s, $\underline{X}(u, v, w)$ is continuous when a second order PDE is used to produce the mapping to the unit cube. If a higher order PDE is used, then higher order continuity is required across C_s. This course of action is analogous to imposing periodic boundary conditions.

It is instructive to carry this example to its conclusion. We insert the cut on the plane $u = 0$ and then the bounding surfaces can be expressed in terms of parameters u, v and w in the form

$$\underline{X}_{s1} = (a \cos 2\pi u, a \sin 2\pi u, 0),$$

$$\underline{X}_{s2} = [(a + b \cos 2\pi v)\cos 2\pi u, (a + b \cos 2\pi v)\sin 2\pi u, b \sin 2\pi v],$$

$$\underline{X}_{s3}(u, w) = \underline{X}_{s4}(u, w), \tag{14}$$

$$\underline{X}_{s5} = \underline{X}_{s6} = [f(w) + g(w)\cos 2\pi v, 0, h(w)\sin 2\pi v],$$

where

$$f(w) = a(e^{2\pi w} + e^{2\pi(1-w)})/(1 + e^{2\pi}),$$

$$g(w) = b[\sinh(2\sqrt{2}\pi w)]/[\sinh(2\sqrt{2}\pi)], \tag{15}$$

$$h(w) = b[\sinh(2\pi w)]/[\sinh(2\pi)].$$

We have taken the $w = 0$ face of the unit cube to correspond to the degenerate surface consisting of the circle

$$x^2 + y^2 = a^2, z = 0. \tag{16}$$

Also periodicity conditions have been applied on $u = 0, u = 1$ and $v = 0, v = 1$. The parametrization with respect to w is somewhat contrived so that a simple analytic solution to Laplace's equation can be obtained for the purposes of this example. In reality there are much simpler ways of obtaining a satisfactory parametrization of the solid torus but the above acts as an illustrative example of the procedure we adopt for dealing with non-simply connected solids. For completeness we write down the solution to the above problem in the form

$$\underline{X}(u, v, w) = \{[f(w) + g(w)\cos 2\pi v]\cos 2\pi u,$$

$$[f(w) + g(w)\cos 2\pi v]\sin 2\pi u, h(w)\sin 2\pi v]\}. \tag{17}$$

Figure 5 shows the relationship between the faces of the (u, v, w) cube and the corresponding surfaces introduced into the surface of the torus by the cutting process. For more complicated objects, i.e. objects with several handles, the solid is made simply connected by the insertion of cuts which map into faces or parts of faces of the unit cube in parameter space and across which appropriate continuity conditions must be enforced when the governing PDE is solved.

An alternative to the use of cuts which may prove desirable in order to exercise control more readily on the generated mesh, is to think of the object as being made up of distinct simpler objects and then map each individual simple object into a

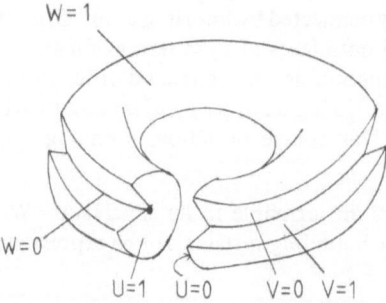

Figure 5. The relationship between the (u, v, w) cube and the faces of the 'cut' torus

separate parameter space with suitable boundary conditions applied so that these objects meet in E^3. This is like thinking of the original object as being made up of rather exotic primitives, defined in terms of bounding surfaces, which are stuck together across common faces. This type of approach is common amongst certain classes of existing mesh generators [7, 8, 9, 11, 12]. An example which will be presented later can be thought of in these terms inasmuch as the solid formed in Plate 4 can be added, in the way just described, to two circular cylinders to create a more complicated solid. As it happens, in this example the resulting solid is simply connected but the principle is nevertheless the same.

6. Control of Parametrization

As has already been mentioned, it is frequently desirable to have a mesh within the solid which is sufficiently refined to represent adequately the geometry, or to facilitate subsequent calculations of physical properties that may have significant variation over short length scales. The particular form of the PDE must therefore be chosen to take account of these considerations. However, we must, in this choice, be aware of the expense of calculating numerical solutions to overly complicated or nonlinear PDEs, and hence it is worthwhile examining what can be achieved from very simple PDEs based on Laplace's equation. In the authors' earlier papers concerned with surface generation, it was found that simple coordinate scaling introduced into the Biharmonic equation was capable of giving a great deal of control [5]. In this paper, we shall consider coordinate scaling via the following equation

$$\left(a^2 \frac{\partial^2}{\partial u^2} + b^2 \frac{\partial^2}{\partial v^2} + c^2 \frac{\partial^2}{\partial w^2} \right) \underline{X} = \underline{F}(u, v, w) \tag{18}$$

in which a, b and c are constants, and u, v and w are confined to the unit cube. In order to understand the roles played by a, b, c and $\underline{F}$ it is more instructive to separate the effects and look at two model systems. Firstly we shall examine the influence on the mesh generated in the solid by the mapping of 'forcing-function' $\underline{F}$ with

$a = b = c = 1$ so that Eq. (18) is simply Poisson's equation. Then the control exercised by the parameters a, b and c can be inferred from a system in which $a = b = 1$, $\underline{F} = \underline{0}$ and variations in c are examined. In both cases, it is possible through an inappropriate choice of parameters or function to produce an unsuitable mapping, parts of which may fall outside the bounding surface. The possibility of this is further discussed in an example given below.

6.1 The Forcing-Function

We can in principle write down the solution of Poisson's equation in terms of a Green's function [13]. If we denote the dependent variable by ϕ, where ϕ may be either x, y or z in the present context, and denote the corresponding component of $\underline{F}$ by f, then the value of ϕ at any fixed point P in the unit cube in parameter space is given by,

$$\phi(P) = -\frac{1}{4\pi} \iint_s \phi_s \frac{\partial G}{\partial n} dS - \frac{1}{4\pi} \iiint_V fG dV \qquad (19)$$

where ϕ_s is the value of ϕ on the surface of the unit cube denoted by S and $\dfrac{\partial}{\partial n}$ denotes the partial derivative in the direction of the outward normal to the surface S. The volume enclosed by S is denoted by V and G is the Green's function for Laplace's equation in the unit cube the value of which, at any point Q in V and on S depends on both Q and P and satisfies

(i) $\nabla^2 G = 0$ in V,
(ii) $G = 0$ on S,
(iii) $G(Q) \to 1/r$ as $r \to 0$, where r is the distance in parameter space from the point P to the point Q,
(iv) G has no other singularities inside V or on S.

The first term on the right hand side of Eq. (19) represents the smoothing property of the Laplacian operator in that at any point p in V, ϕ is the weighted average of its value over the bounding surface, $\dfrac{\partial G}{\partial n}$ being the weighting function. The second term shows the contribution that arises from a non-zero value of f and its dominant influence can be gauged by reference to condition (iii) on the Green's function, i.e. the weighting of its value at any location decreases inversely with the distance from that point.

Let us illustrate this point with a very simple example. We shall consider an analytic solution so that the effect of the source term is readily apparent. With reference to the example considered earlier, for which the boundary conditions are given by Eqs. (7) to (10) and the solution by Eq. (11), it can be seen that the variation of the coordinate z with u is linear. This variation is of course specified over $\underline{X}_{s1}$ and $\underline{X}_{s2}$, and for the purposes of this example may be supposed to represent an appropriate mesh structure for a hypothetical physical problem that needs to be calculated for

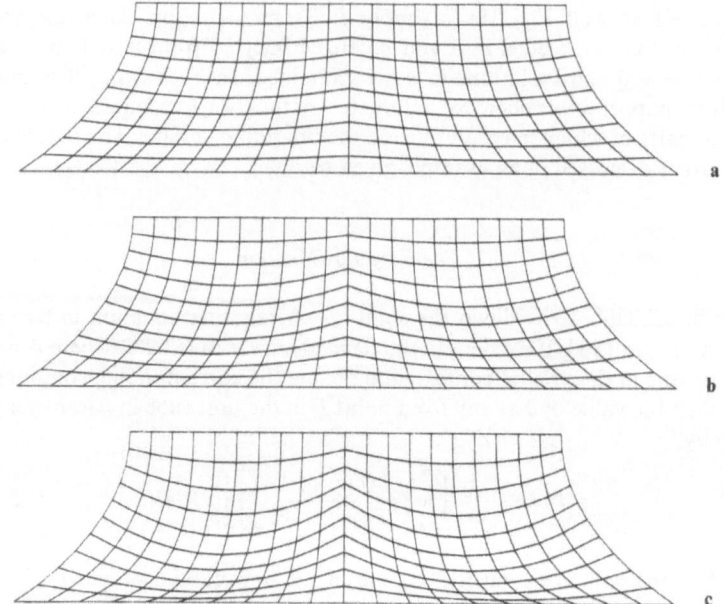

Figure 6a. The effect of a source term: the intersection curves of the $w =$ constant planes and $u =$ constant planes (running left to right) on the v-planes $v = 0$ and $v = 1/2$ for $\alpha = 1$

Figure 6b. The effect of a source term: the intersection curves of the $w =$ constant planes and $u =$ constant planes (running left to right) on the v-planes $v = 0$ and $v = 1/2$ for $\alpha = 2$

Figure 6c. The effect of a source term: the intersection curves of the $w =$ constant planes and $u =$ constant planes (running left to right) on the v-planes $v = 0$ and $v = 1/2$ for $\alpha = 3.5$

the solid e.g. a temperature distribution. Now suppose that the physics indicated that the resolution in the z direction should be refined in a centrally located annular region (around $w = 1/2$) close to the surface $\underline{X}_{s6}$. This can be achieved if z is increased in the region of the parameter space near $w = 1/2$ and $u = 1/2$ so that the mesh adjacent to $u = 0$ is 'stretched' while that near $u = 1$ is refined. Hence we leave the equations for the coordinates x and y unchanged so that the solution is given by the first two component of Eq. (11) but the equation for z is now taken to be

$$\nabla^2 z = -2\alpha[u(1 - u) + w(1 - w)] \qquad (\alpha > 0), \qquad (20)$$

in other words the vector source function $\underline{F}(u, v, w)$ is given by

$$\underline{F} = \{0, 0, -2\alpha[u(1 - u) + w(1 - w)]\}. \qquad (21)$$

The solution for z is now given by

$$z = u + \alpha u(1 - u)w(1 - w) \qquad (22)$$

and elementary analysis shows that this represents a valid transformation provided the parameter $\alpha < 4$. Incidently, this example also illustrates one of the complications of using a source term in that the resulting transformation may carry points outside of the bounding surfaces and thus create an invalid solid. This occurs in the above example when $\alpha > 4$. Figure 6a, b, c shows the intersection curves of the $w = $ constant and the $u = $ constant planes on the v-planes $v = 0$ and $v = 1/2$ for various value of α (cf Fig. 4).

6.2 The Parameter c

In discussing the effect of the parameter c it is useful to draw on our earlier analysis [5] where we dealt with surface generation using a two-dimensional modified biharmonic eqation. An equivalent, essentially coordinate scaling, parameter occurred in that analysis and it was shown how it controlled the relative smoothing between the two parametric coordinate directions. In the present case we have, of course, three independent variables and for the purposes of this discussion we have fixed two of these scaling factors to be unity and, to avoid complicating the issue, $\underline{F} = 0$. The PDE we need to consider is

$$\left(\frac{\partial^2}{\partial u^2} + \frac{\partial^2}{\partial v^2} + c^2 \frac{\partial^2}{\partial w^2} \right) \underline{X} = 0. \tag{23}$$

On the basis of our earlier analysis and bearing in mind that the Laplacian operator is simply an averaging process, we can assert that in this case, the 'smoothing' in the u and v directions is of equal weighting while c controls the weighting of the averaging in the w direction. Thus, if c is small, the averaging takes place primarily over u and v in the $w = $ constant planes, that is to say that the boundary conditions imposed on the faces $u = 0$, $u = 1$, $v = 0$, and $v = 1$ are averaged with uniform weighting over the u, v plane. However, in general the form resulting from this process for the dependent variable on the surfaces $w = 0$ and $w = 1$ would not correspond to the imposed boundary conditions on these faces. The adjustment to these conditions takes place over a length scale in the w direction of $0(c)$, i.e. a 'boundary layer' exists on these faces. In the two dimensional analysis referred to earlier [5], where the problem was concerned with blend generation, this was referred to as a trimlayer.

We will now present a simple example to illustrate these points and again follow our usual practice, wherever possible, of contriving an analytic solution to the problem so that the influence of the parameter c is readily seen. Let us consider the object shown in Plate 3, a circular cylinder with a domed top, which may be defined by the bounding surfaces given by,

$$\underline{X}_{s1} = (u \cos 2\pi v, u \sin 2\pi v, 0),$$
$$\underline{X}_{s2} = (u \cos 2\pi v, u \sin 2\pi v, 1 + h \sin \pi u),$$
$$\underline{X}_{s3}(u, w) = \underline{X}_{s4}(u, w), \tag{24}$$
$$\underline{X}_{s5} = (0, 0, w),$$
$$\underline{X}_{s6} = (\cos 2\pi v, \sin 2\pi v, w).$$

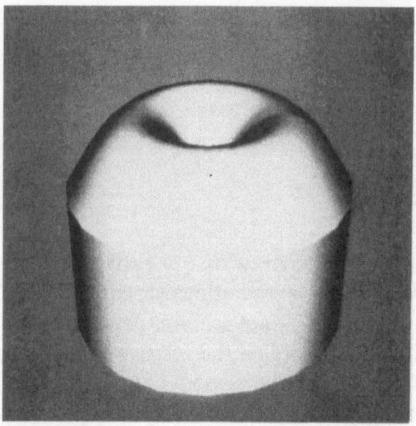
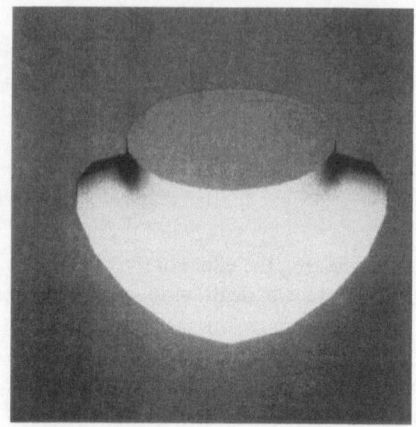

Plate 3. Shaded image of a simple analytic solid (see Eq. (25) used to demonstrate the effect of the smoothing parameters

Plate 4. Shaded image of the exterior of a numerically generated solid which forms a blend between two perpendicular cylinders

Using these as the boundary conditions for Eq. (23), after a little analysis, it can be seen that the solution for $\underline{X}$ is given by

$$\underline{X}(u, v, w) = (u \cos 2\pi v, u \sin 2\pi v, w + f(w) \sin \pi u), \tag{25}$$

where

$$f(w) = h[\sinh(\pi w/c)]/[\sinh(\pi/c)]. \tag{26}$$

Now, for large values of c, the above expression for f may be approximated, to $0(\pi^3/\underline{c}^3)$, by

$$f(w) = hw \tag{27}$$

showing that the influence of the bump on the top of the cylinder is spread throughout the range of w and the magnitude of the bulge on the parametric surfaces $w = $ constant varies linearly from zero at the bottom of the cylinder to h at the top. A more interesting result occurs if c is small when f may be approximated by

$$F(w) = h(e^{\pi(w-1)/c} - e^{-\pi(w+1)/c}) \tag{28}$$

This is of course zero at $w = 0$, to satisfy the boundary condition, but remains close to zero as w increases until w is very close to unity, whereupon it rapidly rises to the value h over a distance of $0(c)$ in order to satisfy the boundary condition on $w = 1$. This example illustrates very clearly the way in which the smoothing parameter c affects the solution and in particular the fact that when c is small boundary layers form at a bounding surface, $w = 1$ in this case, across which the solution adjusts rapidly to the imposed boundary condition. Figure 7a, b, c shows the intersection curves of the $w = $ constant and $u = $ constant planes on the v-planes $v = 0$ and $v = 1$ for various values of c.

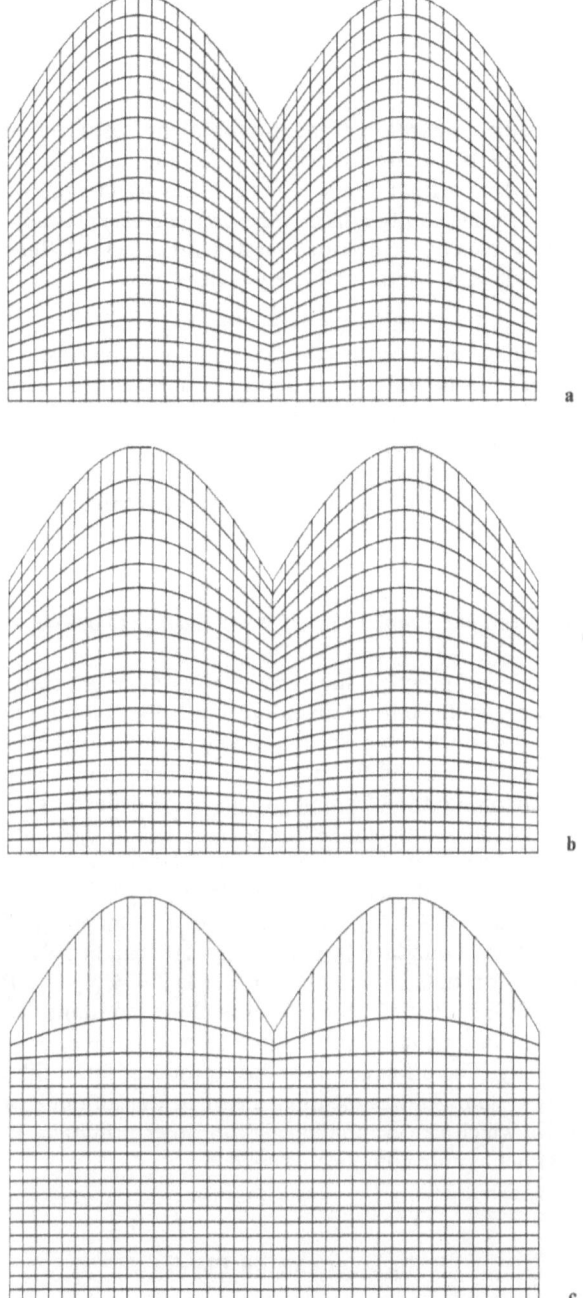

Figure 7a. The effect of the smoothing parameter: the intersection curves of the w = constant planes (running left to right) and u = constant planes on the v-planes $v = 0$ and $v = 1/2$ for $c = 0.1$

Figure 7b. The effect of the smoothing parameter: the intersection curves of the w = constant planes (running left to right) and u = constant planes on the v-planes $v = 0$ and $v = 1/2$ for $c = 1$

Figure 7c. The effect of the smoothing parameter: the intersection curves of the w = constant planes (running left to right) and u = constant planes on the v-planes $v = 0$ and $v = 1/2$ for $c = 10$

7. Higher Order PDES

It can be seen from the description of the smoothing process given above that the parametrization of the bounding surfaces plays a dominant role in determining the isoparametric surfaces within the solid. For this reason, it is important that when the mapping from a surface on to a face of the unit cube is determined, account is taken of the requirements of the eventual mesh arrangement in the solid, bearing in mind, where necessary, the physics of the problem.

Moving to a higher order PDE to determine the mapping for the solid can offer certain attributes which enhance the control on the parametrization of the solid. This is because the extra boundary conditions required to complete the solution allow for the specification of normal derivatives (in (u, v, w) space) to the bounding surfaces and thus control the mesh spacing adjacent to the boundaries. Clearly this may be a useful tool in the physical analysis of certain problems when, for example, it is expected that rapid rates of change of physical quantities may occur near the boundaries.

8. Example: From a Simple Blend Surface to a Solid

In most circumstances of practical importance, the simple illustrative analysis of the previous sections will have to be abandoned in favour of numerical solutions. We shall consider a solid which acts as a blending bridge between two cylindrical bars (see Plate 4). This is a blend that we have considered in an earlier paper [5] and therefore we shall take the blending surface to be defined as a solution to a suitably chosen PDE and this will be the surface $w = 1$. The surface $w = 0$ is degenerate and is the line along the axis of cylinder 2 from $z = R$ to $z = R + h$. The surface $u = 0$ is that part of the surface of cylinder 1 for which $x^2 + y^2 \leq R_2^2$, and $u = 1$ is the flat surface $x^2 + y^2 \leq R_1$, $z = R + h$. The mapping of the solid is periodic in v. With these boundary conditions, Eq. (18) can be solved for specified values of a, b and c. The results for $a = 1, b = 1, c = 1, R = 4, R_1 = 2, R_2 = 3, h = 1$, are shown in Plates 5a, b, c&d which show selected $w = $ constant planes, $u = $ constant planes, $v = $ constant planes, and an impression of the associated mesh within the solid, respectively. These images have been produced on a Silicon Graphics 4D/VGX graphics system. As was mentioned earlier, we can think of this solid as an exotic primitive which can be combined with other primitives in the manner indicated to construct a more complicated solid.

9. Incorporating the Physics

9.1 Mass Properties

In this section we shall give examples of the way in which the method we have used to map a solid on to a simple unit cube in the parameter space may be used to

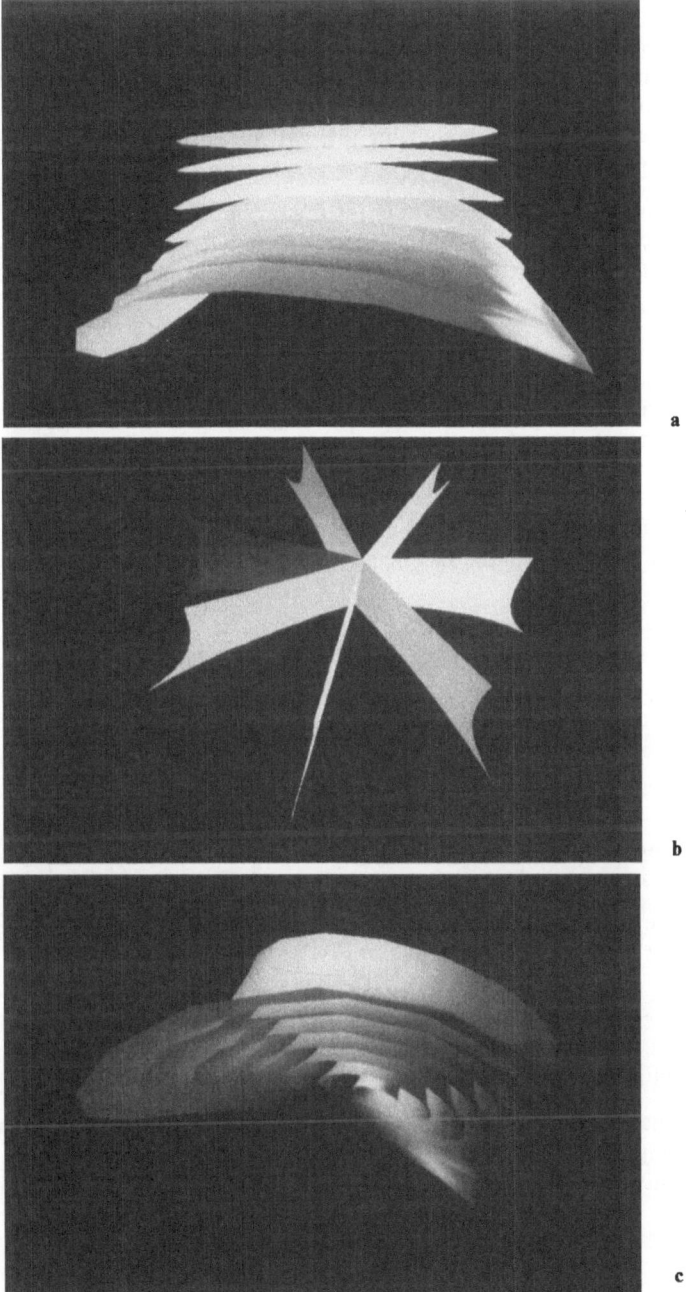

Plate 5a. Shaded image of selected u = constant planes for the solid shown in Plate 4
Plate 5b. Shaded image of selected v = constant planes for the solid shown in Plate 4
Plate 5c. Shaded image of selected w = constant planes for the solid shown in Plate 4

reformulate a physical problem and facilitate its solution. We shall confine ourselves
to problems involving mass properties as this avoids the quite extensive algebra,
straightforward in principle, that is normally required in the transformation of
partial differential equations. We shall try to make the problems as simple as
possible yet consistent with showing the way in which the method might be employed in practice.

Let us consider the solid we used in the first example which is defined by Eq. (6).
We shall calculate its mass and its centre of mass assuming that it is of uniform
density, which we take to be unity for convenience. The mass is given by the volume
integral

$$M = \iiint_V dx\,dy\,dx. \tag{29}$$

Rather than calculate this in the physical space we shall transform it to parameter
space where an element of volume is given by

$$dV = (\underline{X}_u \wedge \underline{X}_v) \cdot \underline{X}_w du\,dv\,dw \tag{30}$$

and the subscripts denote partial differentiation. The triple scalar product in Eq.
(30) is just the inverse of the Jacobian of the transformation J defined by

$$J^{-1} = \begin{vmatrix} x_u & x_v & x_w \\ y_u & y_v & y_w \\ z_u & z_v & z_w \end{vmatrix} \tag{31}$$

thus Eq. (28) becomes

$$M = \int_0^1 \int_0^1 \int_0^1 J^{-1} du\,dv\,dw. \tag{32}$$

Substituting from Eq. (6) for $\underline{X}$ to obtain J the integral in Eq. (31) is evaluated to
give $M = 2$.

The centre of mass $\underline{X}_G$ for this solid can now be found in much the same way by
evaluating another simple integral

$$\underline{X}_G = \frac{1}{M} \int_0^1 \int_0^1 \int_0^1 \underline{X} J^{-1} du\,dv\,dw \tag{33}$$

The result is $\underline{X}_G = (13/12, 11/12, 47/45)$.

Clearly the same approach can be used to calculate other mass properties but in
more complicated circumstances the calculations would have to be carried out
numerically with J being evaluated numerically at suitable points in the unit cube.

9.2 Analysis of Function

A similar procedure can be used for more complicated physical problems which are
governed by a PDE, for example the heat conduction equation. In this case the

governing PDE must be transformed by changing from x, y and z as independent variables to u, v and w and then solved in the simple geometry of the unit cube using finite differences. Frequently, the most convenient way of transforming the PDE governing the physics is to express the physical problem in variational form in terms of the Cartesian coordinates x, y and z. Transformation of the integral to parametric coordinates is a good deal simpler than transforming the original PDE due to the lower order derivatives in the integrand. The problem can now be solved directly as a variational problem in parameter space or by expressing the variational problem as an equivalent PDE using parametric coordinates as the independent variables. Whichever approach is used, the enormous advantage of solving the problem in the simple geometric parameter space, typically a unit cube, is evident. Furthermore, for numerical solutions, the relative ease with which the mesh within this simple geometry can be adapted either statically or dynamically to suit the physical problem is a distinct advantage over similar procedures in physical space.

To illustrate this approach we shall calculate the temperature distribution within the solid shown in plate (2) with $R = \cosh 1$. Denoting the steady state temperature by T, the equation that T must satisfy within the solid is,

$$\left(\frac{\partial^2}{\partial x^2} + \frac{\partial^2}{\partial y^2} + \frac{\partial^2}{\partial z^2} \right) T = 0. \tag{34}$$

The boundary conditions are that $T = 0$ over the whole of the surface of the solid except the disc corresponding to $u = 1$ where T is given in terms of v and w for convenience by,

$$T = \cos(2\pi v). \tag{35}$$

Notice that, although the body has axial symmetry, the problem is not axially symmetric but fully three-dimensional owing to the boundary conditions imposed. To set up the problem in parameter space we express the physical problem in variational form as,

$$\delta \iiint_{\text{solid}} (T_x^2 + T_y^2 + T_z^2) dxdydx = 0. \tag{36}$$

The volume integral can be transformed into an integral over parameter space as done previously for the calculation of solid properties. The resulting variational problem is then converted to an equivalent PDE to be solved using finite differences in the unit cube in parameter space subject to the boundary conditions. The procedure is straightforward but nothing is gained from presenting the detailed algebra here. Plate 6 shows the results of the calculation of the temperature distribution through the solid.

An alternative approach is to employ a finite element method in physical space using the mesh derived from the isoparametric surfaces in the solid to form the basis of the finite element mesh. In other words, the PDE method introduces a (u, v, w) parametrisation of the solid which could be used in certain circumstances as a finite element mesh in its own right, or alternatively used to give a set of points from

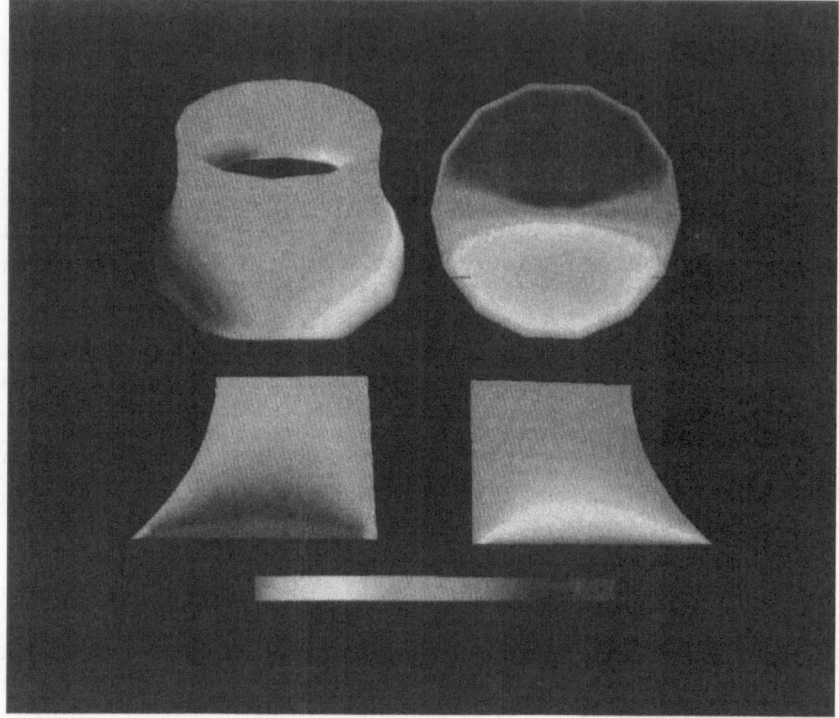

Plate 6. Shaded image of temperature distribution in the solid shown in Plate (2). Blue and red indicate the cold and hot ends of the temperature spectrum respectively

which a mesh could be produced by Delaunay triangulation, rather like certain existing techniques [12, 14].

10. Conclusions

The method presented in this paper allows a solid defined in terms of its boundary surfaces, which are given parametrically, to be mapped on to the unit cube in a new three dimensional parameter space. The use of a partial differential equation with parameters and source functions which can be specified arbitrarily allows a large degree of control of the final transformation. The rules governing the choice of these parameters and source functions to meet various requirements have been explained and the examples give some insight into the way the choice might be made intuitively. The detailed form of the transformation is particularly important in engineering applications when an analysis of physical properties of the object is required and this can be facilitated by the design of a suitable parametrization within the solid. This can be thought of as providing a system of boundary fitted coordinates which can be used in subsequent calculations involving the solid. Thus the parametric coordinate mesh used in the design system is employed in the subsequent functional analysis and so a link between design and analysis is established.

We have seen that, although the physical solid is mapped onto the unit cube in (u, v, w) space, the range of solids that can be handled by this technique is not limited to hexahedral objects. Furthermore, we have indicated how, an object having more than six discernible faces, with an appropriate choice of parameterisation to amalgamate certain adjacent faces into one composite face, can be put into a 'format' suitable for this method. This is not to say that a gross volume decomposition of a complex object into simpler hexahedral units is completely ruled out. One could also envisage the design of solids proceeding in a manner whereby a complex object was built by progressively adding 'exotic' (although perhaps topologically simple) PDE solids together. Unlike conventional systems, where data from the geometric modeller must be processed before analysis can be carried out, the approach outlined here avoids any such hiatus and allows analysis to be linked directly to the geometry. This will greatly facilitate the process of design which in many cases will start with the object's surface, and pass from there to a parametrisation of the enclosed volume and subsequent analysis. Furthermore, it is hoped that an analysis of the object's physical properties will provide the basis for the same sort of automated design process that has already been carried out for the surface design of some objects [15].

A final application of this method for solid modelling is its use in layer manufacturing techniques such as StereoLithography. In such techniques an object is fabricated by building it up layer by layer—in the case of SLA from a bath of polymer which is irradiated by laser light. The relevance of the PDE approach is that a solid model of an object is naturally divided up into layers by the three families of isoparameteric surfaces within the oject. It should be possible to use these as the layers from which SLA builds up an object.

Acknowledgements

The authors would like to thank M. Susan Bloor for commenting upon the manuscript, and acknowledge the SERC for financial support.

References

[1] Piegl, L.: Key Developments in computer-aided geometric design. CAD *21*, 262–273 (1989).
[2] Requicha, A. A. G., Voelcker, H. B.: Solid modelling: A historical summary and contemporary assessment. IEEE Computer Graphics and Applications *2*, 9–24 (1982).
[3] Requicha, A, A. G., Voelcker, H. B.: Solid modelling: A historical summary and contemporary assessment; IEEE Computer Graphics and Applications *3*, 25–37 (1983).
[4] Bloor, M. I. G., Wilson, M. J.: Generating blend surfaces using partial differential equations. Computer Aided Des. *21* (3), 165–171 (1989).
[5] Bloor, M. I. G., Wilson, M. J.: Blend design as a boundary-value problem. In: Straßer, W., Seidel, H.-P. (eds.) Theory and practise of geometric modeling, pp. 221–234. Berlin, Heidelberg: Springer 1989.
[6] Bloor, M. I. G., Wilson, M. J.: Using partial differential equations to generate free-form surfaces. Comput.-Aided Des. *22*, 202–212 (1990).

[7] Thompson, J. F., Warsi, Z. U. A., Mastin, C. W.: Boundary-fitted coordinate systems for numerical solution of partial differential equations—A review. J. Comp. Phys. 47, 1–108 (1982).
[8] Thompson, J. F.: A General three-dimensional elliptic grid generation system on a composite block structure. Comp. Methods in Applied Mech. and Eng. 64, 377–411 (1987).
[9] Thompson, J. F.: Some current trends in numerical grid generation. In: Morton, K. W., Baines, M. J. (eds.) Numerical methods for fluid dynamics III, pp. 87–100. Oxford: Clarendon Press.
[10] Bloor, M.I.G., Wilson, M.J.: Generating N-sided patches with partial differential equations. Earnshaw, R. A., Wyvill, B. (eds.) New advances in computer graphics, pp. 129–145. Berlin, Heidelberg: Springer 1989.
[11] Wordenweber, B.: Finite-element analysis for the naive user. In: Pickett, M. S., Boyse, J. W. (eds.) Solid modeling by computers, pp. 81–102. Plenum Press 1984.
[12] Cavendish, J. C., Field, D. A., Frey, W. H.: An approach to automatic three-dimension finite element mesh generation. Int. J. Num. Meth. Eng. 21, 329–347 (1985).
[13] Zauderer, E.: Partial differential equations of applied mathematics. New York: Wiley 1983.
[14] Schroeder, W. J., Shepard, M. S.: A combined octree/delaunay method for fully automatic 3-D mesh generation. Int. J. Num. Meth. Eng. 29, 37–55 (1990).
[15] Lowe, T., Bloor, M. I. G., Wilson, M.J.: Functionality in blend design. CAD, 22, 655–665 (1990).

M. I. G. Bloor
Department of Applied Mathematical Studies
University of Leeds
Leeds, LS2 PJT
United Kingdom

Computing Suppl. 8, 43–57 (1993)

Featuremodelling with an Object-Oriented Approach

W. Brandenburg and **B. Wördenweber,** Lippstadt

Abstract. The notion of 'features' is an integral part of the design methodology for vehicle lights at Hella and the system implementation supporting it. Object-oriented mechanisms are utilised to define part structure and behaviour models. The mechanism is part of an 'application specific modelling platform' or 'framework', which allows application software to interact with the CAD/CAM system on a high-level and to capture, evaluate and store the design intent.

The paper describes the modelling approach which leads to the use of features. It outlines the implementation of features in an existing CAD/CAM system using an object-oriented paradigm. The paper finally illustrates feature-based modelling techniques in industrial application and lists practical advantages.

Key words: Feature, object-oriented programming, modelling.

1. Platform for Application Specific Modellers

The concept of 'feature modelling' is part of the design methodology used at Hella to develop vehicle lighting systems. The first chapter provides historical overview and leads into design, implementation and use of 'feature modelling'.

In 1989 Hella was desperate for more intelligent modelling tools for the following reasons:

i. Hella had begun technical computing in 1971 and since then invested a good deal of company know-how in proprietary software for technical computation, simulation and engineering support,

ii. Hella had changed from a proprietary modelling system [9] to commercially available CAD/CAM software and

iii. did not want to reinvest the man-decades of programming to reimplement the company know-how.

The first approach was to establish neutral interfaces for interaction, graphics and data storage/retrieval as a lifeline between the CAD/CAM system and the proprietary application software (Fig. 1). This approach was successful in so far as many application programs were then grafted onto the CAD/CAM system with a minimum amount of modification once the interfaces were established. It did, however, not work for all application software.

A number of application programs used inside knowledge of design or manufacture to simplify the geometric complexity of parts and behave seemingly more intelli-

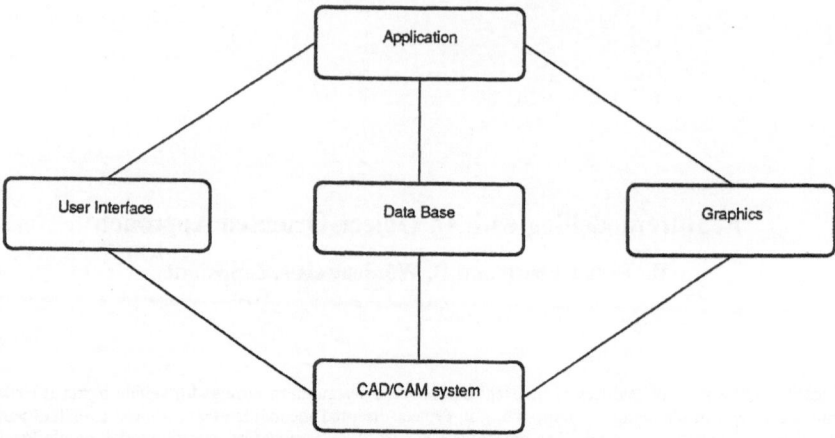

Figure 1. Unintelligent graft onto system software

gently to the user. In fact, the application software dealt in terms of 'features' which:

- linked geometry with engineering function,
- incorporated knowledge of manufacturing technology,
- were available to undergo consistency checks, validation or simulation and
- could exchange and elaborate their representation from implicit to explicit geometry.

The second approach (Fig. 2) permitted high-level modelling information to be communicated, evaluated and stored. It required a more powerful interface between application and CAD/CAM systems software capable of handling structure defini-

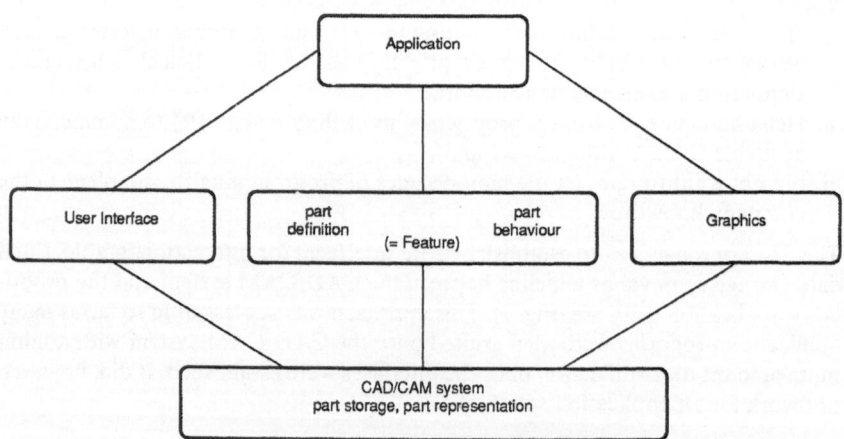

Figure 2. Intelligent graft onto system software

tions and part dependent behaviour. The choice of an 'object-oriented mechanism' for feature representation became apparent.

2. Design Methodology and 'Features'

A conventional headlight consists of light source, reflector and refraction lens. Modern products use free-form surface reflectors and projection lenses to become increasingly compact and powerful. Tail- and indicator lights principally contain the same basic optical elements but are often combined with colour filters, fresnel and other lenses for stylistic or optical reasons (see figure).

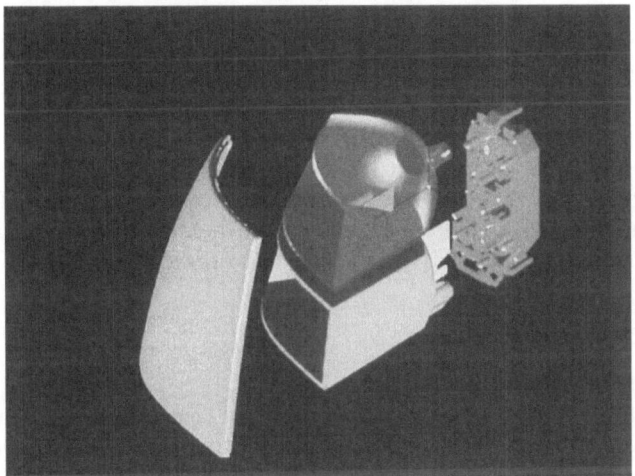

Figure 3. Exploded assembly of a typical Hella product

Vehicle lights have become important safety devices. A headlight for example should illuminate the road in fore- and farfield for normal driving without blinding oncoming traffic. A brakelight should show a large and evenly lit signal surface and be visible from a wide range of directions. The optical properties of head- and taillights are typical examples of **Function-Features** of the assembled product, which in turn are composed out of the function-features of the individual components such as light source and reflector.

The function-features describe functional properties of a product or its components. **Manufacturing-features** mark properties relevant for production and assembly. For example, an optic for a taillight is made up of narrowly spaced rows of refraction lenses. The resulting large matrix (see figure) is milled out of an injection moulding die.

Storing and manipulating function and manufacturing features within the CAD-model enables us to simulate the function of complex assemblies and plan their

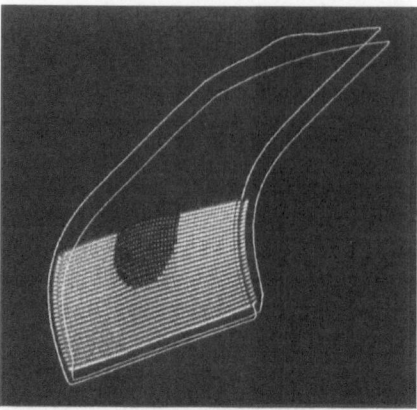

Figure 4. Optic elements of a refraction lens surface

manufacture with greater certainty. A feature relates to one or more topological entites of the CAD-model and brings them into a semantic context. This article describes how features are implemented using the tools of object-oriented programming. A practical application is illustrated and the advantages listed (see also [10]).

3. The Requirements for a Feature Tool

Programs for modelling of a product feature must be available to the designer for use in his usual working environment, the CAD system. For this reason, the tool used to develop these kinds of programs must be **integrated** within the CAD system, and aspects of the system like programming languages and data store, should be considered and used.

The next hurdle for such a programming tool is to overcome the conditions imposed on it by the design of modelling programs in general. One job of a modelling program is to define, according to **design rules**, which feature and which structure will be used to describe the product. Between features and within individual features there exist **dependencies** like equal length, tolerance for mating parts or limits.

Naturally, besides the design rules and dependencies, the feature itself must be described. A **prototype** defines the attributes and construction of a feature, like length, materials and the part hierarchy. Additionally, the prototype has associated functions which define the behaviour of the feature. The behaviour describes the processing of requests such as geometric conversion or part list retrieval. Features are, at the time of design, not completely describable and, for that reason, they are continually being adapted. This should not affect existing programs and data.

Features are often variants of other parts (e.g. standard part libraries) and should be **derivable** from them.

The object-oriented (abbreviated "oo") answers to these requirements are:

Integration into the CAD system → oo-data design in the CAD database

Design Rules	→ rules, active values/demons and methods
Dependencies	→ delegation/propagation, active values/demons and constraints
Feature-Prototype ·	→ objects, classes, methods and dynamic bindings
Feature-Derivation	→ class hierarchy and inheritance

The following chapter will give more information on these 'oo-answers' in oosCA.

4. The Tool oosCA

The system oosCA (object-oriented shell on CATIA database) is a pragmatic approach which enables the use of object oriented data design and programming techniques for the development of modelling programs within the CATIA system, an environment which itself is not tailored for oo-systems. oosCA is realised on an IBM mainframe and integrated in CATIA as a programming shell on top of the access routines (CATGEO [3]) for the geometric kernel of CATIA (see also [5]). To program in oosCA, FORTRAN, C or any other procedural language can be used.

With that approach, one may flexibly define the behaviour and structure of all CAD objects in CATIA and, independently of the application program, may modify and extend these objects. This applies to **all** CAD objects from the simple native CATIA elements like LINE, SURFACE and SOLID to new objects defined by the user (e.g. features) (see also[7]).

oosCA is part of HooD (Hella's object-oriented Data and Programming Environment), a tool family for working with objects in heterogeneous object spaces. An object space is defined as a collection of objects in different oo-systems, oo-tools or tasks (e.g. CAD, Dialog, Graphic, DB, etc.). In the context of HooD, cosCA is one object space. Through the HooD interface (using common message passing), oosCA objects are accessed and may access foreign objects. This allows for crossing of system boundaries. This conversation is realized by, for example standardized message keys and basic system functions with HooD.

4.1 Object-Oriented Functionality in oosCA

4.1.1 Objects

The base unit in oosCA is the object. An object is identified by a system created ID and consists of a sequence of properties. It has an inner state which is defined by the values associated with its properties. Requests to and operations on an object may affect this inner state or may trigger requests/operations on other objects, all of which are defined by the object's behaviour.

4.1.2 Classes: Instance_of Relation

Useful properties and behaviour are collected in classes, and objects are called instances of these classes. Each object is an instance of a single class (**instance_of relation**), with the inner state serving to distinguish between these instances. The object's behavior is defined by the methods of the class to which they belong.

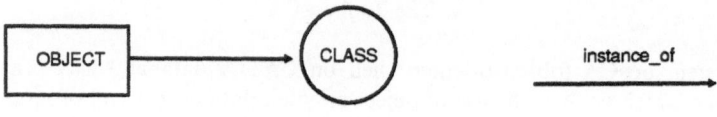

Figure 5. Object/class relation

4.1.3 Class Hierarchy: Is_a Relation

One may build subclasses which inherit properties and methods from their super-class. Inside the subclass, one may add new properties and methods or overwrite those defined in the superclass. This allows a stepwise specification of object prop-erties and behaviour in class hierarchies. A class is a specification of its superclass (**is_a relation**). Instances of a class contain not only the properties and methods of their class, but all properties and methods of the superclass as well. To simplify implementation, classes are allowed to have only one superclass (single inheritance). The root of oosCA's class hierarchy is the predefined class **OBJECT**.

4.1.4 Metaclass

In oosCA classes are treated in the same way as objects. They are accessible by messages and have their own inner state and behaviour, and classes are merely instanced to form metaclass/class relationships. The class object is described with class properties and class methods, the same as any other object.

Like a shadow to the class hierarchy the metaclasses build a hierarchy which uses the predefined class CLASS as it's root. And again, each metaclass is an instance of the single class METACLASS. The top classes are similar to those used in Smalltalk [6].

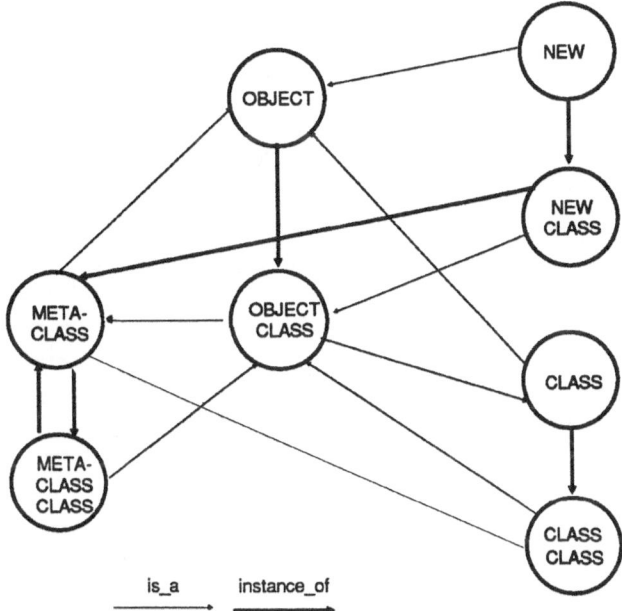

Figure 6. Predefined top of system classes

4.1.5 CATIA-Native Objects

Normal CATIA-elements are treated exactly as other objects are and for each native CATIA element a class is predefined. Their inner structures are predefined (CATIA math.block) and their behaviour is freely redefinable.

4.1.6 Requests/Operation: Messages, and Methods

Each request to, or operation on an object is done by sending a message to that object. This message usually consists of a selector and an optional argument(s). The sender may have values returned to it through arguments. The method filter of the class then processes the selector and an associated method body is looked up and executed. The method body is written procedurally, usually in FORTRAN or C. The object which received the message is called **"self"**.

4.1.7 Inheritance Technique

An object's structure and behaviour are defined in the object's class. If there is no method- or property-description in that class, inheritance takes over, and the information is looked up in the superclasses along the class hierarchy.

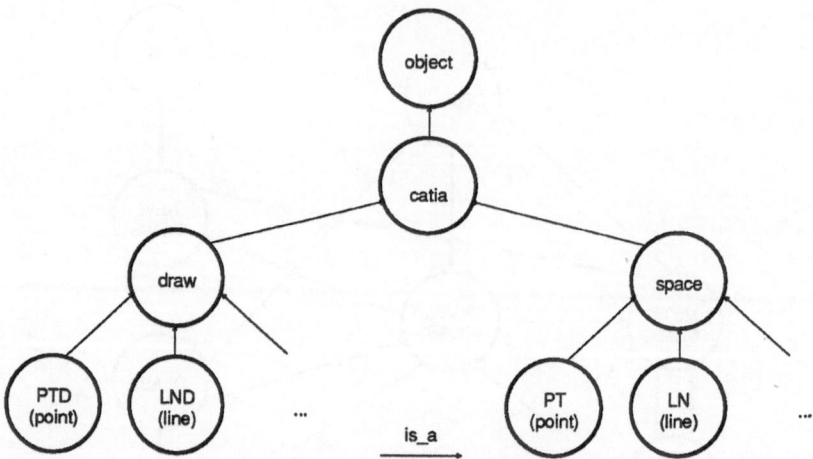

Figure 7. Predefined CATIA-native classes

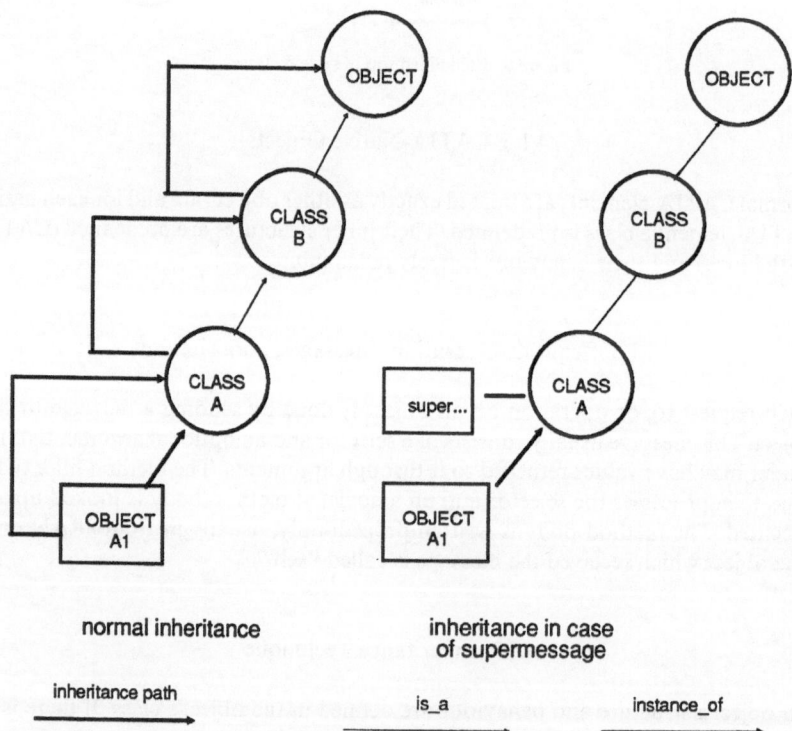

Figure 8. Inheritance technique

4.1.8 Supermessage

The class where the currently executing method was found is known as **"owner"**. If from within a method body a so called supermessage is sent, the method search starts at the superclass of "owner". The "self" object remains the same. This mechanism allows a subclass which overwrites a method to trigger the "old" method from within the "new" extended method. (See also Smalltalk [6]).

4.1.9 Properties: Client_of Relation

A property is identified by a property key (selector) which is associated with a property value (PV).

Usually property values are object identifiers (or lists), allowing an object to use other objects to set up its own inner state (**client_of relation**). An assembly of parts, for example, may be defined with client_of relations (part structure).

This kind of property value may be restricted to a property value class (PVC). In this case, only an instance or subclass instance of the given class may be used as property values.

For more flexible processing, name data, instead of (numerical) object identifiers, may be used, since the data types INTEGER, REAL or STRING are possible.

4.1.10 Property Access

Within a method body only properties of the local object "self" can be accessed. Properties of other objects must be accessed through their object's methods (data hiding). As an exception to this rule, class methods (see metaclass) may access the properties of their instances directly. In Smalltalk this is usually done by overwriting the "self"-variable. The binding of "self" PVs to a method, which is the way the inner state of an object is accessed by a method body, is done in a FORTRAN or C program via a function call.

4.1.11 Property Description: Defaults, Restrictions

If, at initialisation, a property value remains undefined, default values are used. Defaults are stored in the class description. As an additional restriction, minimum and maximum length of a property value list may also be set.

4.1.12 Delegation/Propagation

Some important features of an oo-system include delegation of operation and property value access, and the propagation of the property values themselves. In the first case the information is located elsewhere in the object hierarchy (an object accesses the information through requests to other objects) and in the second case the information is immediately propagated to other objects. In either case, information may be transformed.

The normal inheritance technique (delegation to superclasses) can be switched over to a client_of relation by modifying delegation rules.

Relations between properties may be defined by delegation/propagation. Simple relationships are **coreferences**, where several different properties have the same value. More complex are **dependencies** and **constraints** between properties, which must include a set of rules to determine valid values. Delegation of requests may be done in oosCA by generating a message to other objects.

Property delegation and propagation are achieved through "active" properties, with attached demons which are automatically triggered when property values are accessed. A demon is triggered by a message to the "self" object. The two forms of activation are:

a) activation on get access "GET_ACTIVE" which allows delegation.
b) activation on set access "SET_ACTIVE" which allows propagation.

4.2 Programming Technique and Implementation in CATIA

Classes defined by a class description language and converted to Data Structures are statically linked into the Application Program. Objects are stored in the CATIA-datastore and are handled through CATGEO-routines (CATIA Data Access [3]) which take care of reference integrity and the generation of Object ID's (surrogates).

Methods are traditionally programmed in FORTRAN or C and triggered by message passing through calling a message handler routine. The property access, including restriction checks and demon trigger is done by SET/GET-routines.

5. The Reflector Modeller REFMOD

For headlight or taillight design, the automotive manufacturer predefines the outer surfaces and the maximum extents of the lighting assembly. In order to optimize the lighting characteristics using these guidelines, the reflector is divided into sectors with different reflection surfaces. These so-called "step-reflectors" often have very complex configurations of reflection surfaces.

The objective was to create a modelling program which allowed the step reflector to be defined interactively in CATIA on a logical and very high level and the information to be stored within the CAD system for later extraction for manufacture or lighting simulations. To meet this objective, a class structure had to be defined with which all possible reflectors may be generated and operations like calculation of luminous flux or face generation are allowed.

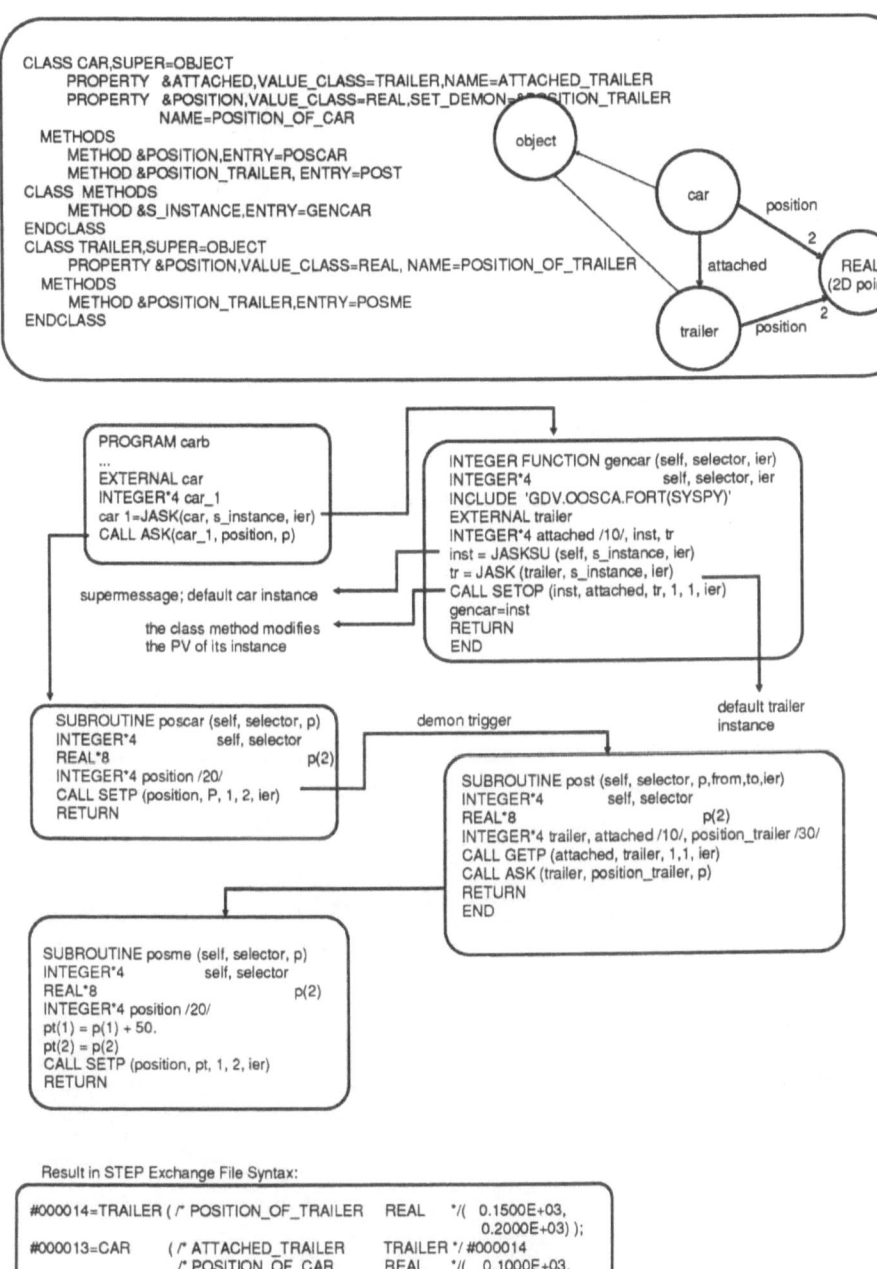

Figure 9. Example

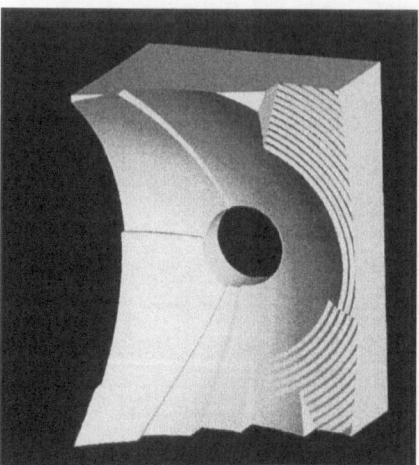

Figure 10. A tail light reflector

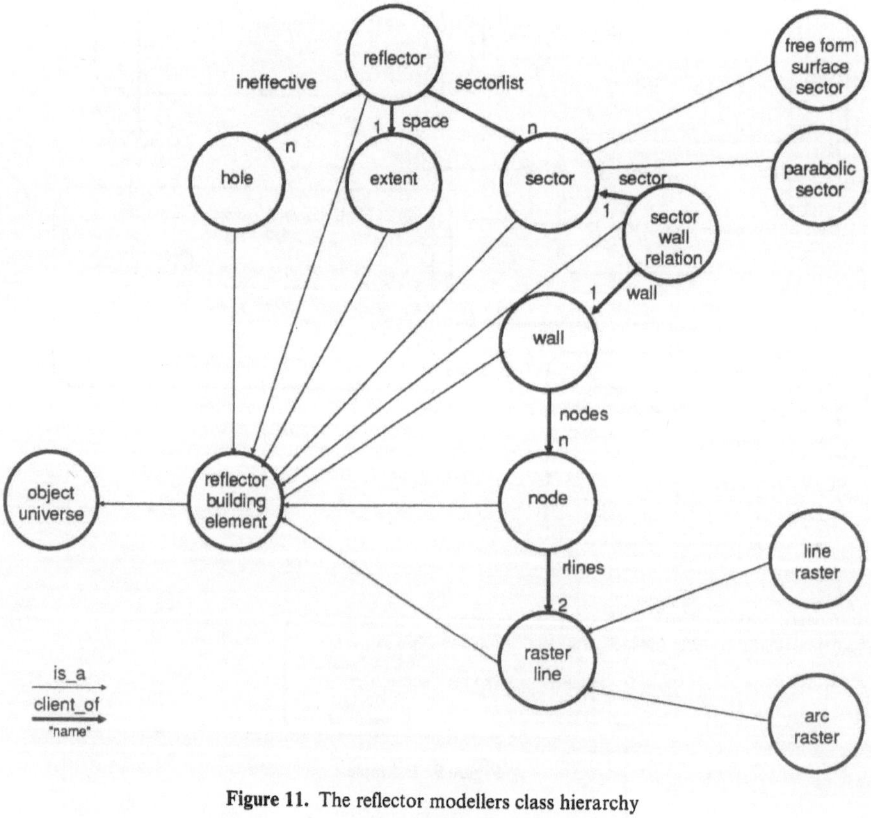

Figure 11. The reflector modellers class hierarchy

The class structure shows the structure of a reflector which is made up of sectors, housing extents and bulb sockets. The sector is described by bounding walls and an interior surface. The bounding walls sit on nodes which are described by crossing raster lines (in this case, raster lines lie on a polar grid).

Within the interactive dialogue, all objects which are used to generate the reflector may be instanciated stepwise. First, raster lines are created on which the walls are placed and connected to each other. Sectors are created by indicating an area. Next, a search algorithm collects all walls which form the boundary around the indicated area. For each sector, a reflection surface must be chosen. All generated sectors are associated with the current reflector. The housing extents and the bulb sockets are

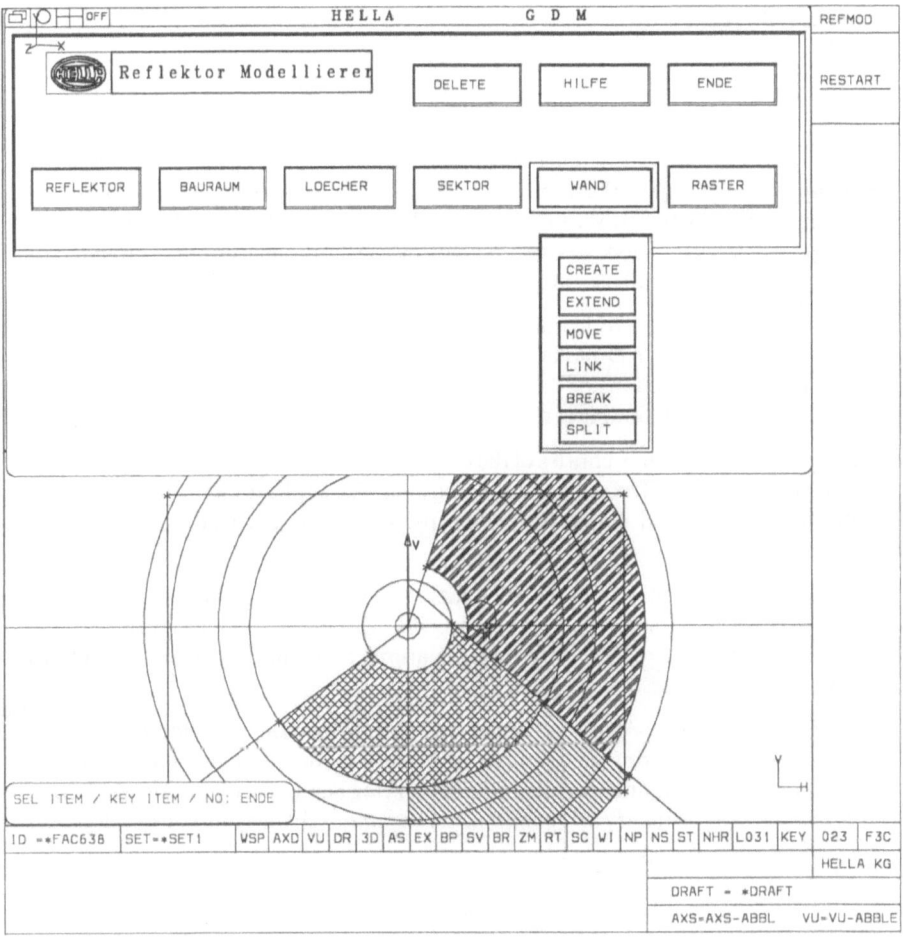

Figure 12. Snapshot of a modelling session

associated similarly. A 2D representation graphic shows the actual development state.

Each operation is based on message exchange between individual objects. In the following, we will show this with an example.

For later work on the reflector, a complete boundary representation (B-Rep) must be derived. For each sector and each step between the sectors, a face (a trimmed surface) must be generated. When the reflector gets a request to generate faces, it delegates this request to the sectors and sector bounding walls. The sector asks the bounding walls for their step surfaces with which it cuts its own reflection surface. Next, the walls ask all sectors which use the wall to report the reflection surface in order to cut their own step surface with these surfaces. Finally, the reflector receives all of the generated surfaces, bundles them, fits them to the housing extents and cuts out the bulb socket.

6. Conclusion

The implementation of the reflector modeller alone proved the importance of the new approach. Despite the need to develop an object-oriented shell within an existing CAD-system, the set timeframe for building the modeller did not have to be adjusted. The function and manufacturing features encorporated provide advantages beyond a database interface.

By attaching properties to features it is now possible to hide information behind the geometry of the CAD-model. The information may concern functional properties, tolerance definitions or data protection. The CAD-model carries and maintains the information until it is required by further programs.

Features relate modelling entities of the CAD-system. These relations present a new, "semantic topology". The semantic topology enables application programs to evaluate the relations and derive further properties or use the feature inherent behaviour. [1].

It is only too easy today to reach the limits of geometric modelling when one tries to model real life objects. It seems impossible to imitate woodgrain or the perforation of, say, a sieve. Only by using features as a shorthand are we able to model these properties.

The above advantages play an important role in product development today. Simultaneous engineering demands that the partners in a development are able to define components simultaneously but with varying degrees of detail. Using feature-based modelling, geometry and components become exchangeable and developable.

References

[1] Anderson, D. C., Chang, T.C.: Automated process planning using object-oriented feature based design, International GI-IFIP Symposium 89 Berlin, FRG.

[2] Barth, P. S.: An object-oriented approach to graphical interfaces. ACM Transaction on Graphics 5 (2), 142–172 (1986).
[3] CATIA Base Geometry Interface Reference Manual, Dassault Systemes 1988.
[4] Chambers, C., Unger, D., Lee, E.: An efficient implementation of SELF, a dynamically-typed object-oriented language based on prototypes, Proceedings of OOPSLA'89.
[5] Dietrich, W. C., Nackman, JR., L., Gracer, F.: Saving a legacy with objects, Proceedings of OOPSLA'89.
[6] Goldberg, A., Robson, D.: Smalltalk-80: The language and its implementation. Reading, MA: Addison Wesley 1983 (Series in Computer Science).
[7] Moreau, P.: Applying object-oriented architecture to industrial geometrie modelers, GI-IFIP Symposium 89 Berlin, pp. 365–374.
[8] Rathke, C., ObjTalk.: Repräsentation von Wissen in einer objekt-orientierten Sprache, Dissertation, Universität Stuttgart, 1986.
[9] Roth, F. K., Menke, H.: Entwerfen und Gestalten am Bildschirm. VDI-Berichte 261, 79–87 (1973).
[10] J. J. Shah, M. T. Rogers,: Expert form feature modelling shell. Computer Aided Design 20 (9), 515–524 (1988).

W. Brandenburg
Technische Datenverarbeilung
CAD/CAM, Abteilung GDV-E
Hella KG
Postfach 2840
Rixbeckerstraße 75
D-W-4780 Lippstadt
Federal Republic of Germany

Computing Suppl. 8, 59–73 (1993)

Computing
© Springer-Verlag 1993

Best Approximations of Parametric Curves by Splines

W. L. F. Degen, Stuttgart

Abstract. A normal distance $d_N(C, C')$ for pairs of plane parametric curves C, C' (with given C and some restrictions for C') is introduced. For certain classes of Bézier curves $\mathscr{B}'$ we get a differentiable manifold of deviation functions and can apply the non-linear approximation theory. As main result we obtain that the error of the best approximant $C_0' \in \mathscr{B}'$ has the alternant property like in the case of Chebyshev approximation. An algorithm to calculate C_0' explicitly and some examples are included.

Key words: Parametric curves, best approximation, normal distance, alternant theorem, Remcz-like algorithm.

1. Introduction

In CAGD the question of how to approximate a given curve within a certain tolerance by a spline of a prescribed type arises frequently and various error estimates have been developed ([De Boor/Höllig/Sabin '87], [Hoskins/Ponzo '74], [Hanna/Evans/Schweitzer '86], [Sakai/Usmani '90]). In particular, it may be required to approximately represent the *offset* of a spline which is, in general, not a spline, or to *convert* a given spline of another kind, say of some higher degree or a rational one, into one of a more restricted class at disposal ([Dannenberg/Nowacki '85], [Hoschek '87])

In this paper, we concentrate on Bézier curves as approximants, however we neither cover the problems of what number of segments is necessary, nor how the subdivision points have to be distributed. Rather we focus on the subsequent problem: to approximate a single segment in the best possible way.

At least at this stage the need arises to measure the rate of approximation. It would be tempting to use the well known Hausdorff distance for this purpose. However, being based only on the point sets, it does not reflect the parameter distributions of the two curves C, C'. Therefore, we introduce the *normal distance $d_N(C, C')$* defined for a certain class $\mathscr{A}(C)$ of "admissible" curves for C (section 2.3). This notion not only overcomes the disadvantages of the Hausdorff distance $d_H(C, C')$ but provides (besides much better convergence properties) simultanously a *deviation function $\rho: [a, b] \to \mathbb{R}$* for each $C' \in \mathscr{A}(C)$, connected with $d_N(C, C')$ by $d_N(C, C') = \|\rho\|$ (maximum norm). We will show that the normal distance is related to the Hausdorff distance by the inequality $d_H(C, C') \le d_N(C, C')$ (proposition 3) and equality holds under an additional condition (theorem 1).

Denoting by $\mathscr{B}$ a certain set of Bézier curves with fixed degree n and specified end conditions and restricting to admissible ones ($\mathscr{B}' := \mathscr{B} \cap \mathscr{A}(C)$), we get a set $\mathscr{M}$ of corresponding deviation functions. We will show (theorem 2) that $\mathscr{M}$ is a differentiable manifold of dimension m (depending on $\mathscr{B}$) and that each $\rho \in \mathscr{M}$ is differentiable.

By this approach, the approximation problem for parametric curves is reduced to one for functions: to select $\rho \in \mathscr{M}$ in such a way that $\|\rho\|$ is minimal. Just at this point the theory of non-linear approximation, which has made great progress in the last decades (see e.g. [Braess '86]), in particular the theory of [Meinardus/Schwedt '64] can be applied. Thus we obtain the result (theorem 4) that the deviation function ρ of the best approximation behaves as in the case of classical Chebyshev approximation, expressed by his famous alternant theorem.

The application of that theory requires to prove its assumptions: the global and the local Haar conditions. As to the former, we gave the proof for some special class $\mathscr{B}$. The general case would be implied by a conjecture on Bézier curves. E. Eisele [Eisele '91], an assistant of mine, succeeded in proving the general global Haar condition (at least for some neighborhood). The local Haar condition is not so difficult, so a proof is included in this paper.

In the final paragraph we will develop an algorithm to compute explicitly the best approximant. It is based directly on the alternant property written as a non-linear system of m equations which will be solved by a quasi Newton method (replacing the Jacobian by a finite difference approximation). Though this method requires some efforts in computing, it yields highly accurate results. Some examples confirm the theory in all details and show that the algorithm is suitable for practical computations.

2. Distances in the Set of Plane Parametric Curves

2.1 Differentiable Parametric Curves in the Euclidean Plane

We denote by $\mathscr{C}^1$ the set of all compact parametric differentiable curves in the Euclidean plane $\mathbf{E}^2$. As usual in differential geometry (see e.g. [Do Carmo '86]) a *curve* $C \in \mathscr{C}^1$ is an equivalence class of parametric representations $\mathbf{x}: [a, b] \to \mathbb{R}^2$. Hereby $\mathbf{x}$ denotes a differentiable mapping of some compact real interval $[a, b]$ ($a < b$) into $\mathbb{R}^2$ (with tangent vector $\mathbf{x}'(t) \neq \mathbf{0}$ for all $t \in [a, b]$); furthermore two such parametric representations are called *equivalent* (giving rise to the same curve C) if there is a diffeomorphism $\phi: [a, b] \to [c, d]$ between their parameter intervals (with $\phi' > 0$ anywhere), called *parameter transformations*. A *geometric object or property*, described in terms of a special parametric representation, must be invariant under parameter transformations.

2.2 The Hausdorff Distance

In order to supply the family of all compact sets of an Euclidean space $\mathbf{E}^n$ with a metric and making this family of point sets into a metric space, Hausdorff introduced [Hausdorff, 1914] the following definition of a *distance* between two compact sets $A, B \subset \mathbf{E}^n$ which is now called the *Hausdorff distance*:

$$d(A, B) := \max \left(\max_{p \in A} \left(\min_{q \in B} d(p, q) \right), \max_{q \in B} \left(\min_{p \in A} d(p, q) \right) \right) \qquad (2.1)$$

($d(p, q)$ denoting the Euclidean distance between the points p and q). This notion has a great importance in the theory of convex sets which is based on Blaschke's famous selection theorem [Blaschke, 1914] saying that the family of all compact, convex sets in $\mathbf{E}^n$ is sequentially compact.

Yet it can serve as a metric in $\mathscr{C}^1$ as well, taking for A and B the point sets $\{C\}, \{D\}$ of two parametric curves; it may easily be expressed by two parametric representations for C, D in the form

$$d_H(C, D) = \max \left(\max_{t \in [a,b]} \min_{s \in [c,d]} \|\mathbf{x}(t) - \mathbf{y}(s)\|, \max_{s \in [c,d]} \min_{t \in [a,b]} \|\mathbf{x}(t) - \mathbf{y}(s)\| \right). \qquad (2.2)$$

Nevertheless, this Hausdorff distance is not an appropriate means to deal with parametric curves because it does not reflect the parameter distribution nor the orientation. Furthermore, there is no convergence property similar to the Blaschke's one: It is easy to construct a sequence $(C_n)_{n \in \mathbb{N}}$ $C_n \in \mathscr{C}^1$ such that $\lim_{n \to \infty} d_H(C_n, C) = 0$ but, for no reparametrization ϕ, $\mathbf{x}_n(t)$ converges pointwise to $\mathbf{x}(\phi(t))$.

2.3 The Normal Field and Admissible Curves

To avoid these disadvantages of the Hausdorff distance, we suggest another approach which is more appropriate to approximate parameter curves. It is based on two main ideas: First, we start from a *fixed* curve C, given by any of its parametric representations $\mathbf{x}: [a, b] \to \mathbb{R}^2$, to be approximated. Second, we remove from a certain (topological) neighbourhood of C those curves that are "undesired". Calling the remaining ones "admissible" and denoting the set of all admissible curves with respect to C by $\mathscr{A}(C)$, it is sufficient to define the distance $d(C, C')$ for all $C' \in \mathscr{A}(C)$. To give an idea of what we called "undesired", we mention curves $\tilde{C}$, lying in a neighbourhood of C (even very close to C), but going back and forewards, including the case of loops.

To be more precise, we assume C being at least of class C^2. Then we associate with each point $\mathbf{x}(t) \in C$ that part of the *normal* $\{\mathbf{x}(t) + \rho\mathbf{n}(t)|\rho \in \mathbb{R}\}$ for which $\rho\kappa(t) < 1$, (ρ unrestricted if $\kappa(t) = 0$) $\kappa(t)$ being the curvature of C and $\mathbf{n}(t)$ the normal unit vector (with such an orientation that $\det(\mathbf{x}'(t), \mathbf{n}(t)) > 0$). Then the region, swept out by these parts of the normals (in general half rays, for inflection points full lines), is called the *normal field* of C and will be denoted by $\mathscr{N}(C)$.

From the Frenet equations one derives immediately that the mapping $(t, \rho) \mapsto$ $\mathbf{x}(t) + \rho\mathbf{n}(t)$ from $\{(t, \rho) \in \mathbb{R}^2 | t \in [a, b], \rho\kappa(t) < 1\}$ onto $\mathcal{N}(C)$ is locally one-to-one, i.e. this mapping is a regular parametrization of $\mathcal{N}(C)$.

Now we can explain the precise meaning of "admissible":

Definition 1. A curve $C' \in \mathcal{C}^1$ is called *admissible* with respect to C if the following conditions hold (C' represented by y: $[c, d] \to \mathbb{R}^2$):

1. For each $s \in [c, d]$ the curve point $\mathbf{y}(s)$ is contained in $\mathcal{N}(C)$; i.e. real values $t \in [a, b]$ and ρ exist which satisfy the equations:

$$\mathbf{y}(s) = \mathbf{x}(t) + \rho\mathbf{n}(t) \tag{2.3}$$

$$\rho\kappa(t) < 1. \tag{2.4}$$

2. For each $t \in [a, b]$ there is exactly one $s \in [c, d]$ for which (2.3) is satisfied.
3. For corresponding values s, t in (2.3), the tangent vector $\dot{\mathbf{y}}(s)$ is not parallel to $\mathbf{n}(t)$.
4. The end points of C and C' are the same, i.e. $\mathbf{x}(a) = \mathbf{y}(c)$, $\mathbf{x}(b) = \mathbf{y}(d)$.

Denoting by $\mathcal{B}$ a given set of Bézier curves with fixed degree n and end conditions specified, our approximation problem consists of selecting from $\mathcal{B}' := \mathcal{A}(C) \cap \mathcal{B}$ that one (if it exists), say C_0, for which $d(C, C_0) = inf\{d(C, C')|C' \in \mathcal{B}'\}$.

The admissibility conditions (no. 1 to no. 4) have further consequences which we will investigate now. Obviously, by (no. 1) and (no. 2), the assignment $t \mapsto s$ defines a *function* σ: $[a, b] \to [c, d]$, $s = \sigma(t)$ in (2.3). Since $\mathbf{y}(s)$ is unique and $\mathbf{n}(t) := \vec{0}$, ρ is unique too. Thus, there is a second function ρ: $[a, b] \to \mathbb{R}$ (with values $\rho(t)$ such that $\rho(t) \cdot \kappa(t) < 1$) satisfying

$$\mathbf{y} \circ \sigma = \mathbf{x} + \rho \cdot \mathbf{n}. \tag{2.5}$$

On the other hand, the values s and ρ can be thought to be the solution of (2.3) as a system of two equations (nonlinear with respect to s) for the unknowns, s, ρ. Its functional determinant is $det(\dot{\mathbf{y}}(s), \mathbf{n}(t))$, (using a dot to indicate derivation with respect to s) hence, by (no. 3), not vanishing. Then the implicit function theorem implies that σ, ρ are *differentiable*. Furthermore, by differentiation of (2.5), we obtain

$$\dot{\mathbf{y}}\sigma' = (1 - \rho\kappa)\mathbf{x}' + \rho'\mathbf{n}. \tag{2.6}$$

By (2.4) the vector at the right hand side of (2.6) does not vanish ($\mathbf{x}'$, $\mathbf{n}$ being linearly independent) and we conclude $\sigma'(t) \neq 0$ for any $t \in [a, b]$. Since by (no. 4), $\sigma(a) = c$, $\sigma(b) = d$, we have $\sigma'(t) > 0$ for all $t \in [a, b]$, hence σ is a *diffeomorphism* and therefore (2.5) is an equivalent representation for C'. Thus we have proved:

Proposition 1. *For any admissible curve C' with respect to a given curve C (of class C^2) exist two differentiable functions σ, ρ: $[a, b] \to \mathbb{R}$ such that σ is an orientation preserving diffeomorphism and (2.5) is an equivalent representation for C'.*

2.4 The Deviation Function and the Normal Distance

The principal idea of the representation (2.5) can now be seen at once. The function ρ measures the distance of corresponding points of C and C' directly:

$$|\rho(t)| = \|\mathbf{y}(\sigma(t)) - \mathbf{x}(t)\|, \, t \in [a, b]. \tag{2.7}$$

So ρ can serve as an error function for the approximation and we can define:

Definition 2. For any admissible curve C', we call ρ the *deviation function* of C and

$$d_N(C, C') := \max_{t \in [a, b]} |\rho(t)| \tag{2.8}$$

the *normal distance* of C' from C.

These notions of admissible curves and their normal distances yield much more than only to exclude some "undesired" aspirants for approximation. From (2.7) one derives immediately:

Proposition 2. *Let* $(C_n)_{n \in \mathbb{N}}$ *be a sequence of curves, admissible to* C, *such that* $\lim_{n \to \infty} d_N(C, C_n) = 0$. *Then* C_n *converges uniformly to* C. *Furthermore, this convergence respects the parameter distribution, i.e. for each* C_n *there is a reparametrization* $\sigma_n: [a, b] \to [c_n, d_n]$ *such that the point* $\mathbf{y}_n(\sigma_n(t))$ *converges to* $\mathbf{x}(t)$ *for each* $t \in [a, b]$.

2.5 Relations Between the Normal and the Hausdorff Distances

Let C, C' be two different curves $\mathscr{C}^1$, C of class C^2 and C' admissible to C. We want to compare the Hausdorff distance with our normal distance. Let t_0, s_0 be that pair of parameter values where $d_H(C, C')$ is attained. Certainly $d_H(C, C') > 0$ because otherwise for each point $\mathbf{x}(t)$ of C there would be a point $\mathbf{y}(s)$ coincident with $\mathbf{x}(t)$; this implies $s = \sigma(t)$, $\rho(t) = 0$ because of the uniqueness in (2.5); hence $d_N(C, C') = 0$ and $C = C'$, contrary to the assumption. Observing (4), we conclude that the extremal distance will be attained at an *interior* point of each curve. Differentiating the distance function $F(s, t) := \|\mathbf{x}(t) - \mathbf{y}(s)\|$ we obtain that $\mathbf{x}'(t_0)$ and $\dot{\mathbf{y}}(s_0)$ are both *orthogonal* to $\mathbf{x}(t_0) - \mathbf{y}(s_0)$. Therefore $\mathbf{y}(s_0)$ lies on the normal of C at t_0, hence $s_0 = \sigma(t_0)$, which implies $d_H(C, C') = \|\mathbf{x}(t_0) - \mathbf{y}(\sigma(t_0))\| = |\rho(t_0)| \leq d_N(C, C')$. Thus we have proved:

Proposition 3. *For any pair of curves* C, C', *the first one being of class* C^2 *and the second admissible to* C, *the following inequality holds*

$$d_H(C, C') \leq d_N(C, C'). \tag{2.9}$$

Some simple counterexamples show that equality in (2.9) does not hold in general. The reason is that the normals of C may cover some regions of $\mathscr{N}(C)$ more than once. We exclude this possibility by the following

Definition 3. C is called to have a *non-overlapping normal field* if each of its points is lying on exactly one normal of C (in this case, the parametrization (t, ρ), described in section 2.3 is one-to-one).

With this additional assumption, we can indeed prove the equality of both distances:

Theorem 1. *For any pair of curves C, C', C being of class C^2 and having a non-overlapping normal field and C' admissible for C, the normal distance is equal to the Hausdorff distance.*

$$d_N(C, C') = d_H(C, C'). \tag{2.10}$$

Proof: Without loss of generality, we can assume that the parameter t in our representation $t \mapsto x(t)$, $t \in [a, b]$ is the arc length and consequently the Frenet formulas are valid.

We start again with the special value $t_1 \in [a, b]$ where $\rho(t_1) = d_N(C, C')$ is attained. As in the previous case, t_1 is in the interior of $[a, b]$, hence $\rho' = 0$, which implies by (2.6) that the tangent vectors of C and C' at $x(t_1)$ and $y(s_1)$ (with $s_1 := \sigma(t_1)$) are parallel, thus the joining line being a common normal of C and C'.

Now we consider the circle B around $y(s_1)$ with radius $r_1 := \rho(t_1) = d_N(C, C')$ and investigate the distance function $r(t) := \|x(t) - y(s_1)\|$ of $x(t)$ (running on C) from the middle point $y(s_1)$ of B. From the Frenet formulas one derives $r(t_1) = r_1$, $r'(t_1) = 0$, $r''(t_1) = (1 - \kappa(t_1)r_1)/r_1$. In particular, by (2.4), $r''(t_1)$ is positive. Hence, these equations have the geometric meaning that C is tangent to the circle B at the point $x(t_1)$ and runs outside of B, both for $t < t_1$ and $t > t_1$ in some neighbourhood.

Well, if $d_H(C, C')$ would be less than $d_N(C, C')$, then, by (2.2), there would be a point $x(t_2)$ closer to $y(s_1)$ than $x(t_1)$, and therefore a point $x(t_3)$ with t_3 between t_1 and t_2, where C enters into B. But then there would be t_4 between t_1 and t_3 where $r(t)$ attains its *maximum* $> r_1$. At this point the tangent to C would be parallel to the circle's tangent at the corresponding point (on the line from $y(s_1)$ to $x(t_4)$). This implies that the normal of C at $x(t_4)$ would pass through $y(s_1)$ which contradicts to the assumption contained in definition 3 because we had two different values t_1, t_4 where the normals of C would pass through the same point $y(s_1)$ of C'. ∎

3. Non-Linear Approximation Theory for Bézier Curves

3.1 The Differential Manifold of Deviation Functions

We start with the set $\tilde{\mathcal{B}}$ of regular Bézier curves $\tilde{\mathcal{B}} \subset \mathcal{C}^1$ of some fixed degree n, for which n is the minimal algebraic degree. Then, for each $C' \in \tilde{\mathcal{B}}$, we have a *unique* parameter representation

$$y(s) = \sum_{k=0}^{n} B_k^n(s)\mathbf{b}_k, \qquad s \in [0, 1] \tag{3.1}$$

with the Bernstein polynomials B_k^n of degree n and a sequence of control points $\mathbf{b}_0, \ldots, \mathbf{b}_n$. Comprising the latter to a vector $\mathbf{b} \in \mathbb{R}^{2n+2}$ and excluding the degenerate cases by requiring certain inequalities, we get an open subset $\tilde{D} \subset \mathbb{R}^{2n+2}$ and a bijective mapping $\tilde{\Lambda}: \tilde{D} \to \tilde{\mathcal{B}}$, assigning to each $\mathbf{b} \in \tilde{D}$ the corresponding $C'_b \in \tilde{\mathcal{B}}$

given by (3.1). Thus we have a parametrization which induces a *differentiable structure* on $\tilde{\mathscr{B}}$.

Next we impose some specified end conditions, which must include (condition 4) of definition 1 in each case, i.e.

$$\mathbf{b}_0 = \mathbf{x}(a), \qquad \mathbf{b}_n = \mathbf{x}(b). \tag{3.2}$$

There may be further conditions to be required, e.g. to guarantee a certain order k of contact with C at the end points. We assume that the end conditions reduce the number of free parameters to m with $1 \le m \le 2n - 2$ and that there is an open subset $D \subset \mathbb{R}^m$ together with an injective and differentiable mapping $\psi \colon D \to \tilde{D}$ such that these end conditions are satisfied if and only if $\mathbf{b} = \psi(\mathbf{p})$ with some $\mathbf{p} \in D$.

Composing both mappings and denoting $\bigwedge := \tilde{\bigwedge}|_{\psi(D)} \circ \psi$, $\mathscr{B} := \bigwedge(D)$ we get a differentiable structure $\bigwedge \colon D \to \mathscr{B}$ on our set of regular Bézier curves of minimal degree n with specified end conditions.

Examples. If no further end conditions beyond (3.2) are given, then, of course, we have $m = 2n - 2$ and the $2n - 2$ coordinates of $\mathbf{b}_1, \ldots, \mathbf{b}_{n-1}$ themselves can serve as parameters. The regularity condition can be obtained by $\mathrm{res}(\dot{y}_1(s), \dot{y}_2(s)) \neq 0$ (res being the resultant of the two coordinate polynomials). Thus, for $n = 2$ we get $D = \{\mathbf{b}_1 \in \mathbb{R}^2 | \det(\mathbf{b}_1 - \mathbf{b}_0, \mathbf{b}_2 - \mathbf{b}_1) \neq 0\}$, In the case of tangent cubics we have $n = 3$, $m = 2$; $\mathbf{b}_0$, $\mathbf{b}_3$ by (3.2), $\mathbf{b}_1 = \mathbf{x}(a) + p_1 \mathbf{x}'(a)$, $\mathbf{b}_2 = \mathbf{x}(b) - p_2 \mathbf{x}'(b)$; $D = \{\mathbf{p} = (p_1, p_2) \in \mathbb{R}^2 | p_1 > 0, p_2 > 0, \mathrm{res}(\dot{y}_1(s), \dot{y}_1(s)) \neq 0\}$.

Now, we take the admissibility conditions (see definition 1) into account. Since no. 4 follows from (3.2) and no. 1, and since no. 3 is given by $\det(\dot{\mathbf{y}}(s), \mathbf{n}(t)) \neq 0$, the vector $\mathbf{b}$ satisfying no. 1, 3 and (3.2) build up an open subset of $\mathbb{R}^{2n+2}$. Thus it remains to show that the same is true for condition no. 2 too. If, for an admissible C', in each of its neighbourhoods there would be a curve C'' not satisfying (no. 2), we had at least for one $t_1 \in [a, b]$ a branching from exactly one s for C' to more than one s for C'' in (2.3); but this would imply that C' is tangent to the normal in contradiction to (no. 3). Thus the restriction $\mathscr{B}' = \mathscr{B} \cap \mathscr{A}(C)$ is equivalent to the restriction of $\bigwedge$ to an open subset $M \subset D$:

$$\bigwedge' := \bigwedge|_M \qquad \bigwedge' \colon M \to \mathscr{B}' \tag{3.3}$$

Thus, the differentiable structure (and the dimension m) are maintained.

Using again the implicit function theorem for the solutions σ, ρ of (2.5) not only for one single C' (as in the proof of proposition 1) but now for the *entire set $\mathscr{B}'$ at once*, we get differentiable functions $\hat{\rho}, \hat{\sigma} \colon M \times [a, b] \to \mathbb{R}$ and their restrictions $\rho_{\mathbf{P}}, \sigma_{\mathbf{P}}$ to $\{\mathbf{p}\} \times [a, b]$ for just one $\mathbf{p} \in M$ are the deviation and reparametrization functions, respectively, for the corresponding Bézier curve $C'_{\mathbf{P}} = \bigwedge'(\mathbf{p}) \in \mathscr{B}'$. We can summarize these results as follows:

Theorem 2. *Each set $\mathscr{B}'$ of admissible Bézier curves as described above is equipped with a differentiable structure (3.4). Furthermore, there is a global differentiable function $\hat{\rho} \colon M \times [a, b] \to \mathbb{R}$ such that the mapping*

$$\Gamma \colon M \to \mathscr{M}, \tag{3.4}$$

assigning the deviation function ρ_P of the corresponding Bézier curve C'_P to each $\mathbf{p} \in M$, is the restriction of $\hat{\rho}$ to $\{\mathbf{p}\} \times [a, b]$; therefore $\mathcal{M}$ is a differentiable manifold (with just one chart (3.4)).

3.2 The Application of the Theory of Meinardus/Schwedt

The theory of Meinardus/Schwedt is concerned with exactly the situation which we found in the previous section: Given $\hat{\rho}: M \times [a, b] \to \mathbb{R}$ as in theorem 2 and $f \in C[a, b]$, find $\rho_{\mathbf{p}_0} \in \mathcal{M}$ such that $\|\rho_{\mathbf{p}_0} - f\| = \inf\{\|\rho_{\mathbf{p}} - f\| \,|\, \mathbf{p} \in M\}$. Of course, in our case, f is zero.

Definition 4.

(a) $\mathcal{M}$ is said to satisfy "the local Haar property at $\mathbf{p}$", if the linear tangent space $T_{\mathbf{p}}\mathcal{M}$ satisfies the classical Haar property, i.e. if each function $\tau \in T_{\mathbf{p}}\mathcal{M}$, $\tau \neq 0$, has at most $m - 1$ zeros.

(b) $\mathcal{M}$ is said to satisfy "the global Haar property", if, for every pair $\mathbf{p}, \mathbf{q} \in M$, $\rho_{\mathbf{p}} - \rho_{\mathbf{q}}$ has at most $m - 1$ zeros.

Remark: In our case, the endpoints are *fixed* by (3.2) and the dimension m of $\mathcal{M}$ is already reduced. Therefore we have $\rho_{\mathbf{p}}(a) = \rho_{\mathbf{p}}(b) = 0$ for all $\mathbf{p} \in M$ and *these zeros have not to be counted* either in the local Haar property or in the global one.

Theorem 3. *(Meinardus/Schwedt): Let $f \in C[a, b]$ and $\mathcal{M}$ be induced by a global differentiable function $\hat{\rho}: M \times [a, b] \to \mathbb{R}$ (as in theorem 2) and satisfy the global Haar property. If there is $\mathbf{p} \in M$ such that the local Haar property is satisfied at $\mathbf{p}$ and $\rho_{\mathbf{p}} - f$ is an alternant with $m + 1$ extremal points, then $\rho_{\mathbf{p}}$ is the best approximation to f and $\rho_{\mathbf{p}}$ is unique.*

Proof: (of the local Haar property). The proof will be given for general Bézier curves without additional end conditions (in the other cases only minor modifications have to be made). Denoting by $D_{\mathbf{v}}$ the directional derivative, we get from (2.5)

$$D_{\mathbf{v}}\mathbf{y} \circ \hat{\sigma} + (\dot{\mathbf{y}} \circ \hat{\sigma})D_{\mathbf{v}}\hat{\sigma} = D_{\mathbf{v}}\hat{\rho}\mathbf{n}$$

hence, at $\mathbf{p}$, observing $\tau = D_{\mathbf{v}}\hat{\rho}(\mathbf{p}, -)$, we obtain

$$\det(D_{\mathbf{v}}\mathbf{y} \circ \sigma_{\mathbf{p}}, \dot{\mathbf{y}} \circ \sigma_{\mathbf{p}}) = \tau \det(\mathbf{n}, \dot{\mathbf{y}} \circ \sigma_{\mathbf{p}}) \tag{3.5}$$

By admissibility assumption (definition 1, (3)), $\det(\mathbf{n}, \dot{\mathbf{y}} \circ \sigma_{\mathbf{p}}) \neq 0$ in $[a, b]$; therefore, the zeros of τ are the same as those of the determinant to the left of (3.5). $\sigma_{\mathbf{p}}$ being a diffeomorphism it can be omitted, thus we are going back to the original parameter s. Now, $\mathbf{y}$ depends linearly on the parameters $p_1, \ldots, p_m$, which are, in our case, the $2(n - 1)$ components of the control points $\mathbf{b}_1, \ldots, \mathbf{b}_n$. Therefore, splitting $\mathbf{v}$ correspondingly into $n - 1$ vectors $\mathbf{v}_k \in \mathbb{R}^2$, $D_{\mathbf{v}}\mathbf{y} = \sum_{k=1}^{n-1} B_k^n(s)\mathbf{v}_k$ and $f(s) := \det(D_{\mathbf{v}}\mathbf{y}, \dot{\mathbf{y}}) = \sum_{k=1}^{n-1} \sum_{j=0}^{n-1} B_k^n(s)B_j^n(s) \det(\mathbf{v}_k, \Delta\mathbf{b}_j)$ is a polynomial of (maximal degree $2n - 1$. Since k is running only from 1 to $n - 1$, every B_k^n contains the factor $(1 - s)s$. Therefore f has at most $m - 1 = 2n - 3$ zeros in the interior of $[0, 1]$. ∎

On the global Haar property. The global Haar property is immediately obtained by convexity arguments for $n = 2$. It becomes much more complicated for $n \geq 3$. For the case of Bézier cubics, tangent to C at $x(a)$ and $x(b)$, we found a proof using the special representation of C' in this case. But it is based on tedious technical details, so it will not be reproduced here. A general proof for arbitrary n and all cases of kth order contact with C at both the end points was recently given by E. Eisele [Eisele '91] using topological arguments combined with an analysis for the resolvent of the intersection condition; however, this proof is valid only for some neighborhood of a fixed $C' \in \mathscr{B}'$.

On the other hand, while trying to obtain a general algebraic proof, I came to the following.

Conjecture: For each pair of Bézier curves there are no more intersection points (where they actually cross and are not just tangent) than those of their control polygons.

This would be a very significant generalization of the variation diminishing property. It might be that some restrictions are necessary, e.g. the control polygons, after being closed by a line from the end point $\mathbf{b}_n$ to the beginning point $\mathbf{b}_0$, should be simple in order to apply Jordan's curve theorem.

In our case, the validity of the conjecture would imply the global Haar property because the intersection points must be *monotone* with respect to the parameters (for both curves) because they intersect the normals of C only once. Then, the same is valid for the control polygons and thus the number is less than or equal to $2n - 3$ (the common end points not counted).

Summarizing, we get:

Theorem 4. *Let $\mathscr{B}'$ be as in theorem 2 (or a submanifold of it) such that the global Haar property holds for it; then, if there is $\mathbf{p} \in M$ such that $\rho_{\mathbf{p}}$ is an alternant with $m + 1$ external points, the corresponding Bézier curve $C_{\mathbf{p}}'$ is the best approximation to C with respect to the normal distance and $C_{\mathbf{p}}'$ is unique in $\mathscr{B}'$.*

4. Numerical Methods and Examples

4.1 Calculating the Best Approximation

For the classical Chebyshev approximation, the algorithm of Remez (cf. [Remez '1934]) is a standard method to calculate the best approximant. However it is based on the *linear* space of polynomials and the possibility to solve the discrete approximation problem (on a finite set of knots) by a linear system of equations, it can not be used directly in our non-linear case.

Instead, we consider the condition for $\rho_{\mathbf{p}}$ to be an alternant directly as a non-linear system of equations, which will be solved by a quasi Newton method. But this procedure requires an initial curve $C_{\mathbf{p}}'$ which is rather close to the solution; at least it must have already the maximum number of $m + 1$ alternating local extremal values, say at the knots $t_1, \ldots, t_{m+1}$, strictly increasing in the interior of the interval $[a, b]$.

So our algorithm is divided into two steps: The first of it is a preliminary one to provide $\mathbf{p} \in M$ so that the deviation function $\rho_{\mathbf{p}}$ of the corresponding curve $C'_{\mathbf{p}}$ has $m + 1$ alternating extremal values as described above. For this first step one can use optimization methods (going down-hill the amount of $\|\rho_{\mathbf{p}}\|_\infty$). The second step is its main part and will be described in the subsequent (we refered to it as a "quasi Newton method").

For given $\mathbf{p} \in M$ we calculate first $C'_{\mathbf{p}}$, its reparametrization function $t \to s = \sigma(t)$ and its deviation function $t \to \rho_{\mathbf{p}}(t)$ as solutions of (2.3) (for a suitable set T of discrete values $t_i \in [a, b]$). This can be done easily by Newton's method applied to the non-linear system (2.3) for each $t_i \in T$. Then we determine the extremal values $\rho_{\mathbf{p}}(\tau_j)$ and the corresponding knots τ_j which can be arranged in their natural monotonic order: $a < \tau_1 < \tau_2 < \cdots < \tau_{\mu+1} < b$.

The input vector $\mathbf{p}$ will be refused, if there are less than $m + 1$ local extreme values $\rho_{\mathbf{p}}(\tau_j)$ or if they do not alternate in sign, i.e. for an appropriate input vector we require

$$\mu = m, \qquad \mathrm{sign}\, \rho_{\mathbf{p}}(\tau_j) \cdot \rho_{\mathbf{p}}(\tau_{j+1}) = -1 \quad \text{for } j = 1 \ldots \mu. \tag{4.1}$$

After having found such an appropriate vector $\mathbf{p}$ we proceed to the second step. Our goal is to find $\mathbf{p}$ which is appropriate in this sence and with the property that its deviation function $\rho_{\mathbf{p}}$ is an *alternant*. Obviously, this is true if all the values

$$f_j(\mathbf{p}) := \rho_{\mathbf{p}}(\tau_j) + \rho_{\mathbf{p}}(\tau_{j+1}) \qquad j = 1, \ldots, m \tag{4.2}$$

are zero. Now we consider the *multivariate function* $\mathbf{f} \colon \tilde{M} \subseteq \mathbb{R}^m \to \mathbb{R}^m \mathbf{f}(\mathbf{p}) := (f_1(\mathbf{p}), \ldots, f_m(\mathbf{p}))$ assigning the vector of components (4.4) to each $\mathbf{p} \in \tilde{M}$, whereby $\tilde{M}$ the set of admissible and appropriate vectors $\mathbf{p}$ $(\tilde{M} \subseteq M)$. Thus the problem of calculating the best approximation is reduced to the solution of the non linear system of equations

$$\mathbf{f}(\mathbf{p}) = \mathbf{0}. \tag{4.3}$$

However this system is not only highly non linear but also we are far away from the knowledge of the function $\mathbf{f}$ itself. The only thing that really can be done is to calculate the *values* (4.2).

Thus we can not apply Newton's iteration, given by

$$\mathbf{p}_{n+1} := \mathbf{p}_n - (D\mathbf{f}(\mathbf{p}_n))^{-1}\mathbf{f}(\mathbf{p}_n), \tag{4.4}$$

starting with some appropriate initial vector $\mathbf{p}_1$ and converging (under additional assumptions on $\mathbf{f}$) quadratically to the solution $\mathbf{p}_0$ of (4.3), because the Jacobian matrix $D\mathbf{f}(\mathbf{p})$ (the elements of which are the partial derivatives $\partial f_j(\mathbf{p})/\partial p_k$) occurs in it. But this difficulty can be avoided by replacing the partial derivatives by suitable estimates. For this purpose we take the divided differences

$$q_{jk} := (f_j(\mathbf{p} + \delta_k \mathbf{e}_k) - f_j(\mathbf{p}))/\delta_k \tag{4.5}$$

(denoting by $\mathbf{e}_k$ the standard unit vectors in $\mathbb{R}^m$ in the kth coordinate direction). The real quantities δ_k must be small enough to keep $\mathbf{p} + \delta_k \mathbf{e}_k$ within $\tilde{M}$ but not too small to avoid cancellation. Now we put the quantities (4.5) as the elements of a matrix $\mathbf{Q}$ and write instead of (4.4) the iteration formula

$$\mathbf{p}_{n+1} := \mathbf{p}_n - \mathbf{Q}^{-1}\mathbf{f}(\mathbf{p}_n) \qquad (4.6)$$

assuming that $\mathbf{Q}$ is regular. This is what we called a quasi Newton method.

In practice, at least so far as our experience, the changement from (4.4) to (4.6) has only minor influence on the rapidity of convergence. Once a good initial vector $\mathbf{p}_1$ was found, only very few steps were necessary until the norm of the correction vector

$$\mathbf{c}_n := \mathbf{Q}^{-1}\mathbf{f}(\mathbf{p}_n)$$

became less than given tolerance ε (some more oders of magnitude than machine accuracy).

To summarize we write down the described method as an algorithm using a (non existing) high level language which includes all mathematical manipulations with vectors, matrices and real-valued functions (though, of course, the latter ones must be restricted to finite sets of arguments and further numerical methods have to be applied to calculate zeros, extremal values etc.).

Algorithm:

Input: $\mathbf{x}: [a,b] \to \mathbb{R}^2$ {param. repres. C}; $D \subseteq \mathbb{R}^m$ {domain for C'_p}; $\varepsilon \in \mathbb{R}^+$
calculate tangent normal vectors $\mathbf{v}, \mathbf{n}: [a,b] \to \mathbb{R}^2_{\|..\|=1}$ and curvature function $\kappa: [a,b] \to \mathbb{R}$ for C; determine $\mathcal{N}(C)$ {preparation step}
Repeat {search of initial vector}
choose $\mathbf{p} \in D$ using strategies of optimization (gradient methods, subdivision subdivision of D etc.) for the total error function $\mathbf{p} \mapsto \|\rho_\mathbf{p}\|_\infty$ {see subr.}
Until $\mathbf{p} \in \tilde{M}$ {admissible & appropriate, Def. 1 & (4.1)}
$n := 1$; $\mathbf{p}_1 := \mathbf{p}$ {start iteration}
Repeat {iteration loop}
For $k := 0$ to m do {displacement vectors}
if $k = 0$ then $\mathbf{p} := \mathbf{p}_n$ else $\mathbf{p} := \mathbf{p}_n + \delta_k \mathbf{e}_k$ calculate deviation function $\rho_\mathbf{p}: [a,b] \to \mathbb{R}$ {see subroutine} calculate its extremal points $\{(\tau_j, \rho_\mathbf{p}(\tau_j)) \mid j = 1\ldots\mu+1\}$ if (4.1) not satisfied then exit {not appropriate} set vector $\mathbf{f}_k := (f_1(\mathbf{p}),\ldots,f_m(\mathbf{p}))$ with $f_j(\mathbf{p})$ by (4.2)
$Q := $ matrix of columns $(\mathbf{f}_1,\ldots,\mathbf{f}_m)$; $\mathbf{Q}^{-1} := $ inverse (Q); $\mathbf{f} := \mathbf{f}_0$ $\mathbf{p}_{n+1} := \mathbf{p}_n - \mathbf{Q}^{-1} * \mathbf{f}$; corr $:= $ norm $(\mathbf{Q}^{-1} * \mathbf{f})$; $n := n + 1$
Until corr $< \varepsilon$
Output: If exit then error message else return $\mathbf{p}_0 := \mathbf{p}_n$

Subroutine: {calculate deviation function ρ_p}

Input:	$\mathbf{p} \in D$
Import: from preparation	$\mathbf{v}, \mathbf{n}: [a, b] \to \mathbb{R}^2_{\|..\|=1}$ {tangent and normal vectors for C} $\kappa: [a, b] \to \mathbb{R}$ {curvature function}
Import: from representation of approximants $\mathbf{p} \to \mathbf{y}_p$ for $\mathbf{p} \in D$ $\mathbf{y}_p$: $\overline{[c, d]} \to \mathbb{R}^2$ {parametric representation of C'_p}	
for each $t \in [a, b]$ calculate the intersection point of the curve $s \mapsto \mathbf{y}_p(s)$ with the normal of C at $\mathbf{x}(t)$ given by $\{\mathbf{x}(t) + \rho\mathbf{n}(t) \mid \rho \in \mathbb{R}\}$; use Newton's iteration applied to (2.3) $$\mathbf{y}(s) = \mathbf{x}(t) + \rho\mathbf{n}(t)$$ considered as a non linear system of two equations for the unknowns s and ρ; check for admissability {see Def. 1}; if so, assemble corresponding values of t and ρ to the desired deviation function $\rho_p: [a, b] \to \mathbb{R}$ and set admissible := true else admissible := false	
Output: $\rho_p: [a, b] \to \mathbb{R}$, admissible: boolean	

Clearly, some more exits have to be made for numerical robustness and stability or limits in computing time but also for mathematical reasons in cases when no solution exists or the matrix Q becomes singular.

4.2 Examples

The following examples show, that the algorithm described in the previous section is robust and converges rapidly, once a good estimate is given to start with.

In the first two examples, we take Bézier parabolas ($n = 2$, $m = 2$) to approximate given curves. Rather than being practical, the reason for these examples is to show that approximants with exactly three extremal points exist and what the best approximant looks like.

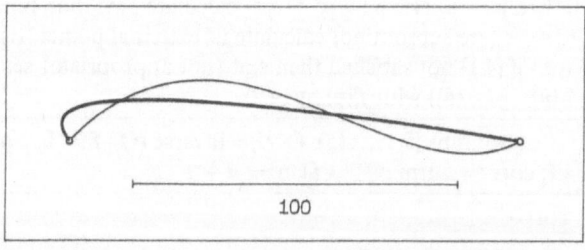

Figure 1. C with inflexion point and its best approximating Bézier parabola

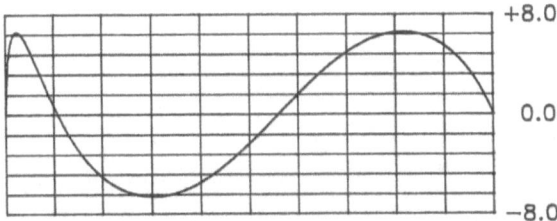

Figure 2. The error function (deviation function) for example 1

The situation in the first example is one of the worst we can imagine, because the given curve has an inflexion point, which cannot be modeled by a parabola: the normal distance remains rather large (see Figs. 1 and 2).

In the second example, the given curve is convex and the approximation is much better (see Figs. 3 and 4).

In the third example we calculate the best approximation of a transcendental curve C (part of the exponential function rotated and stretched) by a cubic Bézier curve tangent to C at the end points (see the examples in sectional 3.1); we have $n = 3$, $m = 2$. Now the best approximation is no longer to be distinguished from the original curve (see Fig. 5).

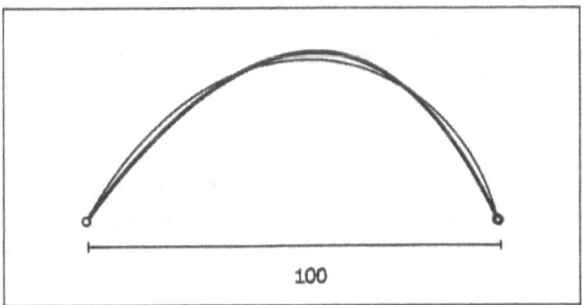

Figure 3. A convex curve and its best approximation by a Bézier parabola

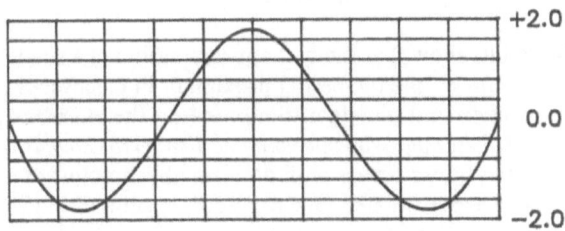

Figure 4. The error function (deviation function) for example 2

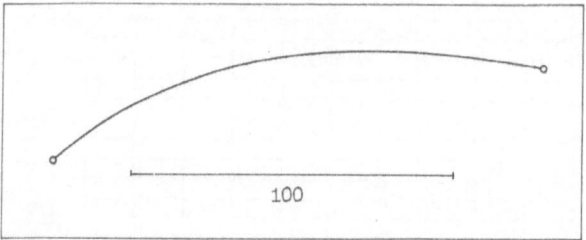

Figure 5. A best approximating tangent cubic to a transcendental curve

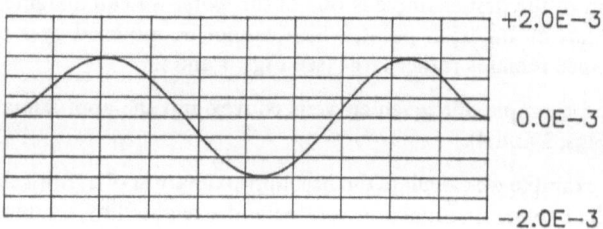

Figure 6. The error function (deviation function) of example 3

The plot of the error function (see Fig. 6) reveals the alternant property. Furthermore one can realize the *high accuracy* of this best approximant: The maximal error (=normal distance) is about 10^{-5} times the length of the curve. We compared this error with the Chebyshev approximants (polynomials of 3rd order) of some real valued functions and found that our tangent Bézier cubic yields, in general, better results.

Finally, H. Ruf, a student at our institute investigated the best approximants of circle arcs in his master's thesis (see also [Dokken et al. '87]). He compared the best approximants with the ones obtained by Hoschek's conversion method [Hoschek '87] and found that the error varies from 10^{-2} to 10^{-3} times the error of the conversion method with the same degree and a comparable number of intermediate points.

4.3 Conclusion

Our theoretical results show that the best approximation of a single curve segment by a Bézier curve behaves like the best approximant in Chebyshev approximation of a univariate function and the algorithm based on the alternant property of its error function *yields best approximants with excellent error bounds* especially in the case of tangent cubic Bézier curves with a reasonable amount of calculation. If a larger curve is to be approximated, one can apply a subdivision procedure previously and take informations from known error estimates ([De Boor et al. '87], [Hanna et al. '86], [Sakai/Usmani '90]) about the number of necessary subdivision points and then calculate the best approximation for each segment separately. Fitting together the tangent Bézier cubics, one get a geometric spline with G^1-continuity; higher order of continuity can be obtained analogously by the corresponding order of contact at the segment's end points.

However, there is one crucial point to be noticed: The region in the parameter domain, where the maximal number of $m + 1$ extremal values is attained, may be very small (in our example 3 it is about 1% of the entire extension in each direction). Thus one has to be careful in finding a starting parameter vector for the quasi Newton algorithm described in section 4.1.

References

[1] Blaschke, W.: Kreis und Kugel (Antrittsrede). Jahresbericht DMV 24, 195–207 (1915).
[2] De Boor, C., Höllig, K., Sabin, M.: High accuracy Hermite interpolation. Computer Aided Geometric Design 4, 269–278 (1987).
[3] Braess, D.: Nonlinear approximation theory. Berlin, Heidelberg, New York: Springer 1986.
[4] Do Carmo: Differential geometry of curves and surfaces. Englewood Cliffs, NJ: Prentice Hall 1986.
[5] Chebyshev, P. L.: Sur les questions de minima qui se rattachent à la représentation approximative des fonctions. Oeuvres, Tome 1, 273–378 (1899).
[6] Dannenberg, L., Nowacki, H.: Approximate conversion of surface representation with polynomial bases. Computer Aided Geometric Design 2, 123–132 (1985).
[7] Dokken, T., Daehlen, M., Lyche, T., Mørken, K.: Good approximation of circles by curvature continuous Bézier curves. Computer Aided Geometric Design 7, 33–41 (1990).
[8] Eisele, E.: Chebyshev approximation of planar curves by splines. Computer Aided Geometric Design (submitted).
[9] Haar, A.: Die Minkowskische Geometrie und die Annäherung an stetige Funktionen. Math. Annalen 78, 294–311 (1918).
[10] Hanna, M. S., Evans, D. G., Schweitzer P. N.: On the approximation of plane curves by parametric cubic splines. BIT 26, 217–232 (1986).
[11] Hausdorff, F.: Grundzüge der Mengenlehre. Leipzig: Veit & Comp. 1914, New York: Reprint Chelsea 1949.
[12] Hoschek, J.: Approximate conversion of spline curves. Computer Aided Geometric Design 4, 56–66 (1987).
[13] Hoschek, J.: Spline approximation of offset curves. Computer Aided Geometric Design 5, 33–40 (1988).
[14] Hoskins, W. D., Ponzo, J. P.: Some approximation properties of periodic cubic splines. BIT 14, 152–155 (1974).
[15] Meinardus, G.: Approximation von Funktionen und ihre numerische Behandlung. Berlin, Göttingen, Heidelberg, New York: Springer 1964.
[16] Meinardus, G., Schwedt, D.: Nicht-lineare Approximationen. Arch. nat. Mech. Analysis 17, 297–326 (1964).
[17] Mørken, K.: Best approximations of circle segments by quadratic Bézier curves. In: Laurent, P.-J., la Méhauté, A., Schumaker, L. L. (eds.) Curves and surfaces. New York: Academic Press 1991.
[18] Remez, E. J.: Sur la détermination des polynômes d'approximation de degreé donnée. Comm. Soc. Math., Kharkov 10, 41–63 (1934).
[19] Rice, J. R.: The characterization of best nonlinear Chebyshev approximation. Transact. Amer. Math. Soc. 96, 322–340 (1960).
[20] Sakai, M., Usmani, R. A.: On orders of approximation of plane curves by parametric cubic splines. BIT 30, 735–741 (1990).

Prof. Dr. W. L. F. Degen
University of Stuttgart
Pfaffenwaldring 57
D-W-7000 Stuttgart 80,
Federal Republic of Germany
E-Mail: LBAA @ DS0RUS1I

Computing Suppl. 8, 75–90 (1993)

© Springer-Verlag 1993

A Modelling Scheme for the Approximate Representation of Closed Surfaces

P. Brunet, I. Navazo, and A. Vinacua, Barcelona

Abstract. An approximate octree representation for closed surfaces is presented, namely face octrees. Face Octrees are based on a hierarchical representation of the subdivision of the space, until either homogeneous or face nodes are reached. Face nodes contain a connected, sufficiently planar part of the surface, within a tolerance ε. The face octree of a surface S defines a thick surface TS(S), union of all bands defined by face nodes, and TS(S) contains S; a band in a face node spans ε to both sides of the plane π approximating S in the node. An algorithm for the generation of the octree, based on the clipping of the surface patches and a planarity test, is presented. On the other hand, algorithms are proposed for point-solid classification, line and plane intersection tests, and interference detection. Space complexity of the proposed representation is discussed, and some bounds are presented.

Key words: Surface representation, octrees, surface-surface intersection.

1. Introduction

Geometric modelling must provide efficient and powerful tools for the representation and manipulation of 3D objects. In fact, geometric modelling systems implement an unambiguous 3D geometric model together with the tools required to perform these manipulations [11]. In unambiguous representations, every valid internal representation corresponds to a single real object. On the other hand, basic geometric manipulations include creating new objects, editing the shape, performing geometric transformations or boolean operations that generate new shapes, interrogating the model and rendering it.

Solid models are used for the representation of closed regions of the space. Solid models can be either evaluated or non-evaluated. In the first case, an explicit representation of the geometry is kept in the model. This is not the case in non-evaluated models like constructive solid geometry [11], where some interrogations can require the explicit computation of a corresponding evaluated model.

Concerning evaluated solid models, they can either model the geometry of the surface of the solid, or the enclosed volume. In the first case, we have the boundary representation schemes. A number of existing systems only allow plane and simple quadric surfaces—cylinder, cone, sphere, etc—in the representation of the object's surface. Geometric interrogations are quite simple in this case, because of the algebraic representation of the oriented faces; on the other hand, a limited number of shapes can be modeled. More sophisticated systems include parametric surfaces, and describe the object's surface through a set of adequately stitched polynomial—or rational—patches [6]. Although very general shapes can be represented, interrogation and shape operations become complex [4] and not always robust. On the other hand, some schemes based upon the representation of surfaces by locally algebraic surfaces have also been proposed, [4].

Among the solid models that explicitly represent the enclosed volume, octrees are the best known decomposition schemes. Classical octrees [7] represent solid objects through the recursive subdivision of a finite cubic universe. The subdivision process is represented by the tree structure of the octree. The root of the tree represents the universe cube. This cube is divided into eight identical octants (cubes), each one being represented by one of the direct descendant nodes of the root. If an octant contains too complex a part of the solid (Grey node), it is also divided into eight identical cubes which are represented as its descendants in the tree. This process is repeated recursively until valid terminal nodes are obtained [7] or a minimum cube size is reached. In the simplest octree model, named classical octrees, terminal nodes represent homogeneous cubes in the spatial decomposition [7]; these homogeneous cubes can be completely outside the solid (White nodes) or inside it (Black nodes). On the other hand, classical octrees represent as Grey nodes all non-homogeneous cubes being partially filled with the solid, Fig. 1. Being a hierarchical representation, the size and location of a cubic octant can be determined from the path to the associated node in the octree. Classical octrees are approximate representations, the accuracy being related to the minimum cube size. On the other hand, the greater the required accuracy, the larger the storage requirements for the octree structure.

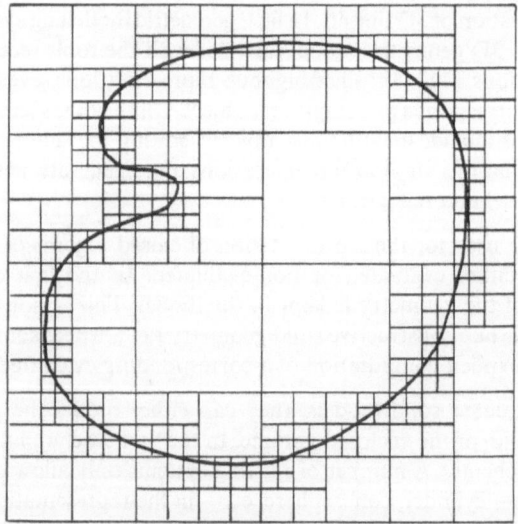

Figure 1. Classical octree representation of a closed region (2D example)

In this paper, an approximate octree representation for closed surfaces is presented, namely face octrees. Face octrees are based on a hierarcical representation of the subdivision of the space, until either homogeneous or face nodes are reached. Face octrees can be used as an auxiliary model of the surface, as simple algorithms can be derived for the most significant solid interrogations (point-solid classifications, line and plane intersections, interference detection between objects, and volume interrogations), and they provide surface bounds in the surface-surface intersection problem. After introducing face octrees in the next section, sections 3 and 4 present the main generation and interrogation algorithms, while the last two sections are devoted to the space complexity analysis and the discussion of several examples.

2. Face Octrees

Although approximate, classical octrees are a suitable model for representing volumes of very complex shape, as is the case in natural or medical applications—modelling the brain, lungs, bones, etc—. Alternative boundary representations can be very verbose in those cases, [2]. On the other hand, octree models that incorporate new terminal nodes containing parts of the object surface have also been proposed. They can be called vector octrees, [13]. In objects of simple shapes as polyhedra, both polytrees [5] and extended octrees [1] yield exact representations which are much more compact than classical octrees [3], by using face, edge and vertex nodes in addition to the classical white and black terminal nodes. Face octrees are halfway between classical octrees and extended octrees: they only allow face nodes in addition to the usual octree nodes [14] [2]. Face nodes contain a sufficiently planar portion of the surface of the object. Face octrees is an approximate representation for surfaces which, as it will be shown in section 5, has less memory requirements than both classical and extended octrees provided that the surface is sufficiently smooth. A related scheme, based on the recursive subdivision of a tetrahedral universe, can be found in [8]. We now give a more precise definition:

Definition. A face octree is an octree with white, black, face and grey nodes together with a tolerance ε. White, black and grey nodes are defined as in classical octrees. Face nodes contain a connected part of the boundary of the object, and have associated with them the equation of a plane π such that $dist(P, \pi) \leq \varepsilon$ for every point P of the object surface S in the cube C associated to the face node, and $dist(Q, S) \leq \varepsilon$ for every point $Q \in \pi \cap C$, Fig. 2.

Thus, the part of the surface of the object within a face node must be sufficiently planar. The value ε associated with the face octree controls the degree of approximation of the representation and the depth of the tree. The encoding of face nodes must include both the corresponding node type and geometric information on the associated plane π. This can be either the explicit plane equation or a pointer to a table of face plane equations. Figure 3 shows the face octree representation (2D) of the same curved closed region as in Fig. 1.

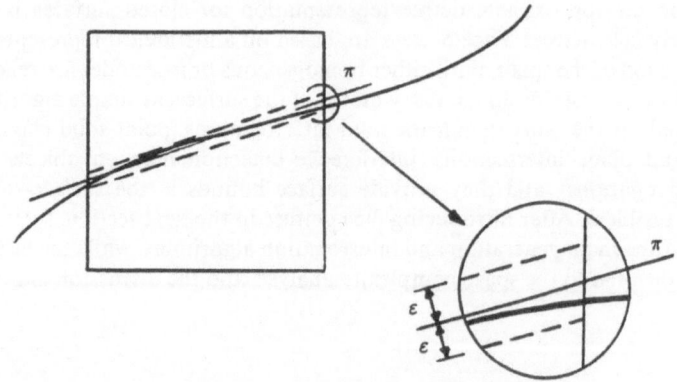

Figure 2. A face node and its associated band

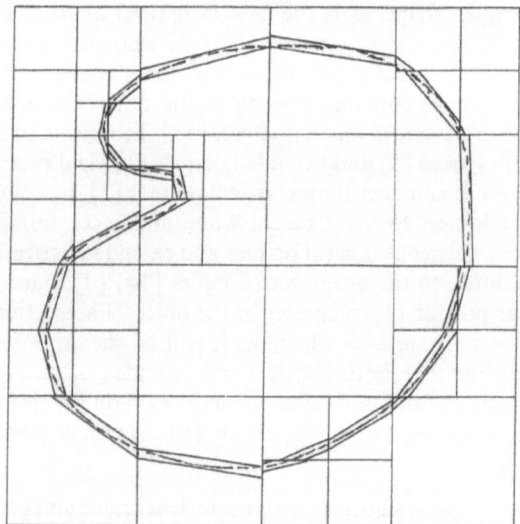

Figure 3. Face octree representation of the same region as in Fig. 1

We will note the face octree representing the closed region bounded by a smooth surface S with a tolerance ε as $FO_\varepsilon(S)$. The band b_i of a face node F_i with associated cube C_i and plane π_i, is defined as the closed set of points P,

$$b_i = \{P, P \in C_i, dist(P, \pi_i) \le \varepsilon\}$$

As a consequence of the face octree definition, the surface S restricted to the cube C_i is contained in b_i, $S \cap C_i \subset b_i$. Then, defining the thick surface of S, $TS_\varepsilon(S)$ as,

$$TS_\varepsilon(S) = \bigcup_{\forall F_i} b_i$$

it follows that

$$S \subset TS_\varepsilon(S)$$

The face octree representation of a closed VC^1 surface S fulfils the following properties:

- they can be considered as an approximate piecewise linear representation of S, $\{\pi_i \cap C_i, i = 1 \ldots n_{fn}\}$, n_{fn} being the number of face nodes in the octree. This approximation is obviously non continuous, but bounded by the tolerance ε.
- they contain a hierarchical, explicit volume representation of the inside region closed by S, defined by black nodes and the inside part $C_i \cap \pi_i^+$ of face nodes. Here π_i^+ represents the halfspace inside the solid defined by the plane π_i of the face node F_i.
- they also define an ε-band representation of S, $TS_\varepsilon(S)$, that contains the represented surface S, Fig. 3. $TS_\varepsilon(S)$ is a connected region that can be used as a bound for S in geometric tests, see section 4: a point P can only be **on** S if it is **in** $TS_\varepsilon(S)$.
- they perform a subdivision of the universe cube to a level that depends on the local curvature of the surface S. On the other hand, the subdivision is independent of the patch boundaries existing in the boundary representation of S.
- face octrees can be considered as an auxiliary data structure, and stored together with the surface model. The composite model $\{S, FO_\varepsilon(S)\}$ is specially well suited for fast geometric interrogations, geometric operations and volume computations, as it will be presented in section 4.
- face octrees, considered as an auxiliary representation for S, give robust region bounds for set operations, as $S_1 \cap S_2 \subset TS_\varepsilon(S_1) \cap TS_\varepsilon(S_2)$, and both thick surfaces are piecewise linear and easy to intersect.
- they can be refined through further clipping and testing present face nodes:

$$\{S, FO_\varepsilon(S)\} \rightarrow FO_{\varepsilon'}(S), \qquad \varepsilon' < \varepsilon$$

The two following sections present and discuss proposed algorithms for the generation of face octrees and basic solid geometric interrogations.

3. Face Octree Generation Algorithm

A set of patches defining a closed surface can be converted onto the corresponding face octree representation, by means of a recursive clipping of the surface and a planarity test. The corresponding algorithm is,

```
procedure build_FO (patch_list,x,y,z,scale)
    clipping (patch_list,x,y,z,scale,sublist)
    if no patches and node_outside then
        output (White)
    elseif no patches and node_inside then
        output (Black)
```

```
        elseif flat_enough then
            compute discrete plane
            output (Face)
        else
            output (Grey)
            for each of the eight subnodes do
                compute_node_location (xn,yn,zn)
                build_FO (sublist,xn,yn,zn,scale/2)
            end for
        end if
    end procedure
```

The algorithm first clips the surface elements against the node, and obtains the list of relevant patches, **sublist**. If this list is empty, then the octree node is either White or Black, depending on the node being outside or inside the surface of the object. Otherwise, a Face node is generated when the patches in **sublist** are connected and flat enough. In the remaining cases the procedure generates a Grey node in the octree, and a recursive call is performed in order to treat descendant nodes. The procedure **output** includes new nodes in the output octree, which is generated in preorder.

The three basic steps in the algorithm are the clipping process, the test for determining if the part of surface inside a node is flat enough, and the computation of the associated discrete plane. Clipping is a procedure that receives a list of patches together with an octree node (location xc, yc, zc and scale of the associated cube), prunes it, and returns a sublist of patches. It works in a way such that it can be asserted that every patch in **patch_list** not in **sublist** is not intersecting the octree node. Clipping can work on general models and topologies of surfaces, including rectangles and triangles together with parametric or algebraic surfaces. In the case of parametric surfaces and assuming a Bezier representation of the patches, it is interesting to maintain information on parametric intervals for every patch in the lists. The structure of both the **patch_list** and the **sublist** for rectangular patches is,

$$(S_i(u, v), u0_i, u1_i, v0_i, v1_i), \qquad i = 1 \ldots n\_patches$$

$$S_i(u, v) = \sum_j \sum_k P_{ijk} B_j(u) B_k(v)$$

Noting by C the cube associated to the present octree node, parametric intervals $u0_i, u1_i, v0_i, v1_i$ restrict the portion of the patch that can intersect the cube C in a way that,

$$S_i(u, v) \cap C = \varnothing, \qquad u < u0_i, \qquad i = 1 \ldots n\_patches$$

and the same for $u > u1_i, v < v0_i, v > v1_i$.

On the other hand, the structure of both the **patch_list** and the **sublist** for triangular patches [6] must keep information on parametric intervals for barycentric coordinates,

$$(S_i^t(u, v, w), u0_i, u1_i, v0_i, v1_i, w0_i, w1_i), \qquad i = 1 \ldots n\_triang\_patches$$

$$S_i^t(u, v, w) = \sum_j \sum_k \sum_l P_{ijkl} B_{jkl}(u, v, w), \qquad j + k + l = degree$$

In both cases, clipping works by performing alternate de Casteljau subdivision of the patch in the different coordinate directions. Let us assume that $P_{ijk}^* = (x_{ijk}^*, y_{ijk}^*, z_{ijk}^*)$ are the control vertices of the significant portion of $S_i(u, v)$ in the **patch_list**,

$$S_i(u, v) \qquad u0_i \le u \le u1_i, \qquad v0_1 \le v \le v1_i$$

Then, the following sequential tests are performed in order to clip S_i in C,

- If $x_{ijk}^* < xc$ or $x_{ijk}^* > xc + scale$ $\forall j, k$, then discard S_i.
- If $y_{ijk}^* < yc$ or $y_{ijk}^* > yc + scale$ $\forall j, k$, then discard S_i.
- If $z_{ijk}^* < zc$ or $z_{ijk}^* > zc + scale$ $\forall j, k$, then discard S_i.
- Increase $u0_i$ while P_{ijk}^{u-} **out** C $\forall j, k$, P_{ijk}^{u-} being the Bezier control points of $S_i(u, v)$, $0 \le u \le u0_i, 0 \le v \le 1$ computed by de Casteljau subdivision, Fig. 4.
- Decrease $u1_i$ while P_{ijk}^{u+} **out** C $\forall j, k$, P_{ijk}^{u+} being the Bezier control points of $S_i(u, v)$, $u1_i \le u \le 1, 0 \le v \le 1$ computed by de Casteljau subdivision, Fig. 4.
- Increase $v0_i$ while P_{ijk}^{v-} **out** C $\forall j, k$, P_{ijk}^{v-} being the Bezier control points of $S_i(u, v)$, $0 \le u \le 1, 0 \le v \le v0_i$ computed by de Casteljau subdivision, Fig. 4.
- Decrease $v1_i$ while P_{ijk}^{v+} **out** C $\forall j, k$, P_{ijk}^{v+} being the Bezier control points of $S_i(u, v)$, $0 \le u \le 1, v1_i \le v \le 1$ computed by de Casteljau subdivision, Fig. 4.

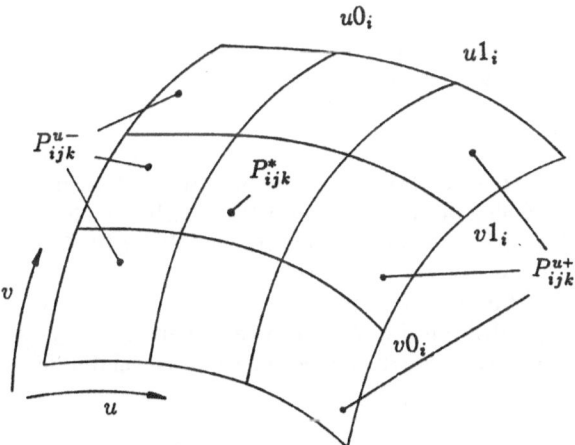

Figure 4. A patch S_i of the sublist, its parametric intervals and the corresponding subpatches and control points

The test for flatness involves all remaining patches in the clipped **sublist**, and it is based on the control points P_{ijk}^* of the associated significant subpatches. The algorithm works in three steps,

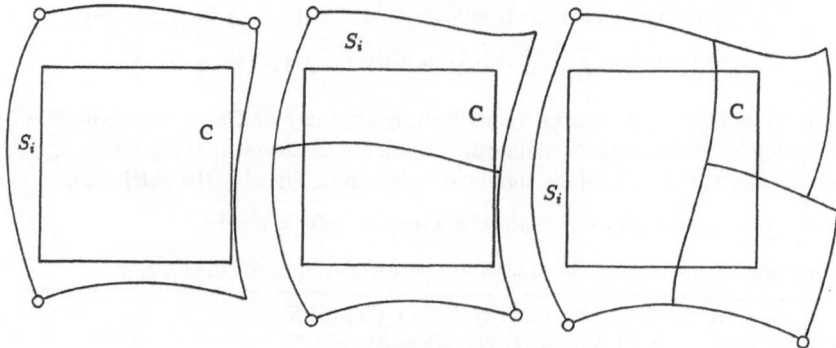

Figure 5. Corner control vertices that define the candidate plane, in the case of a list containing one, two or three patches

- First, three corner control points of the patches are selected, in a way that they are as distant as possible. Corner control vertices opposed to the common edges are chosen in the case of several patches meeting together in C, Fig. 5.
- The plane π defined by these three points is computed as a candidate for a linear approximation of the surface in C.
- Plane π is accepted as the associated plane of the face node if $dist(P^*_{ijk}, \pi) \leq \varepsilon \ \forall j$, $k \ \forall i \in sublist$.

Finally, in the case of affirmative result of the flatness test, the algorithm computes an integer representation of the associated plane [2] and generates a face node encoding both the plane equation and the sublist of concerned patches with their parametric intervals.

The classification of the octree nodes that contain no part of the surface onto White and Black nodes can be stablished based on the terminal Face nodes, provided that a convention assuming outward normals in the patches is adopted. In a first step, an octree containing face nodes and undetermined White/Black nodes can be built; then, inside and outside zones defined by face nodes can be propagated in a germ-like way [12] to the rest of the White-Black terminal nodes of the octree, in order to determine their precise color. This can be in fact a way of testing whether a set of patches bounds a closed volume and defines a valid solid, or not.

4. Interrogation Algorithms

Face octrees, when used as an auxiliary data structure for closed surface representations, allow simple and fast volume interrogation algorithms for most of the usual cases. This section presents and discusses algorithms for point-solid classification and line and plane-solid intersection tests. Other algorithms include interference detection, line (plane)-solid intersection computation, volume interrogations and bounds in surface-surface intersection.

4.1 Point-Solid Classification

This is the basic membership test for solid models. Given a point P and a solid model, the algorithm must ascertain if P is **in**, **on** or **out** with respect to the solid. The face octree representation of S solves the classification with no need of the surface representation in all cases where $P \notin TS_\varepsilon(S)$, by means of a tree traversal driven by the location of P,

```
function class_P_solid (P, FO)
    case of type (FO)
        White: return out
        Black: return in
        Face: case of classification (P, band(FO))
                outside_solid: return out
                inside_solid: return in
                inside_band: refine (P, band(FO))
              end case
        Grey:
              son_node := location (P, FO)
              return class_P_solid (P, son_node)
    end case
end function
```

The classification of P against the band of a face node is very simple, being based on the distance $dist(P, \pi)$. If $dist(P, \pi) \leq \varepsilon$, then P is inside the band. Otherwise, it is classified as outside-solid or inside-solid depending on to which of the halfspaces defined by π P belongs. Function refine is called in the only case when the face octree representation does not suffice. In this case, either P must be classified directly against the portions of patches in the list of the face node, or this node must be refined to a lower tolerance. Refine can obviously return **on** as a result, as well as **in** and **out**. On the other hand, the function **location** returns the son node of FO containing P in its associated cube. It is based on the P coordinates and the son node location and scale.

4.2 Line-Solid and Plane -Solid Tests

Given a straight line (plane) and a solid model, this algorithm must ascertain if it intersects the solid or not. In this case, a tree traversal must be performed, only face nodes being considered. The line test algorithm is,

```
function line_test (line, FO)
    case of type (FO)
        White: return false
        Black: return false
        Face: case of line_band_intersect (line, band(FO))
                intersect no main boundaries of the band: return false
                intersect one main boundary of the band: refine
                intersect two main boundaries of the band: return true
```

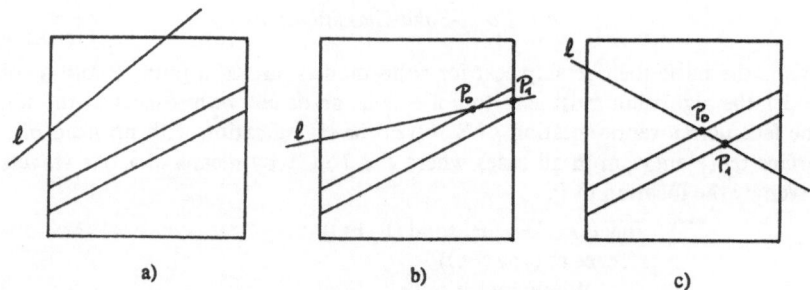

Figure 6. Cases in line-band intersection in a face node. In **a** the line 1 intersects no plane boundaries of the band. In **b** it intersects only one boundary. In **c** it intersects both plane boundaries

```
        end case
    Grey:
        i := −1
        repeat
            i := i + 1
        until i = 8 orcondline_test (line, FO.son [i])
        return i ≠ 8
    end case
end function
```

While cases a and c in Fig. 6 are unambiguous, if the intersection affects only one of the plane boundaries of the band (Fig. 6-b) the line can either intersect or not the surface. In this case, either an approximate solution is sufficient, or the exact surface model must be taken into account. On the other hand, both cases 6-b and 6-c define intervals $[P_0, P_1]$ on the line (plane) which can be used as starting solutions of an iterative numerical computation of the geometric intersection. This is also the reason for not considering black nodes during the tree traversal. On the other hand, the plane test algorithm only differs with the procedure plane-band-intersection, in that it now must compute the number of boundaries of the band intersecting the plane.

4.3 Other Interrogation Algorithms

Interference detection between two face octree encoded volumes can be performed through a simultaneous traversal of both trees operating on homologous face nodes. An interference can exist between both solids only if the band b_i of a face octree of the first octree intersects the band b_j of a face node of the second octree, $b_i \cap b_j \neq \varnothing$. Note that simultaneous traversal ensures that nodes sharing the same spatial location are always operated [3]. It must be also observed that in some limit cases (Fig. 7-c) interference can be concluded in cases of non-intersecting surfaces being at a distance less than 4ε each other, unless exact geometric information on the surfaces is used.

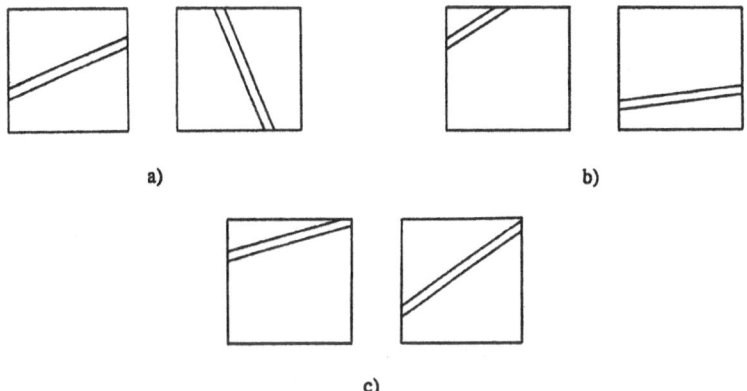

Figure 7. Operation of homologous face nodes. In **a** an interference is detected. Case **b** does not conclude interference. Case **c** will also detect interference; however, a real interference will exist or not depending on the location of the surfaces within the bands

The union of all the polyhedra $b_i \cap b_j$ resulting of the intersection between bands of homologous face nodes is a volume that contains the intersection curve between the surfaces of the objects. Therefore, it can be used as a volume bound in surface-surface intersection algorithms. It also brings information on the topology of the intersection curve with an accuracy depending on ε.

A canonic surface contained in $TS_\varepsilon(S)$ can be obtained from the face octree representation of S, [2], [9]. It can be used for rendering purposes, or whenever a smooth geometry is required yet it is not possible to access the exact geometric information of S.

Finally, the face octree representation can be used for the approximate computation of volume properties (volume, center of mass, inertia momenta, etc). In this case, a tree traversal is required and both black and face nodes must be processed. In the summation, only black nodes and the contribution of the inside part $C_i \cap \pi_i^+$ of face nodes must be considered.

5. Complexity of the Face Octree Representation

By defining the size of an octree as the total number of nodes that it contains, it is known that most of the algorithms related with octree representations show linear complexity with respect to the size of the octrees being interrogated or operated. This section discusses the spatial complexity—size of the octree—of the face octree representation as a function of several geometric parameters of the surface of the object, and compares it with the spatial complexity of classical octrees for the representation of the same solid.

The size of a particular face octree depends on the smoothness of the surface and the thickness of the different parts of the solid, [2]. More precisely, let us define the

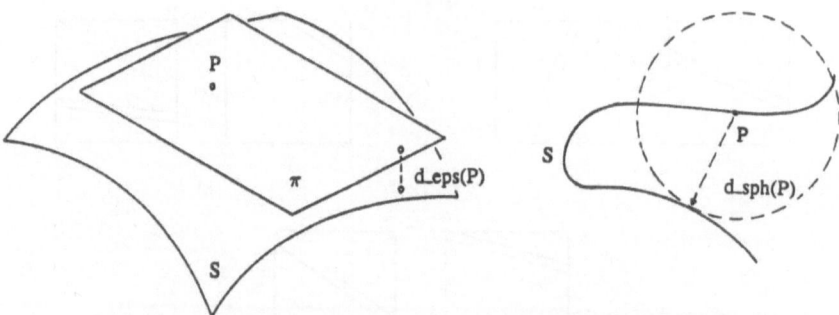

Figure 8. The function d_eps(P) measures the divergence of the surface S with respect to the tangent plane in P. On the other hand, d_sph(P) is a measure of the thickness in P

following point functions d_eps and d_sph, for every point P of the object surface S:

d_eps(P) = min(dist (P,Q)), for all points $Q \in S$ such that $dist(Q, \pi_P) \geq \varepsilon$, ε being the tolerance of the face octree representation and π_P the tangent plane to S in P, Fig. 8-a.

d_sph(P) is the maximum value so that every sphere $B_r(P)$ centered in P with radius $r \leq d\_sph(P)$ intersects S in a part $B_r(P) \cap S$ which is homeomorphic to a disc.

Now, the surface of the object can be divided into zones z_k depending on the value of d_eps and d_sph. If the edge size of the universe cube is 2^N, z_k is defined as the set of points $P \in S$ such that,

$$\sqrt{3}2^{N-k-1} \leq min(d\_eps(P), d\_sph(P)) < \sqrt{3}2^{N-k}$$

Zones z_k with k small represent very flat parts of the object surface. On the other hand, zones corresponding to large values of k are either very curved or thin, Fig. 9. Then, it can be shown that a bound on the size of the face octree can be obtained as a function of the extent of the different zones z_k. In other words, the shape and extent of the zones z_k contains the geometric information on S necessary for the characterization of the face octree complexity. Noting by S_k and P_k the surface area and perimeter of the zone z_k Fig. 9, it can be shown [10] that the number of face nodes n_{face} and the total number of nodes of the face octree n_{nod} fulfil,

$$n_{face} \leq \sum_k (4^{k-N}S_k + 2^{k-N}P_k) * 20$$

$$n_{nod} \leq 9 \sum_k (4^{k-N}S_k + 2^{k-N}P_k) * 20.$$

As $\sum_k S_k$ is the total surface area of the solid, the last equation states that smooth zones have little contribution to the total number of nodes in the face octree whereas abrupt zones generate most of the octree terminal nodes. When comparing with classical octrees, where it is known that the number of nodes n_{nod} is of the order of the total surface area of the object, it can be concluded that the larger the zones z_k

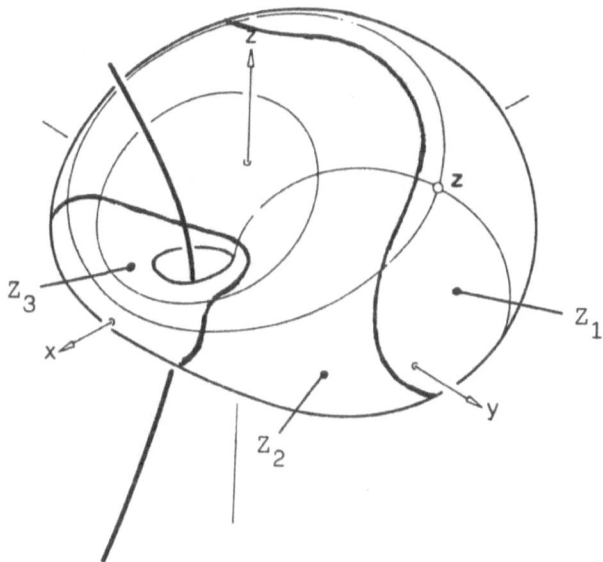

Figure 9. Map of the zones z_k in a simple solid. Observe the characterization of smooth and abrupt zones

with k small, the smaller the number of nodes of the face octree representation related to the number of nodes of the classical octree. This aspect will be further discussed in the next section through particular examples.

6. Discussion and Conclusions

Figure 10 shows the face octree representation of a radius 1000 sphere, in a universe cube with N = 14. Two results with different tolerances ε are shown, $\varepsilon = 2$ and $\varepsilon = 10$. A zoom enlarging the squared area in each case shows the cracks between associated planes π_i in neighbour face nodes, much more important in the case $\varepsilon = 10$, as it was already presented in Fig. 3. The face octree representation of a cyclide is presented in Fig. 11.

The following table shows the complexity of the classical octree and face octree representations as a function of the tolerance ε,

Observe that all the surface S is classified onto the same zone z_k, because of the symmetry of the sphere. Independently of ε, the size of the classical octree is of the order of the surface area of the object which is 12.5×10^6 in this case where the size of the universe cube is 2^{14}. Observe that this value is much higher than both the actual number of face octree nodes and the computed bounds. On the other hand, it can be seen that the bound for n_{face} proposed in the last section is approximately sixty times the actual number of nodes (except for the case $\varepsilon = 20$ which has been classified as being in z_8 although being very close to z_7). Finally, it can be observed

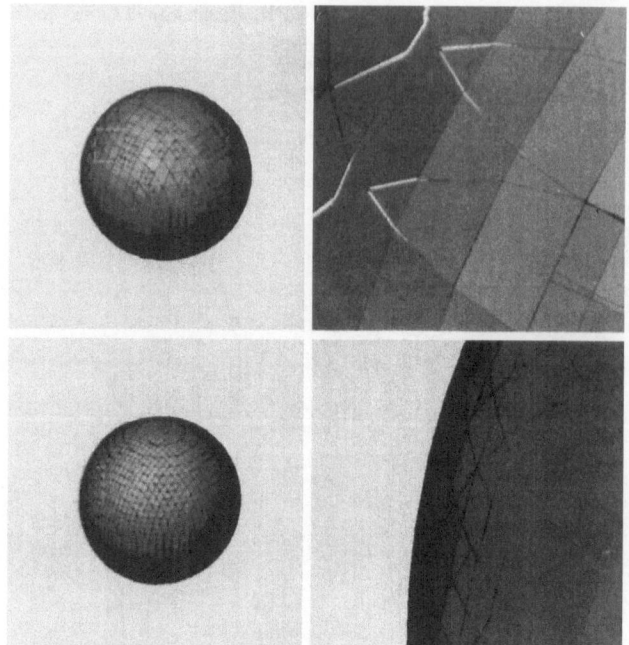

Figure 10. Face octree representation of a sphere

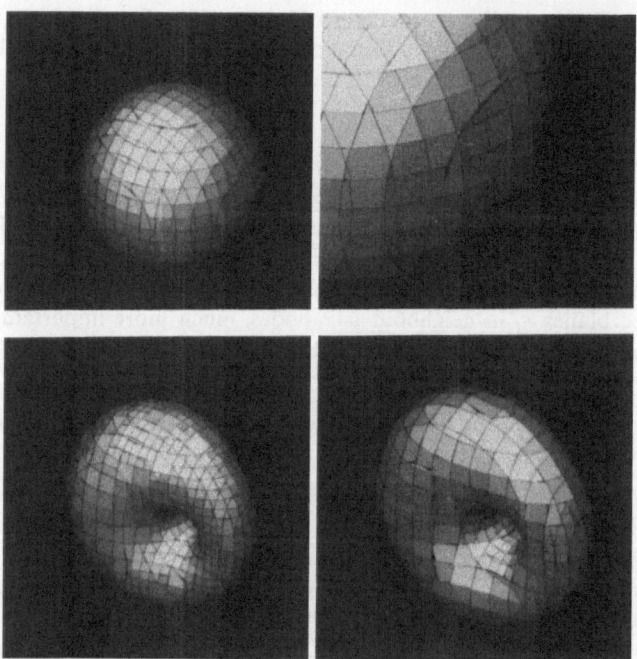

Figure 11. An sphere, and a cyclide represented by two different tolerances

Table 1. Complexity of face Octree representations

Bigskip

ε	d_eps	z_k	n_{face} (actual)	n_{face} (bound)($*1/20$)
2	31.6	z_9	4544	12272
10	70.7	z_8	1085	3068
20	100	z_8	437	3068
50	160	z_7	251	767

that the complexity of the face octree decreases as the tolerance of the approximation ε is less severe.

As a conclusion, we can say that the proposed face octree representation is both an approximate representation of S and a ε-band representation of the boundary of the closed region. Face octrees can be used as an auxiliary representation for S in order to speed up geometric tests, operations and volume interrogations. The complexity of the face octree representation depends on the curvature of the surface and, for sufficiently smooth surfaces it is much smaller than the spatial complexity of the corresponding classical octree representation.

Acknowledgements

The authors would like to thank Nuria Pla for her significative contribution to the derivation of the bounds on the number of face nodes in a face octree. They would also like to thank Marc Vigo for his help in the computation of Table 1 and Fig. 11. This work has been supported in part by the spanish agency CICYT under grant TIC-0574.

References

[1] Ayala, D., Brunet, P., Juan, R., Navazo, I.: Object representation by means of non minimal division quadtrees and octrees. ACM Transactions on Graphics 4, 41–59 (1985).
[2] Brunet, P.: Face Octrees. Involved algorithms and applications, Report LSI-90-14, Software Dept, Polytechnical Univ of Catalonia, Barcelona 1990.
[3] Brunet, P., Navazo, I.: Solid representation and operation using extended octrees. ACM Transactions on Graphics 9 (2), 170–197 (1990).
[4] Brunet, P., Vinacua, A.: Surfaces in solid modeling. In: Hagen, H., Roller, D. (eds.) Berlin, Heidelberg: Springer 1991.
[5] Carlbom, I., Chakravarty, I., Vanderschel, D. A.: A hierarchical data structure for representing the spatial decomposition of 3-D objects. IEEE Computer Graphics and Applications 5, 24–31 (1985).
[6] Farin, G.: Curves and surfaces for computer aided geometric design. San Diego, CA: Academic Press 1990.
[7] Meagher, D.: Geometric modelling using octree encoding. Computer Graphics and Image Proceeding 19, 129–147 (1982).
[8] Moore, D., Warren, J.: Multidimensional Adaptive Mesh Generation, Research Report Rice COMP TR 90-106, Dept of Computer Science, Rice University, febr. 1990.
[9] Pla, N.: Approximate curve representation using edge quadtrees, Eurographics Workshop on Computer Graphics and Mathematics, Genova, Italy 1991.
[10] Pla, N.: Personal communication.
[11] Requicha, A. A. G., Voelcker, H. B.: Solid modelling: A historical summary and contemporary assessment. IEEE Computer Graphics and Applications 2 (2), 9–24 (1982).

[12] Samet, H.: The quadtree and related hierarquical data structures. ACM Computing Surveys *16*, 187–260 (1984).
[13] Samet, H.: The design and analysis of spatial data structures. Reading, MA: Addison-Wesley Publishing Company 1989.
[14] Samet, H.: Data structures to support bezier based modelling. Computer-Aided Design *23*, 162–176 (1991).

Pere Brunet
Polytechnical University of Catalonia
E-08028 Barcelona, Spain

Computing Suppl. 8, 91–100 (1993)

Cross Boundary Derivatives for Transfinite Triangular Patches

T. A. Foley, S. Dayanand, and R. Santhanam, Tempe

Abstract. For interpolation of scattered data using a triangular patch, the cross boundary derivatives on the triangle edges are commonly computed to vary linearly from one vertex to another. Such an approach often yields an interpolant with visual ridges along some of the edges of the triangle patches. The improved approach presented here for computing cross boundary derivatives makes use of the network curves of neighboring triangle patches. These cross boundary derivatives vary as an arbitrary degree polynomial in Bernstein-Bézier form.

Key words: Scattered data, multivariate interpolation, data fitting, triangle patches.

1. Introduction

A method for computing cross boundary derivatives is presented which can be used with transfinite triangular patches to solve the following scattered data interpolation problem. Given N distinct points $V_i = (x_i, y_i)$ and N real values z_i, the problem is to construct a C^1 bivariate function $F(x, y)$ that satisfies $F(x_i, y_i) = z_i$, for $i = 1, \ldots, N$. Several scattered data interpolation methods and many examples are surveyed in [1, 7, 8, 9, 16].

This paper focuses on computing cross boundary derivatives for the side-vertex method in [12], although it applies equally well to the BBG triangular patch in [2]. These patches are commonly referred to as transfinite triangular patches because they interpolate to arbitrary position and first derivative values on the boundary of a triangle. For interpolation of scattered data, these methods are usually discretized so that the boundary curves are cubic polynomials and the cross boundary derivatives are generally selected to vary linearly from one vertex to another. The boundary curves are dependent upon estimates of $F_x(V_i)$ and $F_y(V_i)$, and there are many effective methods for selecting these values in [7, 9, 13, 18]. Instead of using cross boundary derivatives that vary linearly, the approach presented in section 3 computes cross boundary derivatives that vary as an arbitrary degree polynomial in Bernstein-Bézier form. The control points for this polynomial on each edge are cross boundary derivatives estimates based upon quadratic interpolation of three different network curves. The resulting surfaces are significantly smoother than those using linear varying cross boundary derivatives.

Very few techniques have been proposed for selecting the cross boundary derivatives and we briefly discuss three of these methods. Using a hybrid Bézier triangular patch, a method is given in [6] for selecting cross boundary derivatives so that the interpolant has cubic precision. The cross derivatives vary quadratically along each edge and the surfaces are visually smoother than when linear varying derivatives are used. For the Clough-Tocher patch, which splits each triangle into three smaller triangles, significantly smoother results are obtained in [4] than when linear varying derivatives are used. These cross derivatives also vary quadratically over each edge. Since the details are lengthy, we briefly note that these two methods in [4] and [6] reduce to joining two cubic triangular patches in a C^1 manner along only one edge of the triangle. With the derivatives fixed at the vertices, each of these methods has an additional degree of freedom for each edge because the cross boundary derivatives of a cubic patch varies quadratically along each edge. With this degree of freedom, the approach in [4] minimizes the C^2 discontinuity across the edge, while cubic precision is obtained with the approach in [6]. The surface plots of these approaches have similarities to the examples shown in this paper.

The minimum norm network method in [13] was recently generalized in [15] to form a C^2 network that involves minimizing the following quantity

$$\sum \int_{e_{ij}} \left(\left(\frac{\partial^3 F}{\partial e_{ij}^3} \right)^2 + \alpha \left(\frac{\partial^3 F}{\partial n_{ij} \partial e_{ij}^2} \right)^2 + \beta \left(\frac{\partial^3 F}{\partial n_{ij}^2 \partial e_{ij}} \right)^2 \right) de_{ij},$$

where e_{ij} is the edge from V_i to V_j, and n_{ij} is normal to the edge. The network curves are C^2 piecewise quintic polynomials and the first and second order partial derivatives that define the network can be computed by solving a sparse linear system with $5N$ equations and unknowns. With these first and second order partial derivatives, cubic varying cross boundary derivatives are used. The triangle patch method is a C^2 side-vertex procedural method that use quintic Hermite blending functions. This technique yields excellent results on smoothly varying data.

2. Side-Vertex Method

Suppose that we are given a planar triangle $V_1 V_2 V_3$ (labeled counterclockwise) and values of F, F_x and F_y on the boundary of the triangle. The cubic blended side-vertex interpolant in [12] is a weighted combination of three partial interpolants $D_1[F]$, $D_2[F]$ and $D_3[F]$. If (x, y) is a point in the triangle, let (b_1, b_2, b_3) be the barycentric coordinates (see [5, 12]) that satisfy $b_1 + b_2 + b_3 = 1$ and $(x, y) = b_1 V_1 + b_2 V_2 + b_3 V_3$. The point of intersection on the edge $V_j V_k$ with the line through V_i and (x, y) will be denoted by S_i, where

$$S_i = \left(\frac{x - x_i b_i}{1 - b_i}, \frac{y - y_i b_i}{1 - b_i} \right). \tag{1}$$

The operator $D_i[F]$ is defined at (x, y) by

$$D_i[F](x, y) = H_0(t)F(V_i) + H_1(t)F(S_i) + H_2(t)R_i'(0) + H_3(t)R_i'(1), \tag{2}$$

where $H_0(t), \ldots, H_3(t)$ are the standard cubic Hermite basis functions on $[0, 1]$, $t = 1 - b_i$ and

$$R_i'(0) = \frac{(x - x_i)F_x(V_i) + (y - y_i)F_y(V_i)}{1 - b_i}, \tag{3}$$

$$R_i'(1) = \frac{(x - x_i)F_x(S_i) + (y - y_i)F_y(S_i)}{1 - b_i}. \tag{4}$$

Along the line segment from V_i to S_i, $D_i[F]$ is the cubic polynomial interpolant that matches position and the directional derivative of F at the vertex V_i and at S_i. Although several different combinations of the $D_i[F]$ are considered in [12], a commonly used combination is

$$D[F](x, y) = \sum_{i=1}^{3} W_i(x, y)D_i[F](x, y). \tag{5}$$

where (b_1, b_2, b_3) are the barycentric coordinates of (x, y) and

$$W_i(x, y) = \frac{b_j^2 b_k^2}{b_1^2 b_2^2 + b_1^2 b_3^2 + b_2^2 b_3^2}, \tag{6}$$

for $i \neq j \neq k$. If F is C^1 on the triangle and C^2 on each edge, then $D[F]$ is a C^1 function with removable singularities at the V_i, which interpolates both position and first derivatives of F on the boundary of the triangle.

If F is a cubic polynomial along each triangle edge, then we can use the following lower degree weight functions

$$W_i(x, y) = \frac{b_j b_k}{b_1 b_2 + b_1 b_3 + b_2 b_3}. \tag{7}$$

Since all of the boundary curves in the remainder of this paper are cubics, all references to the side-vertex method will use the weight functions in (7) together with (5).

For interpolation of scattered data using the side-vertex method, the points $V_i = (x_i, y_i)$ are first triangulated using some local or global optimal criteria, such as those in [1, 3, 9, 17]. The next step involves estimating partial derivatives $F_x(V_i)$ and $F_y(V_i)$ based upon some or all of the data (x_i, y_i, z_i). Several effective techniques for this derivative estimation are given in [7, 9, 13, 15, 18]. With these derivative values, a piecewise cubic network can be defined over the triangle edges in the following manner.

For the edge V_j to V_k, let $e_{jk} = (V_k - V_j)$ and denote the directional derivative by

$$\frac{\partial F}{\partial e_{jk}}(V) = \langle e_{jk}, (F_x(V), F_y(V)) \rangle,$$

where $\langle ., . \rangle$ is the inner or scaler product. If we represent points on the edge $V_j V_k$ by $E_{jk}(u) = (1 - u)V_j + uV_k$ for $0 \leq u \leq 1$, then the network curve on the edge can be represented by a cubic

$$F_{jk}(u) = H_0(u)F(V_j) + H_1(u)F(V_k) + H_2(u)\frac{\partial F}{\partial e_{jk}}(V_j) + H_3(u)\frac{\partial F}{\partial e_{jk}}(V_k). \tag{8}$$

Figure 1. Contours of the side-vertex interpolant with linear cross boundary derivatives

The value $F(S_i)$ in (2) can be computed as $F_{jk}(\bar{u})$, where $\bar{u} = b_k/(b_j + b_k)$ because S_i in (1) can also be written in the form

$$S_i = E_{jk}(\bar{u}) = \frac{b_j}{1 - b_i}V_j + \frac{b_k}{1 - b_i}V_k.$$

Since we will later need the e_{jk} direction derivative of F at S_i, we observe now that

$$\frac{\partial F}{\partial e_{jk}}(S_i) = F'_{jk}(\bar{u}).$$

To complete the definition of a C^1 composite surface, we need to specify values of F_x and F_y along the triangle edges. A commonly used approach is to have them vary linearly from one vertex to the other. Figure 1 displays contour curves for the side-vertex interpolant using linearly varying cross boundary derivatives applied to the $N = 33$ point sets generated by the test function $F_1(x, y)$ in [7]. The Delaunay triangulation and exact partial derivatives at the vertices V_i are used in this and subsequent figures. Although the composite interpolant is C^1, observe that the contour curves are poorly behaved near many of the triangle edges.

If the cross boundary derivatives vary quadratically, then it is shown in [6] that the side-vertex method can be implemented in a compact hybrid Bézier form. They also give an algorithm for selecting cross boundary derivatives that has cubic precision and it yields significantly smoother plots. The following section uses higher degree cross boundary derivatives to increase the visual smoothness.

3. Cross Boundary Derivatives

In this section, we describe some techniques for selecting cross boundary derivatives that take advantage of the transfinite capabilities of the side-vertex method. The

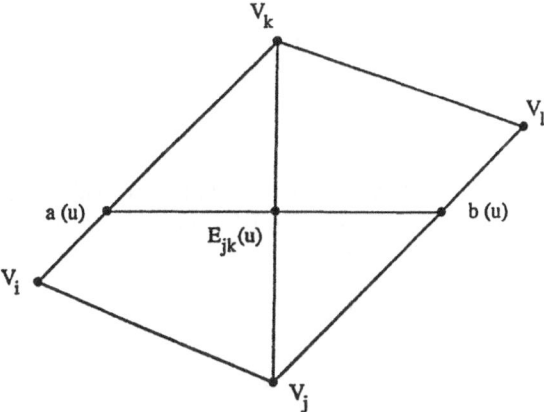

Figure 2. Two adjacent domain triangles

basic approach involves estimating the directional derivatives perpendicular to the edge by using the values of the network curves on two adjacent triangle patches. Let $n_{jk} = (y_k - y_j, x_j - x_k)$ be the normal vector to $V_j V_k$ in the direction away from V_i. It will be computationally efficient to represent $R_i'(1)$ in (4) by directional derivatives in n_{jk} and e_{jk}. Since

$$\frac{\partial F}{\partial n_{jk}}(V) = \langle n_{jk}, (F_x(V), F_y(V)) \rangle, \tag{9}$$

it follows from (4) and (9) that

$$R_i'(1) = \frac{\langle p_i, n_{jk} \rangle}{\|n_{jk}\|^2} \frac{\partial F}{\partial n_{jk}}(S_i) + \frac{\langle p_i, e_{jk} \rangle}{\|e_{jk}\|^2} \frac{\partial F}{\partial e_{jk}}(S_i), \tag{10}$$

where $p_i = (x - x_i, y - y_i)/(1 - b_i)$.

The e_{jk} directional derivative in (10) is simply the derivative of the cubic network curve and the n_{jk} derivative will now be described. Consider the two adjacent triangles in Fig. 2 with the common edge $V_j V_k$. Let $E_{jk}(u) = (1 - u)V_j + uV_k$ be a point on $V_j V_k$ and let $a(u)$ and $b(u)$ be points on the bordering triangle edges such that $a(u)$, $b(u)$ and $E_{jk}(u)$ are colinear and the line through them is perpendicular to $V_j V_k$. Let $t_1(u) = \|E_{jk}(u) - a(u)\| \|n_{jk}\|$, $t_2(u) = \|E_{jk}(u) - b(u)\| \|n_{jk}\|$, and define $Q(t, u)$ to be the quadratic interpolant in t to the three points $(0, F(E_{jk}(u)))$, $(t_1(u), F(a(u)))$ and $(t_2(u), F(b(u)))$. Define

$$G(u) = \frac{\partial Q}{\partial t}(0, u) = F(E_{jk}(u))\frac{-t_1(u) - t_2(u)}{t_1(u)t_2(u)} + F(a(u))\frac{t_2(u)}{t_1(u)(t_2(u) - t_1(u))}$$

$$+ F(b(u))\frac{t_1(u)}{t_2(u)(t_1(u) - t_2(u))}. \tag{11}$$

$G(u)$ is an estimate to the n_{jk} directional derivative of F at the point $E_{jk}(u)$ using the derivative of the corresponding quadratic interpolant to the three points of the network curves. Although $G(u)$ generally yields effective cross boundary derivatives along the edge $V_j V_k$, some modification is required because of the following problems. At the points where $a(u) = V_i$ and $b(u) = V_l$, $G(u)$ is only C^0, thus the interpolant will only be C^0. Another problem occurs if one of the angles in either triangle is greater than or equal to 90 degrees at V_j or V_k. If this occurs at V_j, for example, then $G(0)$ would not necessarily be equal to the value computed by (9) at V_j. A computational problem with using $G(u)$ on the entire edge is that there is no simple closed formula that can be easily stored or evaluated. Thus for each evaluation point, after S_i is computed, it would be necessary to compute the points $a(u)$ and $b(u)$ before calculating $G(u)$.

Instead of using $G(u)$, we use a polynomial approximation to $G(u)$ of degree M, where M could be different for each edge in the triangulation. Since we have partial derivative estimates for F_x and F_y at the vertices, define $G(0)$ by (9) with $V = V_j$, and define $G(1)$ by (9) with $V = V_k$. With $u_m = m/M$, we define the cross boundary derivatives at an arbitrary point $E_{jk}(u)$ by

$$\frac{\partial F}{\partial n_{jk}}(E_{jk}(u)) = \sum_{m=0}^{M} G(u_m)B_m^M(u), \tag{12}$$

where $B_m^M(u)$ is the mth Bernstein basis function of degree M. We also considered polynomial interpolation using Lagrange basis functions in (12), but we ran into the expected stability problems associated with polynomial interpolation.

A minor modification of this method is required for boundary triangles because adjacent triangles are used. For example, if $V_j V_k$ is a boundary edge and $a(u)$ is on an interior edge, then let $b(u)$ be the intersection of the ray $\overrightarrow{E_{jk}(u)a(u)}$ with the edge of the next triangle. Since $a(u)$ and $b(u)$ are on the same side of $V_j V_k$, we change the sign of $t_2(u)$ and then apply (11). If both $E_{jk}(u)$ and $a(u)$ are on boundary edges, then set

$$G(u) = \frac{F(E_{jk}(u)) - F(a(u))}{-t_1(u)}.$$

To implement this approach, after F_x and F_y are computed for each vertex, values $G(u_1), \ldots, G(u_{M-1})$ can be computed and stored for each edge. An alternative that avoids storing the values of $G(u_i)$ for all edges is to evaluate all points in one triangle at a time and temporarily store the $G(u_i)$ only for the three edges of that triangle. The storage/time trade-off in this case is that the $G(u_i)$ are computed twice for each edge. To evaluate $D_i[F](x, y)$ in (2), we compute $F(S_i)$ and the e_{jk} directional derivative at S_i using the discussion following (8). The n_{jk} directional derivative is computed by (12) with $u = b_k/(b_j + b_k)$ and $R_i'(1)$ is then computed by (10).

4. Examples

Figure 3 contains contour plots of the side-vertex interpolant with cross boundary derivatives computed by (12) with $M = 4$. This interpolant uses the same network

Figure 3. Contours of the interpolant using cross boundary derivatives of degree $M = 4$

curves as those in Fig. 1 and it is observed that the contours in Fig. 3 are visually smoother than those in Fig. 1. The observed maximum and RMS errors for these methods do not differ much because the errors are influenced more by the boundary curves, which are identical for the two methods. The smoothly shaded surface plots in Fig. 4 also shows the difference between these two approaches. The linearly varying cross boundary derivatives yields ridges along many of the triangle edges, while the new approach does not. The small squares in Fig. 4 are the given data points.

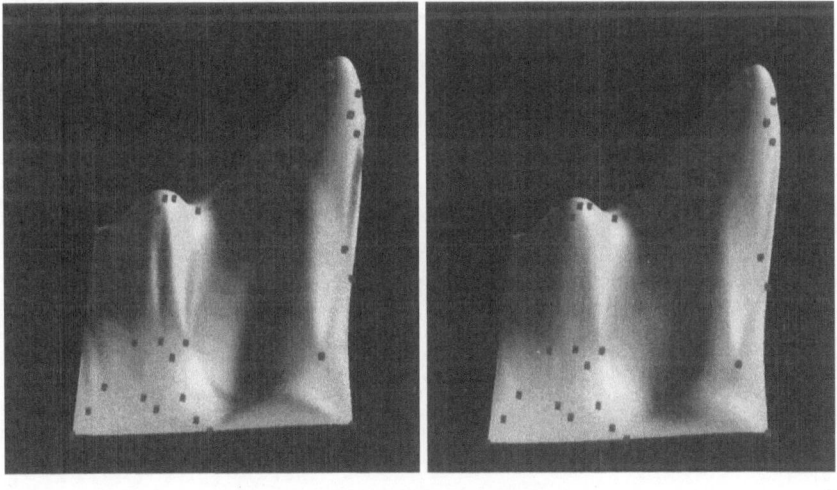

a   b

Figure 4. Smooth shaded surfaces of the interpolants in Fig. 1 in **a** and Fig. 3 in **b**

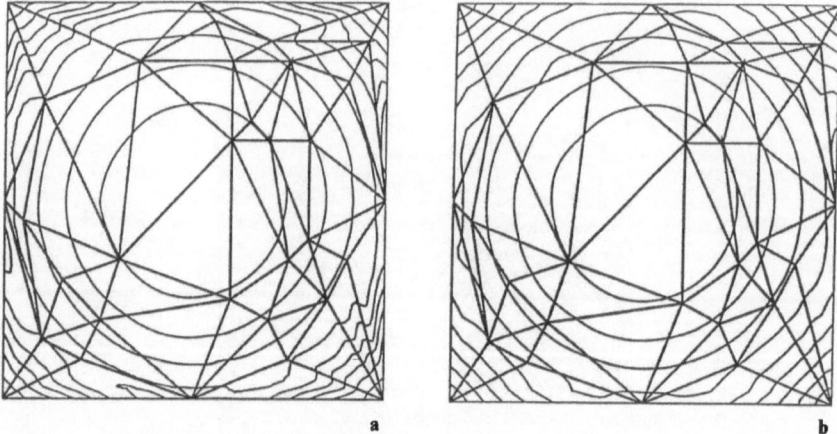

Figure 5. Isophotes for the interpolant to the sphere data whose cross boundary derivatives are **a** linear and **b** of degree $M = 4$

We applied these methods to the other data sets in [7], and the results were consistent with those shown here. On the $N = 33$ point set in [7] generated by the sphere test function $F_6(x, y)$, the contour plots of both methods appeared similar. However, the isophotes of the two methods on the sphere data in Fig. 5 indicate that the new method yields a significantly smoother surface. Displaying isophotes is an effective surface interrogation technique used in [11, 15] for observing the smoothness of the surface normals. Isophotes are points on the surface that have equal diffuse light intensity and they can be computed by contouring the function whose value is the angle between the surface normal and the light direction.

Figure 6. Isophotes for the interpolants to F_1 whose cross boundary derivatives are **a** linear and **b** of degree $M = 4$

Isophotes are similar to the reflection lines in [4], except that the strip lights are concentric circles. If the surface is G^k, then the isophote curves are G^{k-1}.

Figure 6 contains isophote plots for the interpolants in Figs. 1 and 3, and the improved smoothness is apparent. For the $N = 33$ point set in [7] generated by the saddle test function $F_3(x, y)$, the contour plots of both methods appear similar on this smooth data. The isophote plots for the interpolants are shown in Fig. 7, and it is observed that the new method with $M = 4$ is significantly smoother than when linear cross boundary derivatives are used.

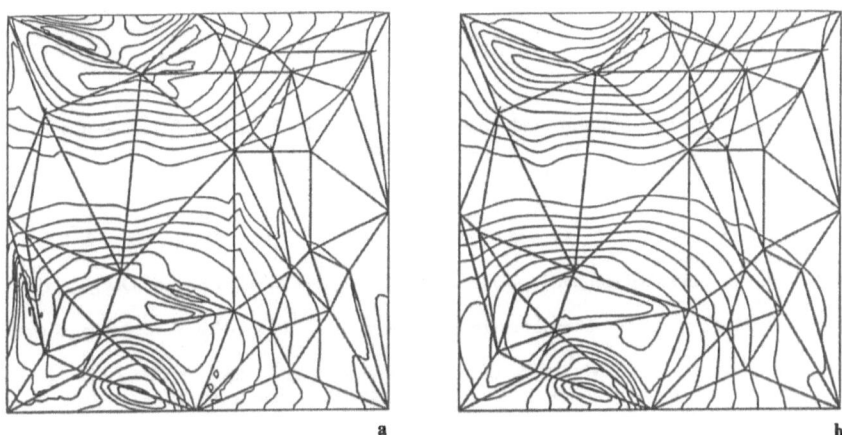

a   b

Figure 7. Isophotes for the interpolant to the saddle data whose cross boundary derivatives are **a** linear and **b** of degree $M = 4$

The optimal choice of M in (12) depends on the data. We have observed very little change for $M \geq 5$, thus an appropriate value should be in the range $2 \leq M \leq 5$. An acceptable default value is $M = 4$. For the case when $M = 2$, the hybrid Bézier patch in [6] can be used to represent the side-vertex patch if cubic boundary curves are used. We also applied these methods using derivatives at the vertices that were generated by the minimum norm network approach in [13], and the results were similar to the examples shown here.

5. Concluding Remarks

Although our examples involved the cubic blended side-vertex method, the method for selecting cross boundary derivatives can be applied to transfinite patches with more general network curves, such as exponential splines in tension. Also, this approach for computing cross boundary derivatives is not restricted to triangle patches. The same basic approach can be used with Coons patches or Gregory patches (see [5]). For transfinite patches that require second derivative boundary information, we can also generate these by taking the second derivative in t of $Q(t, u)$

at discrete points on the edge. With some modification, this basic technique of estimating cross boundary derivatives based on neighboring network curves can also be used for parametric surface patches, such as the transfinite triangular patches in [10, 14].

Acknowledgements

A portion of this research was supported by the NSF and AFOSR grant DMS-9116930 at Arizona State University. This was also supported by an Associated Western Universities sabbatical fellowship at Lawrence Livermore National Lab under the DOE contract W-7405-Eng-48.

References

[1] Barnhill, R. E.: Representation and approximation of surfaces. In: Rice, J. R. (ed.) Mathematical Software III, pp. 69–120. New York: Academic Press 1977.

[2] Barnhill, R. E., Brikhoff, G., Gordon, W. J.: Smooth interpolation in triangles. J. Approx. Th. 8, 114–128 (1973).

[3] Dyn, N., Levin, D., Rippa, S.: Algorithms for the construction of data dependent triangulations. In: Mason, J. C., Cox, M. G. (eds.) Algorithms for approximation II, pp. 185–192. London: Chapman and Hall 1990.

[4] Farin, G.: A modified Clough-Tocher interpolant. Computer Aided Geometric Design 2, 19–27 (1985).

[5] Farin, G.: Curves and surfaces for computer aided geometric design. New York: Academic Press 1990.

[6] Foley, T. A., Opitz, K.: Hybrid cubic Bézier triangle patches. In: Lyche, T., Schumaker, L. L. (eds.) Mathematical Methods in Computer Aided Geometric Design II, pp. 275–286. New York: Academic Press 1992.

[7] Franke, R.: Scattered data interpolation: tests of some methods. Math. Comp. 38, 181–200 (1982).

[8] Franke, R.: Recent advances in the approximation of surfaces from scattered data. In: Chui, C. K., Schumaker, L. L., Utreras, F. (eds.) Topics in multivariate approximation, pp. 175–184. New York: Academic Press 1987.

[9] Franke, R., Nielson, G. M.: Scattered data interpolation and applications: a tutorial and survey. In: Hagen, H., Roller, D. (eds.) Geometric modeling: methods and their applications. Heidelberg: Springer 1991.

[10] Hagen, H., Pottmann, H.: Curvature continuous triangular interpolants. In: Lyche, T., Schumaker, L. L. (eds.) Mathematical methods in computer aided geometric design, pp. 373–384. New York: Academic Press 1989.

[11] Hagen, H., Schreiber, T., Gschwind, E.: Methods for surface interrogation. In: Kaufman, A. (ed.) Visualization '90, pp. 187–193. Los Alamitos, CA: IEEE Press 1990.

[12] Nielson, G. M.: The side-vertex method for interpolation in triangles. J. Approx. Th. 25, 318–336 (1979).

[13] Nielson, G. M.: A Method for interpolation of scattered data based upon a minimum norm network. Math. Comp. 40, 253–271 (1983).

[14] Nielson, G. M.: A transfinite, visually continuous, triangular interpolant. In: Farin, G. (ed.) Geometric modeling: algorithms and new trends, pp. 235–246. Philadelphia, PA: SIAM 1987.

[15] Pottmann, H.: Scattered data interpolation based upon generalized minimum norm networks. Constr. Approx. 7, 247–256 (1991).

[16] Schumaker, L. L.: Fitting surfaces to scattered data. In: Lorentz, G. G. Chui, C. K., Schumaker, L. L. (eds.) Approximation theory, New York: Academic Press 1976.

[17] Schumaker, L. L.: Triangulation methods. In: Chui, C. K., Schumaker, L. L., Utreras, F. (eds.) Topics in multivariate approximation. New York: Academic Press 1987.

[18] Stead, S. E.: Estimation of gradients from scattered data. Rocky Mt. J. Math 14: 219–232 (1984).

T. A. Foley
Arizona State University
Tempe, AZ 85287, U.S.A.

Computing Suppl. 8, 101–115 (1993)

© Springer-Verlag 1993

Reconstruction of C^1 Closed Surfaces with Branching

T. N. T. Goodman, B. H. Ong, and K. Unsworth, Dundee

Abstract. This paper describes a scheme for reconstructing 3D objects from cross-sectional data, where bifurcation is allowed between adjacent heights. The scheme is an extension of an earlier method developed by the same authors, in which a closed C^1 surface, representing the boundary of an object, is generated using a parametric piecewise rational cubic shape preserving curve interpolation scheme and cubic Hermite interpolation, but in which no account of branching was taken. In this method, a 'saddle' surface patch, represented by a hyperboloid, is fitted in a C^1 manner to accomodate the branching of the boundary surface between two heights. The paper gives a detailed account of precisely how this is done, and some numerical examples are given.

Key words: Surface reconstruction, contour data, parametric surface, C^1 interpolation, serial sections.

1. Introduction

In [5], the authors presented a C^1 scheme for reconstructing 3D objects from cross-sectional data. However as noted in that report, the method restricts itself to cases in which no branching occurs between the heights at which the data are given, i.e. each height has precisely one contour. In this paper, we address the problem of branching, and a scheme is developed for extending the method of [5] to deal with such cases while still retaining C^1 continuity. To date, we have considered the case in which one contour branches into two, however the application of the method to cater for more complex situations is to be investigated. As noted in [8], one of the main application areas for 3D reconstruction algorithms is in computerized tomography, and a number of references associated with work in this area are given in [5]. However, Schumaker's review article [8] provides many more references, together with an introduction to the current state of the art in this field.

The particular problem of branching has been addressed in [2], in which the boundary surface of the object comprises triangular surface patches. In [2], to deal with branching, additional triangular facets are introduced with a vertex half-way between the two heights between which branching occurs. This results in a relatively simple approach to the problem which works well if both of the upper contours are convex. In [1] a 3D Delaunay tetrahedrization of the volume of the object is obtained from the 2D Delauney triangulation of the points on adjacent contours. Once this is obtained, the surface of the object may be obtained by consideration of the boundary of this volume. Wang et al. [9] describe a three step approach to the problem. First, an association is built up between adjacent contours. This is followed by the generation of a local surface representation using planar triangular patches between associated contours. The final step involves the generation of a surface representation hierarchy by coalescing certain adjacent patches, thereby obtaining a reduction in the amount of stored data necessary to represent the surface. None of the above schemes however addresses the problem of generating a smooth surface, i.e. each surface is represented by planar polygonal facets, thereby creating surfaces with just C^0 continuity.

A different approach is offered by Lin et al. [7]. Given data on adjacent contours, intermediate contours are generated using an elastic interpolation scheme. This set of contours is then used as an initial approximation to a surface defined over a rectangular grid. A surface value is given by the y-value of a point lying on a contour, the location of this point in the grid being determined from the x- and z-values. The x-, z-values corresponding to the remaining grid points are obtained by a combination of linear and cubic B-spline interpolation. It then remains to generate y-values at all of these grid points, which is achieved using a constrained optimization scheme. The final result comprises a set of discrete points lying on the boundary surface of the original object.

The fundamental idea behind the present scheme is to introduce a surface patch, in the shape of a saddle, between the heights where branching is known to occur. The saddle shape can be suitably represented by a hyperboloid; this particular surface offering an advantage when it comes to blending it with the remainder of the surface, since at a fixed height the hyperbola lying on this surface can be represented in the rational Bézier form. Thus given a set of heights

$$z^1 < z^2 < \cdots < z^l < z^u < \cdots < z^{n-1} < z^n,$$

we assume that the cross-sectional data has been specified at each height. At each of these heights either one or two closed planar curves which interpolate(s) the set(s) of ordered data will be generated, this (these) being known as the contour(s) at each height. A contour is generated using the local convexity preserving parametric interpolation scheme described in [4, 6]. The application of the scheme to this problem is described in [5], and in the present paper we assume that these contours have been obtained. Suppose for each of the heights z^i, $i = 1, \ldots, l$, there is one contour, the remaining heights each containing two contours. At $z = z^l$, the contour must be 'split', in order to associate an appropriate part of this contour with the two contours at the height z^u. This is described in 2. The establishment of a one to one correspondence between points on adjacent contours using the 'minimum distance' criterion as presented in [5] is explained in 3. The main part of the work occupies 4. This describes the generation of the saddle surface, 'located' at $z = z^s$, where $z^s = (z^l + z^u)/2$. It contains a description of the way in which a contour at $z = z^s$ is generated (4.1), the representation of the saddle surface patch (4.2), additional contours introduced at $z = z^{ls} = (z^l + z^s)/2$ and $z = z^{us} = (z^u + z^s)/2$ (4.3), and finally (4.4) the matching of parameters between each contour between z^l and z^u. In order to obtain the final surface, the contours and saddle must be blended in a C^1-manner, e.g. using Hermite interpolation. This is explained in 5. The surface is closed with the introduction of a base point at $z \geq z^0$, and two crown points at $z \leq z^{n+1}$, see 6. In 7 an outline of the entire reconstruction algorithm is presented; with numerical results obtained from an implementation of the algorithm discussed in 8. The paper finishes with a summary and proposals for further work on the scheme.

With regard to notation used in the paper, various references are made to curves which interpolate two points, say **A** and **B**. Unless otherwise indicated, **AB** will refer to the curve which starts at **A** and terminates at **B**, moving in an anti-clockwise direction.

2. Contour Splitting

We consider the case where, for the surface section in which branching has occurred, the lower branching height, z^l, contains one contour, while the upper branching height, z^u, contains two contours. In order to blend the contours at these two heights, the corresponding portions of the lower contour which are to be blended with each of the upper contours need to be identified.

Let C_1, C_2 represent the upper contours, and L the lower contour. Let G_1, G_2 be the centroids of the data interpolated by C_1, C_2 respectively. A positive even integer, M, is chosen and the points P_i, $i = 1, \ldots, M$, which are uniformly spaced parametrically (as in [5]), are chosen from L and are divided into two groups, Λ_1, Λ_2, with Λ_i associated with G_i, $i = 1, 2$. This is done as follows. First, identify a point S_1 such that, from amongst all the P_i, $i = 1, \ldots, M$, S_1 generates the minimum value for d where

$$d = |G_1 - P_i|/|G_2 - P_i|, \qquad i = 1, \ldots, M,$$

i.e. $S_1 = P_j$ for some $1 \leq j \leq M$; for which d is a minimum. Then let $S_1 \in \Lambda_1$. Similarly identify a point S_2 such that $1/d$ is a minimum, and let $S_2 \in \Lambda_2$. The membership of the sets Λ_i, $i = 1, 2$, is then increased by recruiting outwards in both directions along the contour, with points being added to Λ_1 (resp. Λ_2) if they are nearer to G_1 (resp. G_2) than to G_2 (resp. G_1). More precisely, points are added to Λ_1 by moving in a clockwise direction from S_1 with the condition that any point P_i is added to Λ_1 if

$$|G_1 - P_i| \leq |G_2 - P_i|. \qquad (2.1)$$

After such a point is added to Λ_1, its immediate neighbour is considered, with the recruitment terminating as soon as (2.1) is not satisfied. The same operation is then repeated from S_1 moving in an anti-clockwise direction. Λ_2 is extended in a similar manner, the condition being that '$\leq$' is replaced with '$>$' in (2.1). After both operations are completed, there may be some P_i which are not in either Λ_1 or Λ_2; suppose $\{Q_1, \ldots, Q_k\}$ is a set of such points. Then $\{Q_1, \ldots, Q_j\}$ are allocated to Λ_1 and $\{Q_{j+1}, \ldots, Q_k\}$ are allocated to Λ_2 where j ($0 \leq j \leq k$) is the integer which minimizes

$$\sum_{l=1}^{j} |\mathbf{G}_1 - \mathbf{Q}_l| + \sum_{l=j+1}^{k} |\mathbf{G}_2 - \mathbf{Q}_l|.$$

Note that the inclusion of 0 and k as possible values for j is used to indicate that one of the two subsets may be empty. Finally, the points in Λ_1, Λ_2 are ordered, in an anti-clockwise direction; with the first and last points of Λ_i, $(i = 1, 2)$ denoted by $\mathbf{F}_i$, $\mathbf{L}_i$ respectively (see Fig. 1). L is now 'cut' at $\mathbf{F}_1$, $\mathbf{L}_1$, $\mathbf{F}_2$, $\mathbf{L}_2$; dividing it into four curve segments, and segment $\mathbf{F}_i\mathbf{L}_i$ is associated with the upper contour C_i, $i = 1, 2$.

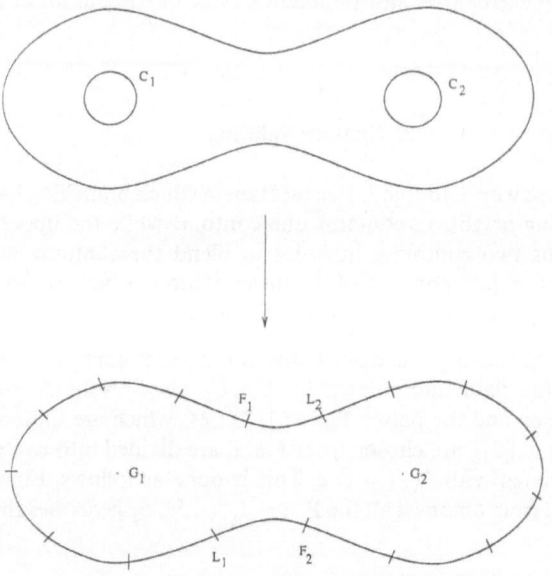

Figure 1. Splitting of contours

From the above discussion it would seem that when building Λ_1 and Λ_2 either one could finish containing only one point, namely $\mathbf{S}_1$ or $\mathbf{S}_2$. To avoid this undesirable situation, the data at z^u first undergo a horizontal translation, so that the centroid of all of the data at height z^u is vertically above the centroid of the data at z^l. Once this has been done, the contour splitting proceeds. To date, this vertical alignment of the centroids has led to satisfactory results when splitting the contours at z^l.

3. Matching Parameters

As explained in [5], in order to blend adjacent contours to form a surface section, a 1-1 correspondence between points with the same parameter values on each of these contours must be established. In this scheme, this operation is done precisely as in [5] for all heights at or below z^l, and at or above z^u, except that there are now clearly two correspondences to be set up at each height $\geq z^u$. Now consider the correspondence between the contours at z^l and z^u. For each of the open curves $\mathbf{F}_i\mathbf{L}_i$

($i = 1, 2$), as generated in 2, M points which are uniformly spaced parametrically and inclusive of $\mathbf{F}_i$ and $\mathbf{L}_i$ are chosen. A 1-1 correspondence between these points and M points chosen from the appropriate upper closed contour C_i ($i = 1, 2$) is then established using the same criterion as in 3 of [5]. We denote the points at height z^u which correspond to $\mathbf{F}_i$, $\mathbf{L}_i$ by $\mathbf{H}_i$, $\mathbf{I}_i$ respectively for $i = 1, 2$. The 'saddle' will be inserted within the region 'enclosed' by these eight points.

It should be noted that before establishing the above 1-1 correspondences, the centroids of adjacent data sets could be 'matched' as described in 2. However this extra operation did not seem to offer any significant benefits.

4. The Saddle Surface Patch

The generation of the saddle surface patch and subsequent inclusion in the entire surface involves a number of different processes. These are described in 4.1–4.4.

4.1 Generation of the 'Figure of 8' Contour

The height of the saddle, z^s, may be anywhere between z^l and z^u. We will assume that $z^s = (z^l + z^u)/2$. Having established the correspondence between the curves $\mathbf{H}_i\mathbf{I}_i$ and $\mathbf{F}_i\mathbf{L}_i$, the curve $\mathbf{H}_i\mathbf{I}_i$ may be blended with $\mathbf{F}_i\mathbf{L}_i$ using Hermite interpolation as described in [5]. Note however, the purpose of this blending operation is not to generate a surface section between z^l and z^u, but to provide a mechanism for obtaining points at height z^s. We denote the set of points C_i^s ($i = 1, 2$) as containing those points obtained by blending $\mathbf{H}_i\mathbf{I}_i$ with $\mathbf{F}_i\mathbf{L}_i$. A contour in the shape of a 'figure of 8', as shown in Fig. 2, is now generated as described below.

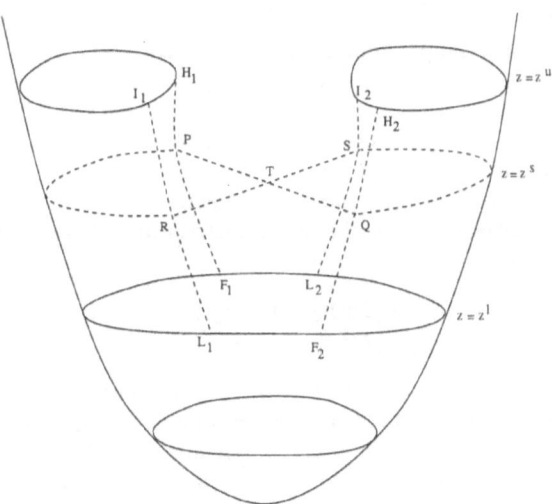

Figure 2. Location of 'figure of 8' contour

From this figure we note that **P** is the point which lies on the Hermite interpolant $F_1 H_1$, similarly **R** lies on $L_1 I_1$, **Q** on $F_2 H_2$ and **S** on $L_2 I_2$. The point **T** denotes the point of intersection of the two straight lines **PQ** and **RS**. This will be referred to as the 'saddle point". Four points **A, B, C, D** are chosen, lying on the straight lines **PT, TQ, RT, TS** respectively such that

$$|AT| = |BT| = |CT| = |DT| = h \min(|PT|, |QT|, |RT|, |ST|), \qquad h \in (0, 1].$$

In the implementation, h typically has a value of 0.5.

First, an open contour segment is generated interpolating **A, P**, the points in C_1^s, **R** and **C**. This is done using the interpolation scheme described in [4, 6], and also used in [5]. This is then repeated for **B, Q**, the points in C_2^s, **S** and **D**. The final contour is obtained by joining **A** to **B** and **C** to **D** by straight line segments.

4.2 Generation of the Saddle

The 'saddle' shape that we wish to include between z^l and z^u may be obtained by representing this surface patch as a hyperboloid. This may be written in the general form:

$$(a_1 x + b_1 y + c_1)(a_2 x + b_2 y + c_2) = kz \qquad (4.1)$$

where, in this application,

$$\left. \begin{aligned} a_1 x + b_1 y + c_1 = 0 \\ a_2 x + b_2 y + c_2 = 0 \end{aligned} \right\}$$

will represent the equations of the two straight lines joining **A** to **B** and **C** to **D** respectively, and where k is an arbitrary real number. Varying the magnitude of k will alter the shape of the saddle.

Let us now consider a simple example of (4.1), namely

$$x^2 - y^2 = z.$$

Consider cross-sections of this surface, for fixed z. When viewed along the z-axis we obtain cross-sectional curves as given in Fig. 3. It can be seen that these curves are in fact hyperbolae, which is also true for the general quadric surface given by (4.1). A conic section may be represented as a parametric rational quadratic curve, which has the Bézier form

$$\frac{b_0 (1 - t)^2 + b_1 2h(1 - t)t + b_2 t^2}{(1 - t)^2 + 2h(1 - t)t + t^2}, \qquad 0 \le t \le 1. \qquad (4.2)$$

As is well known, this curve interpolates b_0, b_2 at its end points, while the straight line joining b_1 to the midpoint of $b_0 b_2$ intersects the curve in the ratio $1 : h$, see Fig. 4. If $h > 1$, this conic section is a hyperbola. A further useful property is that the expression (4.2) can be easily transformed into a rational cubic by degree raising [3]; this will be required later when the curve is considered as a segment of a piecewise rational cubic curve.

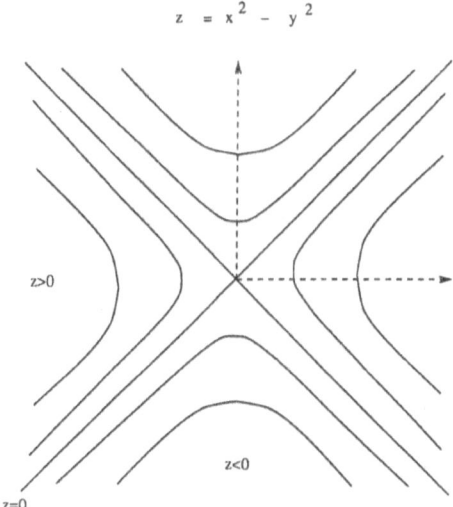

$$z = x^2 - y^2$$

Figure 3. Cross-sectional hyperbolae

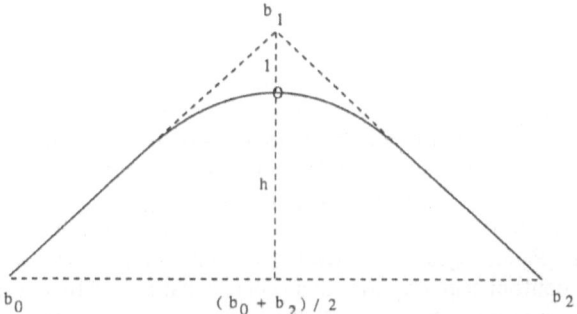

Figure 4. Rational Bézier form of hyperbola

Finally, we need to determine the region which will be occupied by the saddle. We define a lower saddle level by $z^{ls} = (z^l + z^s)/2$; similarly an upper saddle level by $z^{us} = (z^u + z^s)/2$. Consider the four vertical planes, passing through **A**, **B**, **C**, **D**; the first two planes parallel to the vertical plane containing **CD**, and the last two parallel to the vertical plane through **AB**. The intersection of the hyperboloid (4.1) with each of these planes is a straight line, and that portion of the hyperboloid which we require as a surface patch is bounded from the sides by these four planes and from above and below by the planes $z = z^{us}$ and $z = z^{ls}$ respectively. We denote the endpoints of the hyperbola at $z = z^{ls}$ by $\mathbf{D}_{13}, \mathbf{D}_{14}, \mathbf{D}_{23}, \mathbf{D}_{24}$ and the endpoints at $z = z^{us}$ by $\mathbf{U}_{13}, \mathbf{U}_{14}, \mathbf{U}_{23}, \mathbf{U}_{24}$. See Fig. 5. The straight lines denoting the intersection of the planes referred to above are given by $\mathbf{D}_{ij}\mathbf{U}_{ij}$ $(i = 1, 2; j = 3, 4)$. It should

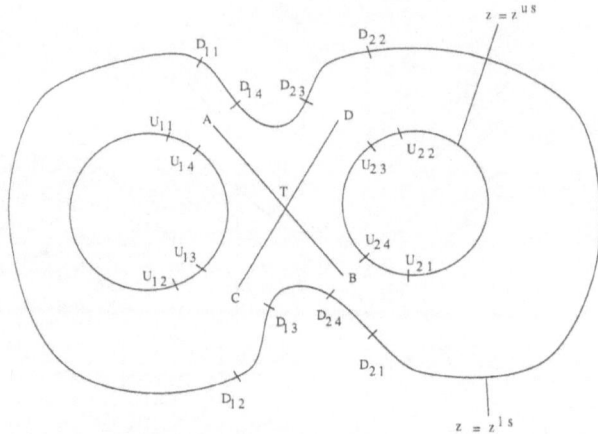

Figure 5. Matching of points on adjacent contours

be noted that this straight line intersection property will be significant when considering the blending of adjacent contours in 5.

4.3 The contours at $z = z^{ls}$ and $z = z^{us}$

It was stated in 4.2 that the saddle is bounded above and below by the planes $z = z^{us}$ and $z = z^{ls}$ respectively. Hence we need to supply contours at these heights, one at z^{ls} and two at z^{us}. As noted in 4.1, Hermite interpolation is used to generate data values at heights between z^l and z^u, and in particular at $z = z^{ls}$ and $z = z^{us}$. Consider the lower saddle level z^{ls}. Let $\mathbf{D}_{11}$, $\mathbf{D}_{12}$, $\mathbf{D}_{21}$ and $\mathbf{D}_{22}$ be the end points of two segments of a contour at z^{ls}, see Fig. 5, obtained from the Hermite interpolants $\mathbf{H}_1\mathbf{F}_1$, $\mathbf{I}_1\mathbf{L}_1$, $\mathbf{H}_2\mathbf{F}_2$ and $\mathbf{I}_2\mathbf{L}_2$ respectively as given in Fig. 2. Two open piecewise rational cubic contour segments are then generated at $z = z^{ls}$ by interpolating data from these two segments, obtained in the same manner as described in 4.1, together with the endpoints of the hyperbola from the saddle surface, i.e. $\mathbf{D}_{14}$, $\mathbf{D}_{13}$, $\mathbf{D}_{24}$, $\mathbf{D}_{23}$. Thus the final contour is represented using the scheme described in [4, 6] for the two open contours between $\mathbf{D}_{14}$ and $\mathbf{D}_{13}$, and $\mathbf{D}_{24}$ and $\mathbf{D}_{23}$ (moving anti-clockwise), and by (4.2), degree raised to a rational cubic, for the hyperbola. Note that this requires tangent directions and curvatures to be specified at $\mathbf{D}_{14}$, $\mathbf{D}_{13}$, $\mathbf{D}_{24}$ and $\mathbf{D}_{23}$. The tangent directions are obtained from the representation of the hyperbola. The curvature values could also be taken from the hyperbola; however when this was done values were generated which seemed too small to produce a 'visually pleasing' result. Hence the curvature values were actually chosen in the same manner as that described in [4, 6], e.g. that of the circle passing through $\mathbf{D}_{11}$, $\mathbf{D}_{14}$ and $\mathbf{D}_{23}$. Finally, as in [5], reparametrization of each segment of the contour is performed, in order to ensure unit magnitude of the tangent vectors at the end points of each segment, followed by normalization of the subsequent global parameter.

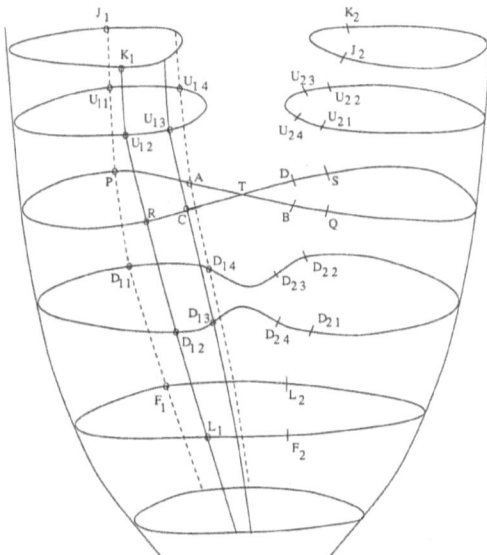

Figure 6. Upper and lower saddle levels

To construct the two contours at $z = z^{us}$, we need to obtain the values of some data points lying on these contours. First, a 1-1 correspondence is established between points lying on the contours at $z = z^u$ and on the corresponding closed part of the contour at $z = z^s$ using the 'minimum distance' criterion as in [5]. Prior to doing this, each closed part of the 'figure of 8' contour must be treated as a separate closed curve, so that reparametrization of each segment and normalization of the corresponding global parameter, must be performed as before. The purpose of this new correspondence is to orientate the saddle surface so that it fits into a natural position with its neighbouring surface sections. Having established the 1-1 correspondence, suppose the points on the contour at height z^u which correspond to the points **P, R** and **Q, S** are $\mathbf{J_1}$, $\mathbf{K_1}$ and $\mathbf{J_2}$, $\mathbf{K_2}$ respectively. See Fig. 6. Then blending is performed between the open curves **PR** and $\mathbf{J_1 K_1}$, and **QS** and $\mathbf{J_2 K_2}$, using Hermite interpolation, thereby allowing data points to be generated at $z = z^{us}$. Let $\mathbf{U_{11}}, \mathbf{U_{12}}, \mathbf{U_{21}}$ and $\mathbf{U_{22}}$ be four such data points lying on the interpolants joining $\mathbf{PJ_1}, \mathbf{RK_1}, \mathbf{QJ_2}$ and $\mathbf{SK_2}$ respectively. Two closed piecewise rational cubic contours are then constructed at $z = z^{us}$ in essentially the same manner as for the contour at $z = z^{ls}$, again degree raising the hyperbolae. Finally reparametrization of each segment, followed by normalization of the global parameter, are performed for each closed curve.

4.4 Matching Parameters

Having constructed the contours at $z = z^l, z^{ls}, z^s, z^{us}$ and z^u, we now need to establish 1-1 correspondences between points lying on adjacent contours. Consider in particular the contours at $z = z^l$ and $z = z^{ls}$. Matching these entire contours using the 'minimum distance criterion' requires, initially, M representative points to be chosen from each contour, these points being uniformly distributed in terms of parametric length. It was found, however, that this could result in a relatively high concentration of such points located within the regions $\mathbf{D}_{12}\mathbf{D}_{21}$ and $\mathbf{D}_{22}\mathbf{D}_{11}$ on the $z = z^{ls}$ contour, one reason for this being the occurrence of a significant 'indentation' of the $z = z^{ls}$ contour in these regions. This subsequently led to undesirable visual results. An alternative approach is to match points on corresponding open contour segments. This has in fact already partially been done; i.e. correspondences have been established between the open contour segments $\mathbf{F}_1\mathbf{L}_1$, $\mathbf{D}_{11}\mathbf{D}_{12}$, $\mathbf{PR}$, $\mathbf{U}_{11}\mathbf{U}_{12}$ and $\mathbf{J}_1\mathbf{K}_1$; and between $\mathbf{F}_2\mathbf{L}_2$, $\mathbf{D}_{21}\mathbf{D}_{22}$, $\mathbf{QS}$, $\mathbf{U}_{21}\mathbf{U}_{22}$ and $\mathbf{J}_2\mathbf{K}_2$. See Fig. 6. We now need to establish correspondences between the following contour segments: $\mathbf{L}_1\mathbf{F}_2$ and $\mathbf{D}_{12}\mathbf{D}_{21}$; $\mathbf{L}_2\mathbf{F}_1$ and $\mathbf{D}_{22}\mathbf{D}_{11}$; $\mathbf{U}_{12}\mathbf{U}_{11}$ and $\mathbf{K}_1\mathbf{J}_1$; $\mathbf{U}_{22}\mathbf{U}_{21}$ and $\mathbf{K}_2\mathbf{J}_2$; $\mathbf{D}_{12}\mathbf{D}_{13}$, $\mathbf{RC}$ and $\mathbf{U}_{12}\mathbf{U}_{13}$; $\mathbf{D}_{24}\mathbf{D}_{21}$, $\mathbf{BQ}$ and $\mathbf{U}_{24}\mathbf{U}_{21}$; $\mathbf{D}_{14}\mathbf{D}_{11}$, $\mathbf{AP}$ and $\mathbf{U}_{14}\mathbf{U}_{11}$; $\mathbf{D}_{22}\mathbf{D}_{23}$, $\mathbf{SD}$ and $\mathbf{U}_{22}\mathbf{U}_{23}$.

(a) $\mathbf{L}_1\mathbf{F}_2$ and $\mathbf{D}_{12}\mathbf{D}_{21}$

Since we must match $\mathbf{L}_1$ with $\mathbf{D}_{12}$, and $\mathbf{F}_2$ with $\mathbf{D}_{21}$, we must find a suitable parametrization for $\mathbf{D}_{12}\mathbf{D}_{21}$ which ensures the matching of these particular points while still retaining the same overall parametric length of unity and the C^1 property of the contour.

Suppose the original parametric length of $\mathbf{D}_{12}\mathbf{D}_{21}$ is L. Let the new parametric lengths of $\mathbf{D}_{12}\mathbf{D}_{13}$, $\mathbf{D}_{13}\mathbf{D}_{24}$ and $\mathbf{D}_{24}\mathbf{D}_{21}$ be l, p, m respectively where

$$l + p + m = L \qquad (4.3)$$

(see Fig. 6).

Let the magnitudes of the tangent vectors at $\mathbf{D}_{12}$ and $\mathbf{D}_{21}$ be M_1, M_2 respectively. The initial representations of the three curve segments comprising $\mathbf{D}_{12}\mathbf{D}_{21}$ are in the 'standard' rational cubic Bézier form, i.e.

$$\frac{\mathbf{A}_i\alpha_i(1-t)^3 + \mathbf{B}_it(1-t)^2 + \mathbf{C}_it^2(1-t) + \mathbf{D}_i\beta_it^3}{\alpha_i(1-t)^3 + t(1-t)^2 + t^2(1-t) + \beta_it^3}, \qquad 0 \le t \le 1, \qquad i = 1, 2, 3.$$
$$(4.4)$$

From (4.4), the magnitudes of the tangent vectors at the respective end points of the three segments; denoted by g and h, d and d, v and w, may be easily obtained. After a reparametrization of the form

$$t_i = \frac{s_i}{a_i(1-s_i) + s_i}, \qquad 0 \le s_i \le 1, \qquad a_i \in \mathbb{R}, \qquad i = 1, 2, 3,$$

and scaling of the unit parametric length by l, p, m respectively, we have, by equating the tangent magnitudes to preserve C^1 continuity,

$$M_1 = \frac{g}{a_1 l} \quad \text{at} \quad \mathbf{D}_{12},$$

$$\frac{ha_1}{l} = \frac{d}{a_2 p} \quad \text{at} \quad \mathbf{D}_{13},$$

$$\frac{da_2}{p} = \frac{v}{a_3 m} \quad \text{at} \quad \mathbf{D}_{24},$$

$$\frac{wa_3}{m} = M_2 \quad \text{at} \quad \mathbf{D}_{21}.$$

On simplifying we obtain

$$l = (ghp/M_1 d)^{1/2}, \qquad m = (vwp/M_2 d)^{1/2}. \tag{4.5}$$

From (4.3) and (4.5) we may obtain values for l, p and m. With the new distribution of these parametric lengths, these three segments are individually matched to the relevant portions of the curve $\mathbf{L}_1 \mathbf{F}_2$ by a linear 1-1 correspondence based upon parametric lengths. The matchings between $\mathbf{L}_2 \mathbf{F}_1$ and $\mathbf{D}_{22} \mathbf{D}_{11}$, $\mathbf{U}_{12} \mathbf{U}_{11}$ and $\mathbf{K}_1 \mathbf{J}_1$, $\mathbf{U}_{22} \mathbf{U}_{21}$ and $\mathbf{K}_2 \mathbf{J}_2$ are performed in a similar manner.

(b) $\mathbf{D}_{12} \mathbf{D}_{13}$, RC and $\mathbf{U}_{12} \mathbf{U}_{13}$

These mappings are straightforward linear mappings of the parameter values which map the corresponding endpoints. The matchings between $\mathbf{D}_{24} \mathbf{D}_{21}$, $\mathbf{BQ}$ and $\mathbf{U}_{24} \mathbf{U}_{21}$; $\mathbf{D}_{14} \mathbf{D}_{11}$, $\mathbf{AP}$ and $\mathbf{U}_{14} \mathbf{U}_{11}$; $\mathbf{D}_{22} \mathbf{D}_{23}$, $\mathbf{SD}$ and $\mathbf{U}_{22} \mathbf{U}_{23}$ are performed in a similar manner.

5. Blending of Contours

Blending between adjacent contours, for the construction of the surface sections, is done using Hermite interpolation as described in [5]. The tangent vector to the surface at a point is, in general, calculated using (5.2) of [5], or an analogous expression, which involves three points lying on adjacent contours. However, the tangent vectors at points on the contour segments $\mathbf{D}_{12} \mathbf{D}_{13}$, $\mathbf{D}_{14} \mathbf{D}_{11}$, $\mathbf{D}_{24} \mathbf{D}_{21}$, $\mathbf{D}_{22} \mathbf{D}_{23}$, and the corresponding points for $z = z^{us}$, require an alternative calculation in order to maintain C^1-continuity in the t direction at $\mathbf{D}_{13}, \mathbf{D}_{24}, \mathbf{D}_{23}$ and $\mathbf{D}_{14}$. For example let $\mathbf{G}_{12}, \mathbf{G}_{13}$ be the tangent vectors at $\mathbf{D}_{12}, \mathbf{D}_{13}$ respectively. $\mathbf{G}_{12}$ can be calculated using (5.1) of [5]. Note, however, that $\mathbf{D}_{13}$ lies on the saddle surface, and recall from 4.2 that the intersection of the hyperboloid with the plane containing $\mathbf{D}_{13} \mathbf{U}_{13}$ and parallel to the vertical plane containing $\mathbf{AB}$ is a straight line. See Fig. 5. Hence we may obtain an expression for $\mathbf{G}_{13}$ by considering the difference of any two points on this line. Consequently, the tangent vector at points in between $\mathbf{D}_{12}$ and $\mathbf{D}_{13}$ are calculated as a linear combination of $\mathbf{G}_{12}$ and $\mathbf{G}_{13}$. The same scheme is used for the other contour segments referred to above. Note that the formula (5.1) of [5] may still be used to calculate the tangent vector at points lying on $\mathbf{AP}$, $\mathbf{RC}$, $\mathbf{BQ}$ and $\mathbf{SD}$ because of the 'straight line intersection' property.

6. The Crowns and the Base

Recall that in [5], a surface was closed by introducing two extra points, the crown point and the base point, located at heights $T \in (z^n, z^{n+1}]$ and $B \in [z^0, z^1)$ respectively. This is also done in this scheme in the same manner, except that clearly there will be one crown, or base, for each branch. These are blended again as described in [5].

7. An Outline of the Algorithm

We now outline the major steps of an algorithm to reconstruct a 3D object using the ideas described in 2–6.

Step 1
Interpolate the data at z^i, $(i = 1, \ldots, n)$ by one or more closed piecewise rational cubic curves, with reparametrization, scaling and subsequent normalization of the global parameters being performed as described in [5].

Step 2
Split the closed curve at $z = z^l$ into four open contour segments as described in 2.

Step 3
Match the parameters of adjacent contours according to the 'minimum distance' criterion as described in [5], with the contours at $z = z^u$ matched with two open contour segments at $z = z^l$.

Step 4
Construct the 'Fig. of 8' contour at $z = z^s$, identify the heights z^{ls} and z^{us}, and construct the saddle surface.

Step 5
Construct the contours at $z = z^{ls}$ and $z = z^{us}$.

Step 6
Match the parameters of the curves at the five heights z^l, z^{ls}, z^s, z^{us} and z^u; redistributing the parametric lengths at $z = z^{ls}$ and $z = z^{us}$ as described in 4.4.

Step 7
Construct the crown point(s) and base point.

Step 8
Blend adjacent contours, and the crowns and base, using Hermite interpolation, to generate surface sections as described in [5], with the tangent vector at a point on the surface calculated using (5.2) of [5], or an analogous expression, or as a linear combination of tangent vectors as described in 5.

Table 1

$z = 1$:	$(x, y) = (1, 0), (0, 1), (-1, 0), (0, -1)$
$z = 2$:	$(x, y) = (2, 0), (0, 2), (-2, 0), (0, -2)$
$z = 3$:	$(x, y) = (3, 0), (0.5, 2.5), (-3, 0), (0, -3)$
$z = 4$:	$(x, y) = (-1, 0), (-2, 1), (-3, 0), (-2, -1)$
	$(x, y) = (3, 0), (2, 1), (1, 0), (2, -1)$
$z = 5$:	$(x, y) = (-1, 0), (-2, 1), (-3, 0), (-2, -1)$
	$(x, y) = (3, 0), (2, 1), (1, 0), (2, -1)$
$z = 6$:	$(x, y) = (-1, 0), (-2, 1), (-3, 0), (-2, -1)$
	$(x, y) = (3, 0), (2, 1), (1, 0), (2, -1)$

Crown points:

$z^7 = 7$	calculated $(x, y, z) = (-2, 0, 6.618)$
	$(x, y, z) = (1.5, 0, 6.3)$
$z^0 = 0$	calculated $(x, y, z) = (0, 0, 0.764)$

8. Numerical Results

Example output, obtained from an implementation of the algorithm outlined above, is shown in Fig. 7–9. The output is generated from the data set given in Table 1. Note that for this data set, one of the two crown points is provided as input, while the other is calculated within the program. Figure 7 presents a view from the side and slightly below the resulting surface. Note that the middle region of the surface, comprising narrow polygons, indicates the region affected by the inclusion of the saddle surface patch. Figure 8 shows a different side view of the surface, giving more detail of the saddle. Finally a top view of the surface is presented in Fig. 9, showing full details of the two crowns and the saddle.

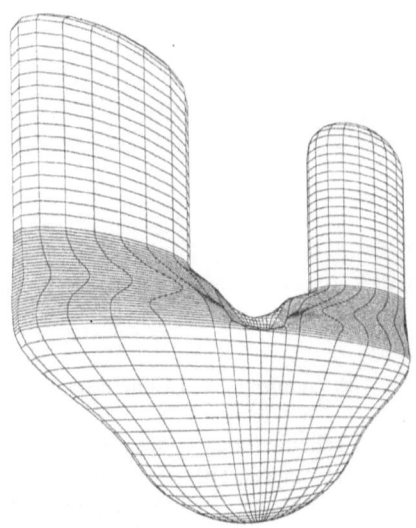

Figure 7. Side view of surface generated from data of Table 1

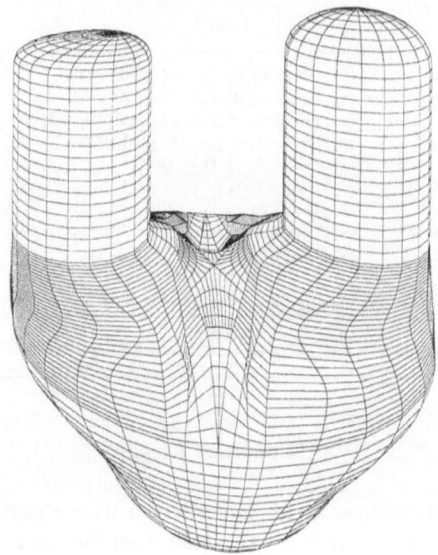

Figure 8. Side view showing construction of saddle

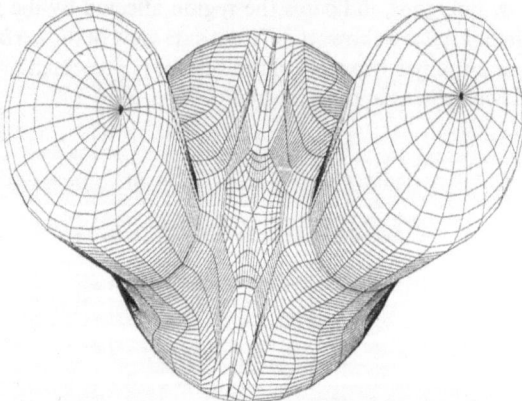

Figure 9. Top view of surface, showing plan of saddle

9. Summary and Conclusions

This paper has presented a scheme for extending the 3D reconstruction method of [5] to be able to reconstruct objects with branching. The fundamental idea is to deal with branching by introducing a saddle shaped surface patch between the heights at which branching is known to occur. A suitable shape is that of a hyperboloid, this surface having the advantage that at a fixed height the resulting hyperbola may be represented in a rational Bézier form, which fits in with the description of the remainder of the surface. The resulting surface has C^1 continuity.

The work of this paper has concentrated upon the particular case in which one contour at a particular height branches into two contours at the next height. A related problem is that in which two contours at one height reduce to one contour at the next height. It would seem that the approach used in this paper could also be applied to this case, either using a direct analogue of the current scheme, or by 'inverting' the regions of data for which this contour 'reduction' occurs so that such a region may be built 'upside down' and be considered as a branching problem. In the particular case in which precisely one contour 'reduction' occurs, and no branching, this would mean constructing the entire surface 'upside down'. This has been implemented and tested. First the sign of the z-coordinate of each of the data points is reversed, and the surface is constructed based upon the resulting set of data points: this surface being constructed as a branching case. Finally the z-coordinates revert to their original signs.

Also under development is a full generalization of the scheme to arbitrary numbers of branches.

References

[1] Boissonnat, J.-D.: Shape reconstruction from planar cross-sections. Comp. Vision Graph. Imag Proc. 44, 1–29 (1988).
[2] Christiansen, H. N., Sederberg, T. W.: Conversion of complex contour line definitions into polygonal element mosaics. Computer Graph. 13(2), 187–192 (1978).
[3] Farin, G. E.: Curves and surfaces for computer aided geometric design, 2nd ed. San Diego: Academic Press 1990.
[4] Goodman, T. N. T.: Shape preserving interpolation by parametric rational cubic splines. Basel: Birkhauser 1988, 149–158 (Intnl. Series of Numerical Mathematics, 86).
[5] Goodman, T. N. T., Ong, B. H., Unsworth, K.: Reconstruction of C^1 closed surfaces from cross-sectional data. University of Dundee Comp. Sci. Report 91/12, 1991 (Submitted).
[6] Goodman, T. N. T., Unsworth, K.: An algorithm for generating shape preserving interpolating curves using rational cubic splines. University of Dundee Comp. Sci. Report 89/01, 1989.
[7] Lin, W.-C., Chen, S.-Y., Chen, C.-T.: A new surface interpolation technique for reconstructing 3D objects from serial cross-sections. Comp. Vision Graph. Imag Proc. 48, 124–143 (1989).
[8] Schumaker, L. L.: Reconstructing 3D objects from cross-section. In: Dahmen, W., Gasca, M., Micchelli, C. A. (eds.) Computation of curves and surfaces, pp. 275–309. Dordrecht: Kluwer 1990.
[9] Wang, Y. F., Aggarwal, J. K.: Surface reconstruction and representation of 3-D scenes. Pattern Recognition 19, 197–207 (1986).

T. N. T. Goodman and K. Unsworth   B. H. Ong
Department of Mathematics and Computer   School of Mathematical Sciences
Science University of Dundee,   Universiti Sains Malaysia
Dundee DD1 4HN,   11800 Penang, Malaysia
Scotland

Computing Suppl. 8, 117–132 (1993)

High Order Continuous Polygonal Patches

J. A. Gregory, V. K. H. Lau, Uxbridge and **J. M. Hahn,** Sindelfingen

Abstract. A polygonal patch method is described which can be used to fill a polygonal hole within a given k'th order continuous rectangular patch complex. The method is relatively easy to implement, since it only requires C^k extensions of the rectangular patch complex defined in terms of the rectangular patch parameterizations. The method is illustrated by reference to C^2 bicubic B-spline surfaces.

Key words: Polygonal patches.

1. Introduction

The parametric representation of surfaces in $CAGD$ is usually based on an assembly of patches with rectangular domains of definition. However, arbitrary surface topolgies cannot be described by a regular rectangular patch framework. Either an arbitrary number of rectangular patches meeting at a vertex has to be allowed, or a polygonal patch has to be filled in. Here we consider the latter approach of constructing a polygonal patch. An n-sided patch will be exhibited which can be used to fill in a hole within a given C^k rectangular patch complex, for any order of k. In particular, the case $k = 2$ (curvature continuity) should be of practical interest in $CAGD$, for example, when filling a polygonal hole within a bicubic B-spline patch complex.

There have been several attempts to construct polygonal patches, see [Gregory, Lau and Zhou'90, Varady'87], but these only achieve C^1 continuous joins with their rectangular patch neighbours. For example, [Charrot and Gregory'84] describe a pentagonal patch defined by a convex combination of parametric surfaces. As pointed out in [Gregory and Hahn'87b], this method cannot be immediately generalized to higher order continuous surfaces, although a particular C^2 solution is given in [Gregory and Hahn'89]. The problem is that the continuity considerations cannot be treated within the given parameterizations, since the patches cannot be considered as being defined in a common parameter plane. The appropriate framework in which to examine continuity is that of k'th order *'geometric continuity'* between the patches, see [Hahn'89], that is C^k continuity under a reparameterization.

A general k'th order solution for the polygonal hole problem is given in an internal report [Gregory and Hahn'87a]. In that report, the rectangular patch data is reparameterized as C^k data around the exterior of a polygonal domain. This data is then extended into the interior of the polygon by a blending function interpolation method. Here, however, we adopt an approach which is much easier to implement. The rectangular patch complex is extended about each corner of the hole in terms of the rectangular patch parameterizations. These extensions are then reparameterized onto the polygonal domain and blended to give the final polygonal patch. Theoretically, these two approaches to the problem are equivalent but practically there is a significant difference in the ease of implementation which leads us to recommend the method proposed here.

The construction of the reparameterizations must be considered with some care and most of the theoretical content of the paper is concerned with this problem. However, given the reparameterization functions (diffeomorphisms) proposed here, together with the C^k extensions of the rectangular patch complex, it is then a simple matter to implement the polygonal patch method.

The polygonal hole problem is described in Section 2 and is followed, in Section 3, by a description of co-ordinate systems (co-ordinate charts) which are defined on the polygonal domain by central projections. These co-ordinate charts then form the basis of the reparameterization method used in the construction of the polygonal patch method in Section 4. This section contains most of the theory of the paper. In particular, the conditions to be satisfied by the reparameterization functions are developed. An alternative construction for the special case of a triangular hole is then considered in Section 5. In the final Section 6, we consider implementation of the polygonal patch schemes for the specific case of bicubic B-spline surfaces which contain polygonal holes.

2. The Polygonal Hole Problem

Assume that $q_j, j = 0, \ldots, n - 1$, describes a given C^k parametric rectangular patch complex around an n-sided hole in $\mathbb{R}^3$, where $n \geq 3$. To make the exposition more concrete, suppose that

$$q_j: \delta \to \mathbb{R}^3, \qquad \delta = [0, 2] \times [-1, 0],$$

where the segment $(s, 0)$, $0 \leq s \leq 1$, is mapped to the j'th boundary segment of the hole in $\mathbb{R}^3$, see Fig. 1. (In practice, q_j will usually be composed of a sub-complex of two, or more, rectangular patches but it is mathematically convenient to represent these as one composite surface patch here.)

The patches are assumed to form a C^k parametric patch complex in the sense that, for two adjacent patches q_{j-1} and q_j, the composite map

$$q_{j-1,j}(u, v) := \begin{cases} q_{j-1}(\sigma(u, v)), & (u, v) \in \sigma^{-1}(\delta) = [-1, 0] \times [-1, 1], \\ q_j(u, v), & (u, v) \in \delta = [0, 2] \times [-1, 0], \end{cases} \qquad (2.1)$$

is C^k continuous on the L-shaped domain $\sigma^{-1}(\delta) \cup \delta$, where

$$\sigma(u, v) := (1 - v, u), \qquad \sigma^{-1}(u, v) := (v, 1 - u), \qquad (2.2)$$

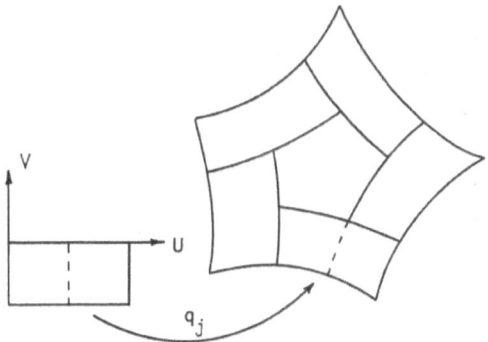

Figure 1. Five patches surrounding a polygonal hole

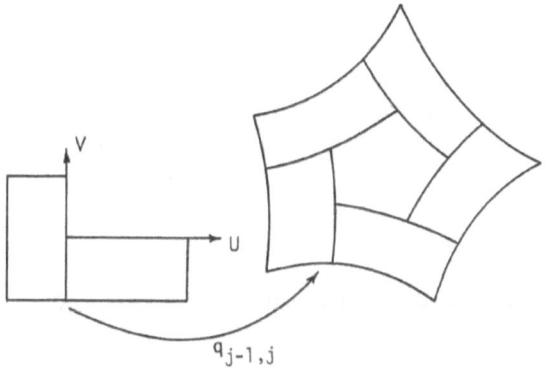

Figure 2. The composite map $\mathbf{q}_{j-1,j}$

see Fig. 2. In addition, the composite map will usually be $C^{k,k}$ continuous, that is, the partial derivatives

$$\partial_{l,m}\mathbf{q}_{j-1,j} := \frac{\partial^{l+m}\mathbf{q}_{j-1,j}}{\partial u^l \partial v^m} \tag{2.3}$$

exist for $0 \leq l$, $m \leq k$, are continuous, and are independent of the order of differentiation.

Our objective is to extend the rectangular patch complex into the n-sided hole in $\mathbb{R}^3$, with a surface patch $\mathbf{p}$ defined on a regular polygon. In order to achieve this goal, we assume that a C^k extension of $\mathbf{q}_{j-1,j}$ into the positive quadrant is supplied for each $j = 0, \ldots, n-1$. In particular, we assume that each $\mathbf{q}_{j-1,j}(u,v)$ is defined for $(u,v) \in [0,1]^2$. These C^k extensions will then be defined on the regular polygon, through appropriate reparameterizations, and be blended together to form the polygonal patch $\mathbf{p}$.

The C^k extensions of $\mathbf{q}_{j-1,j}$ into the unit square $[0,1]^2$ can easily be constructed in practice. For example, if the surrounding patch complex has a $C^{2,2}$ bi-cubic B-spline representation, then it is natural to extend $\mathbf{q}_{j-1,j}$ as a bi-cubic B-spline surface with appropriate additional control points. The important point to note is that the extensions are to be constructed with respect to a rectangular patch parameterization. The theory of the reparameterization of the $\mathbf{q}_{j-1,j}$ with respect to a regular polygonal domain is now considered in the following two sections.

3. The Polygonal Domain

Let Ω be a closed, regular, n-sided polygon in $\mathbb{R}^2$ with centre $\mathbf{0} = (0,0)$ and sides of unit length. Its vertices are denoted by $X_j, j = 0, \ldots, n-1$, and its edges are E_j, parameterized as

$$E_j(s) := (1-s)X_j + sX_{j+1}. \tag{3.1}$$

In order to reparameterize the extensions $\mathbf{q}_{j-1,j}(u,v), (u,v) \in [0,1]^2$, on this regular polygon, we find it necessary to introduce co-ordinate charts $\phi_{j-1,j}, j = 0, \ldots, n-1$, defined on Ω. These co-ordinate charts respectively transform the angles at the vertices $X_j, j = 0, \ldots, n-1$, to $\pi/2$, and are defined here by central projections.

3.1 The Central Projection Co-Ordinate Chart

Let Z_j be the point of intersection of the edge E_{j-1} with E_{j+1} and, for a point $X \in \Omega$, let $u_j = u_j(X)$ be such that

$$E_j(u_j) := (1-u_j)X_j + u_jX_{j+1} \tag{3.2}$$

is the point of intersection of the edge E_j with the ray from Z_j through X, see Fig. 3. Let

$$v_j := 1 - u_{j-1}, \tag{3.3}$$

then

$$E_{j-1}(1-v_j) = (1-v_j)X_j + v_jX_{j-1}. \tag{3.4}$$

The co-ordinate chart $\phi_{j-1,j}$ is now defined by

$$\phi_{j-1,j}(X) := (u_j(X), v_j(X)). \tag{3.5}$$

This chart maps Ω into $[0,1]^2$, where X_j is mapped to $(0,0)$ and the two edges $E_j(u)$, $E_{j-1}(1-v)$ are mapped onto $(u,0), (0,v)$ respectively. Hence, the interior angle of the polygon at X_j is mapped to $\pi/2$.

The co-ordinate chart is conveniently computed as

$$\phi_{j-1,j} := (u_j, v_j) := \left(\frac{d_{j-1}}{d_{j-1} + d_{j+1}}, \frac{d_j}{d_{j-2} + d_j} \right), \tag{3.6}$$

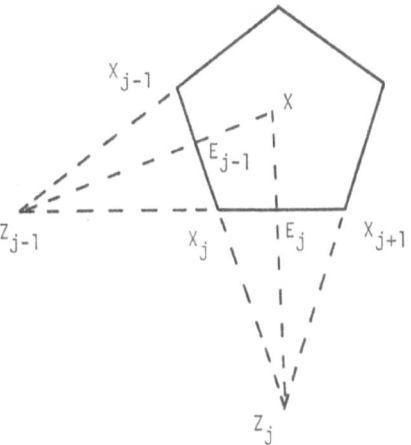

Figure 3. Central projection co-ordinate chart construction

where $d_j = d_j(X)$ is the perpendicular distance of $X \in \Omega$ from the side E_j. In particular, in the case $n \geq 5$,

$$d_j = d_j(X) := \langle X_j - X, Z_j - 0 \rangle / \| Z_j - 0 \|, \qquad (3.7)$$

where $\langle \cdot, \cdot \rangle$ denotes the Euclidean scalar product in $\mathbb{R}^2$.

In the case $n = 3$, the domain Ω is a triangle and it is more convenient to work directly with the barycentric co-ordinate system (b_0, b_1, b_2) of the point $X \in \Omega$. Thus

$$X = b_0 X_0 + b_1 X_1 + b_2 X_2, \qquad b_0 + b_1 + b_2 = 1. \qquad (3.8)$$

In this case the co-ordinate chart representation (3.6) becomes

$$\phi_{j-1,j} := (u_j, v_j) := \left(\frac{b_{j+1}}{b_j + b_{j+1}}, \frac{b_{j-1}}{b_j + b_{j-1}} \right). \qquad (3.9)$$

It should be remarked that the co-ordinate chart is singular at $X = Z_j$ and $X = Z_{j-1}$, but for $n \geq 4$ these points are outside the polygonal domain. (In the case $n = 4$, the central projections are parallel projections with Z_j becoming a point at infinity.) In the case $n = 3$, the central projection introduces singularities at the vertices $Z_j = X_{j-1}$ and $Z_{j-1} = X_{j+1}$. However, these singularities will be removable in the final scheme and will not cause numerical difficulties. Alternatively, a triangular scheme based on a nonsingular parallel projection co-ordinate chart construction can be derived, and this alternative scheme is considered in Section 5.

4. The Polygonal Patch

Let $\Phi_{j-1,j} \colon \Omega \to [0,1]^2$ be a C^k diffeomorphism which maps the vertex X_j to $(0,0)$ and the edges $E_j(u)$, $E_{j-1}(1-v)$ to $(u,0)$, $(0,v)$ respectively. Also, let $\mathbf{p}_{j-1,j} \colon \Omega \to \mathbb{R}^3$

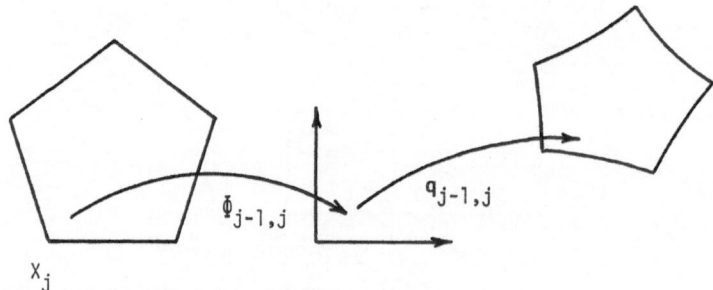

Figure 4. The map $\mathbf{p}_{j-1,j}$

be defined by the composition

$$\mathbf{p}_{j-1,j}(X) := \mathbf{q}_{j-1,j}(\Phi_{j-1,j}(X)), \qquad X \in \Omega, \tag{4.1}$$

where $\mathbf{q}_{j-1,j}(u, v)$, $(u, v) \in [0, 1]^2$, is the C^k extension described in Section 2, see Fig. 4. Then, by definition, $\mathbf{p}_{j-1,j}(X)$, $X \in \Omega$, and $\mathbf{q}_{j-1,j}(X)$, $X \in \sigma^{-1}(\delta) \cup \delta$, see (2.1), form a C^k surface in the sense that there exists a reparameterization in which the surface is C^k. More precisely, $\mathbf{p}_{j-1,j}(X)$, $X \in \Omega$, and $\mathbf{q}_{j-1,j}(\Phi_{j-1,j}(X))$, $X \in \Phi_{j-1,j}^{-1}(\sigma^{-1}(\delta) \cup \delta)$ form a C^k surface, where $\Phi_{j-1,j}$ defines the reparameterization. Two such surfaces patches, which join with C^k continuity under a reparameterization, are said to have a *geometric continuous* GC^k join in the *CAGD* literature, see, for example, [Hahn '87]. Thus, $\mathbf{p}_{j-1,j}$ defines a C^k surface patch on the polygonal domain Ω, which, along the edges E_j and E_{j-1}, has GC^k joins with the rectangular patch complex.

The polygonal patch $\mathbf{p}: \Omega \to \mathbb{R}^3$, is required to have a GC^k join with the rectangular patch complex around the entire boundary of Ω. We thus define

$$\mathbf{p}(X) := \sum_{j=0}^{n-1} w_j(X)\mathbf{p}_{j-1,j}(X), \qquad X \in \Omega, \tag{4.2}$$

where the weights $w_j: \Omega \to \mathbb{R}$ are C^k functions such that

$$\sum_{j=0}^{n-1} w_j(X) = 1, \qquad w_j(X) \geq 0, \qquad X \in \Omega, \tag{4.3}$$

and, for all $i = 0, \ldots, k$, the derivative maps are such that

$$\partial^i w_j|_{E_l} = 0, \qquad l \neq j - 1, j; \qquad l = 0, \ldots, n - 1. \tag{4.4}$$

Hence $\mathbf{p}$ is a convex combination of the $\mathbf{p}_{j-1,j}$ with weights w_j chosen to be zero to order k on those sides where $\mathbf{p}_{j-1,j}$ does not match the surrounding patch complex.

The unwary reader may feel that the problem is now completely solved, since the work of the previous section suggests that the co-ordinate charts $\phi_{j-1,j}$ provide appropriate definitions for the diffeomorphisms $\Phi_{j-1,j}$. However, as observed in [Gregory & Hahn '87] for the case $k = 2$, a convex combination patch of the form (4.2) will not, in general, have a GC^k join with the surrounding rectangular patch complex. This problem occurs because of the different diffeomorphisms $\Phi_{j-1,j}$

defining the reparameterizations of the GC^k joins. Thus more care is needed in the construction of the $\Phi_{j-1,j}$.

To investigate this problem further, we consider the join of $\mathbf{q}_j$ with $\mathbf{p}$ along the edge E_j. From (4.1)–(4.4), we can write

$$\mathbf{p} = w_j \cdot \mathbf{p}_{j-1,j} + w_{j+1} \cdot \mathbf{p}_{j,j+1} + \mathbf{r}_j \tag{4.5}$$

$$= w_j \cdot (\mathbf{q}_{j-1,j} \circ \Phi_{j-1,j}) + w_{j+1} \cdot (\mathbf{q}_{j,j+1} \circ \Phi_{j,j+1}) + \mathbf{r}_j, \tag{4.6}$$

where

$$\partial^i \mathbf{r}_j|_{E_j} = 0, \qquad i = 0, \ldots, k, \tag{4.7}$$

and

$$\left.\begin{array}{l} (w_j + w_{j+1})|_{E_j} = 1, \\ \partial^i (w_j + w_{j+1})|_{E_j} = 0, \qquad i = 1, \ldots, k, \end{array}\right\}. \tag{4.8}$$

Now, by definition, $\mathbf{q}_{j-1,j}$ is a C^k extension of $\mathbf{q}_j$ and $\mathbf{q}_{j,j+1}$ is a C^k extension of $\mathbf{q}_j \circ \sigma$ across $(s, 0)$, $0 \le s \le 1$, see (2.1). Thus $\mathbf{p}_{j-1,j}$ is a C^k extension of $\mathbf{q}_j \circ \Phi_{j-1,j}$ and $\mathbf{p}_{j+1,j}$ is a C^k extension of $\mathbf{q}_j \circ \sigma \circ \Phi_{j,j+1}$ across $E_j(s)$, $0 \le s \le 1$. Hence $\Phi_{j-1,j}$ and $\sigma \circ \Phi_{j,j+1}$ define the reparameterization functions for the GC^k joins. It follows, from (4.5)–(4.8), that if $\Phi_{j-1,j}$ and $\sigma \circ \Phi_{j,j+1}$ agree to order k along E_j, that is if the GC^k joins are identical, then $\mathbf{p}$ has an (identical) GC^k join with $\mathbf{q}_j$. In fact, the following proposition shows that this condition on the diffeomorphisms can be weakened.

Proposition 1. *The polygonal patch $\mathbf{p}$ has a GC^k join with $\mathbf{q}_j$ if*

$$\partial^i \Phi_{j-1,j}|_{E_j(s)} = \partial^i (\sigma \circ \Phi_{j,j+1})|_{E_j(s)}, \qquad i = 0, \ldots, k-1, \tag{4.9}$$

that is if $\Phi_{j-1,j}$ and $\sigma \circ \Phi_{j,j-1}$ agree to order $k - 1$ on $E_j(s)$, $0 \le s \le 1$.

Proof. Since $\mathbf{q}_{j-1,j}$ is a C^k extension of $\mathbf{q}_j$ and $\mathbf{q}_{j,j+1}$ is a C^k extension of $\mathbf{q}_j \circ \sigma$, then differentiating (4.6) along any direction U using Leibniz' theorem gives

$$\frac{\partial^l \mathbf{p}}{\partial U^l}\bigg|_{E_j} = \sum_{i=0}^{l} \binom{l}{i} \left\{ \frac{\partial^{l-i} w_j}{\partial U^{l-i}} \frac{\partial^i \mathbf{q}_j \circ \Phi_{j-1,j}}{\partial U^i} + \frac{\partial^{l-i} w_{j+1}}{\partial U^{l-i}} \frac{\partial^i \mathbf{q}_j \circ \sigma \circ \Phi_{j,j+1}}{\partial U^i} \right\}\bigg|_{E_j}. \tag{4.10}$$

Now, from the hypothesis (4.9),

$$\frac{\partial^i \mathbf{q}_j \circ \Phi_{j-1,j}}{\partial U^i}\bigg|_{E_j} = \frac{\partial^i \mathbf{q}_j \circ \sigma \circ \Phi_{j,j+1}}{\partial U^i}\bigg|_{E_j}, \qquad i = 0, \ldots, k-1. \tag{4.11}$$

Hence, (4.8) and (4.10) give, for $l = 0, \ldots, k$,

$$\frac{\partial^l \mathbf{p}}{\partial U^l}\bigg|_{E_j} = \left\{ w_j \frac{\partial^l \mathbf{q}_j \circ \Phi_{j-1,j}}{\partial U^l} + w_{j+1} \frac{\partial^l \mathbf{q}_j \circ \sigma \circ \Phi_{j,j+1}}{\partial U^l} \right\}\bigg|_{E_j} \tag{4.12}$$

$$= \left\{ \frac{\partial^l}{\partial U^l} \mathbf{q}_j \circ [w_j \cdot \Phi_{j-1,j} + w_{j+1} \cdot (\sigma \circ \Phi_{j,j+1})] \right\}\bigg|_{E_j}. \tag{4.13}$$

This latter result comes after expansion of (4.12) and (4.13), using the chain and product rules, where we again make use of (4.8) and of the fact that $\Phi_{j-1,j}$ and

$\sigma \circ \Phi_{j,j+1}$ agree to order $k - 1$ on E_j. For brevity, we omit the details of these expansions. We have thus shown that $\mathbf{p}$ is a C^k extension of $\mathbf{q}_j \circ [w_j \cdot \Phi_{j-1,j} + w_{j+1} \cdot (\sigma \circ \Phi_{j,j+1})]$ across E_j and hence $\mathbf{p}$ has a GC^k join with $\mathbf{q}_j$. $\square$

4.1 Construction of the Diffeomorphism $\Phi_{j-1,j}$

Proposition 1 shows that $\Phi_{j-1,j}$ must agree with $\sigma \circ \Phi_{j,j+1}$ to order $k - 1$ on E_j and with $\sigma^{-1} \circ \Phi_{j-2,j-1}$ to order $k - 1$ on E_{j-1}. Consider the central projection co-ordinate charts defined in Section 3. Then

$$\phi_{j-1,j} = (u_j, v_j) \tag{4.14}$$

agrees with

$$\sigma \circ \phi_{j,j+1} = (1 - v_{j+1}, u_{j+1}) \tag{4.15}$$

to order zero on E_j, and agrees with

$$\sigma^{-1} \circ \phi_{j-2,j-1} = (v_{j-1}, 1 - u_{j-1}) \tag{4.16}$$

to order zero on E_{j-1}. Thus $\Phi_{j-1,j} := \phi_{j-1,j}$ is an appropriate definition for the diffeomorphism in the case $k = 1$ of Proposition 1, which is the construction used in [Charrot and Gregory'84]. In general, let $\alpha: [0, 1] \to \mathbb{R}$ be a C^k function such that

$$\alpha^{(i)}(0) = \delta_{i,0}, \qquad \alpha^{(i)}(1) = 0, \qquad i = 0, \ldots, k - 1. \tag{4.17}$$

Then the following proposition shows that the diffeomorphism $\Phi_{j-1,j}$ can be constructed by matching a blend of the charts (4.14)–(4.16).

Proposition 2. *Let* $\Phi_{j-1,j}, j = 0, \ldots, n - 1$ *be* C^k *diffeomorphisms which respectively match*

$$\psi_j(X) := \alpha(u_j)(u_j, v_j) + \alpha(v_{j+1})(1 - v_{j+1}, u_{j+1}) \tag{4.18}$$

to order $k - 1$ *on* E_j *and*

$$\sigma^{-1} \circ \psi_{j-1}(X) = \alpha(v_j)(u_j, v_j) + \alpha(u_{j-1})(v_{j-1}, 1 - u_{j-1}) \tag{4.19}$$

to order $k - 1$ *on* E_{j-1}. *Then the diffeomorphisms satisfy the* GC^k *conditions* (4.9) *of Proposition 1.*

Proof. At $X = X_j, u_j = v_j = 0$ and $v_{j+1} = u_{j-1} = 1$. Hence, using (4.17), $\psi_j(X)$ and $\sigma^{-1} \circ \psi_{j-1}(X)$ agree to order $k - 1$ with $\phi_{j-1,j} = (u_j, v_j)$ at X_j. Thus the C^k functions defined by (4.18) and (4.19) are compatible to order $k - 1$ at $X_j = E_j \cap E_{j-1}$. Hence a C^k diffeomorphism $\Phi_{j-1,j}$ can be constructed which matches these C^k functions to order $k - 1$ on E_j and E_{j-1}. (An explicit construction involving the central projection co-ordinate charts is given below and an alternative construction involving parallel projection co-ordinate charts is given in Section 5.) Furthermore, since $\Phi_{j-1,j}$ matches ψ_j and $\Phi_{j,j+1}$ matches $\sigma^{-1} \circ \psi_j$ to order $k - 1$ on E_j, it immediately follows that $\Phi_{j-1,j}$ and $\sigma \circ \Phi_{j,j+1}$ agree to order $k - 1$ on E_j. Thus the conditions (4.9) of Proposition 1 hold. $\square$

Diffeomorphisms $\Phi_{j-1,j}$ must now be constructed which satisfy the conditions (4.18) and (4.19) of Proposition 2. With the central projection co-ordinate charts, this problem has a simple solution. Observe that

$$\alpha^{(i)}(v_{j+1})|_{E_{j-1}} = \alpha^{(i)}(u_{j-1})|_{E_j} = \alpha^{(i)}(1) = 0, \qquad i = 0, \ldots, k-1, \qquad (4.20)$$

and

$$\alpha^{(i)}(u_j)|_{E_{j-1}} = \alpha^{(i)}(v_j)|_{E_j} = \alpha^{(i)}(0) = \delta_{i,j}, \qquad i = 0, \ldots, k-1. \qquad (4.21)$$

It then follows that $\Phi_{j-1,j}$ can be defined by the tensor-product like construction

$$\Phi_{j-1,j} := [\alpha(u_j)\alpha(v_{j+1})] \begin{bmatrix} (u_j, v_j) & (v_{j-1}, 1-u_{j-1}) \\ (1-v_{j+1}, u_{j+1}) & (a_j, b_j) \end{bmatrix} \begin{bmatrix} \alpha(v_j) \\ \alpha(u_{j-1}) \end{bmatrix} \qquad (4.22)$$

Here, $0 \le a_j$, $b_j \le 1$ ensures that $\Phi_{j-1,j} : \Omega \to [0,1]^2$. In practice we find that $(a_j, b_j) = (0,0)$ produces a satisfactory result and prefer to control the shape of the patch through the choice of the extensions $\mathbf{q}_{j-1,j}$.

4.2 The Weight Functions w_j and α

The polygonal patch construction is summarized as being defined by (4.2), where $\mathbf{p}_{j-1,j}$ is defined by (4.1) and $\Phi_{j-1,j}$ is defined by (4.22). Given the weight functions w_j in (4.2) and the weight α in (4.22), then the user has only to supply the C^k extensions $\mathbf{q}_{j-1,j}$ of the rectangular patch complex.

We define the weight functions w_j, which satisfy conditions (4.3) and (4.4), by

$$w_j := \frac{\prod_{i \ne j-1,j} d_i^{k+1}}{\sum_{l=0}^{n-1} \prod_{i \ne l-1,l} d_i^{k+1}}, \qquad (4.23)$$

where $d_j = d_j(X)$ is the perpendicular distance of X from the side E_j, see Section 3. For $n = 3$, these weight functions can be written in terms the barycentic co-ordinates as

$$w_j := \frac{b_j^{k+1}}{\sum_{l=0}^{n-1} b_l^{k+1}}. \qquad (4.24)$$

(The case $n = 3$ also exhibits an alternative polynomial definition for the weights, see Section 5.)

The weight α, in the definition of the diffeomorphism $\Phi_{j-1,j}$, must satisfy conditions (4.17). Hermite two point Taylor interpolation then gives

$$\alpha(s) := (1-s)^k \sum_{j=0}^{k-1} \frac{(k-1+j)!}{(k-1)! j!} s^j \qquad (4.25)$$

as an appropriate definition. Thus, for $k = 2$ (curvature continuity),

$$\alpha(s) := (1-s)^2(1+2s). \qquad (4.26)$$

For $k = 1$ (tangent plane continuity), we have $\alpha(s) = 1 - s$ in the definition (4.22) but, as was observed in the introduction to this section, $\Phi_{j-1,j} = \phi_{j-1,j}$ is also valid in this case. Finally, it can be noted that

$$\alpha(s) + \alpha(1 - s) = 1.$$ (4.27)

Thus, since $v_{j+1} = 1 - u_j$ for the central projection co-ordinate charts, it follows that (4.18) defines a convex combination of the two charts (u_j, v_j) and $(1 - v_{j+1}, u_{j+1})$.

5. An Alternative Triangular Patch

In the triangular domain case $n = 3$, the component $\mathbf{p}_{j-1,j}$ of the patch definition (4.2) has singularities at the vertices X_{j-1} and X_{j+1}. These are introduced by the singularities of the central projection co-ordinate charts in the definition (4.22) of the diffeomorphism $\Phi_{j-1,j}$. These singularities are removable to order k, since the weight w_j in (4.2) has a $k + 1$'st order zero along the edge E_{j+1} which joins X_{j-1} and X_{j+1}. Hence the patch definition is numerically stable to order k. It is, however, possible to totally avoid the introduction of singularities through the use of parallel projection co-ordinate charts on the triangle. In this case, the construction of $\Phi_{j-1,j}$ as in (4.22) is no longer valid and hence an alternative construction is required.

5.1 The Parallel Projection Co-Ordinate Chart

For a point $X \in \Omega$, let $u_j = u_j(X)$ be such that

$$E_j(u_j) := (1 - u_j)X_j + u_j X_{j+1}$$ (5.1)

is the point of intersection of the edge E_j with the ray through X parallel to the side E_{j-1}. Also, let $v_j = v_j(X)$ be such that

$$E_{j-1}(1 - v_j) := (1 - v_j)X_j + v_j X_{j-1}$$ (5.2)

is the point of intersection of the edge E_{j-1} with the ray through X parallel to the side E_j. Then

$$\phi_{j-1,j}(X) := (u_j(X), v_j(X))$$ (5.3)

defines the co-ordinate chart, see Fig. 5. As in the case of the central projection, the

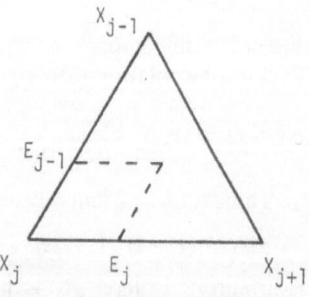

Figure 5. Parallel projection co-ordinate chart construction

co-ordinate chart maps X_j to $(0,0)$ and the two edges $E_j(u)$ and $E_{j-1}(1-v)$ onto $(u,0)$ and $(0,v)$ respectively. Also, for $n=3$, the chart maps the triangle Ω into $[0,1]^2$. However, for $n \geq 5$ there are points $X \in \Omega$ for which $\Phi_{j-1,j}(X) \notin [0,1]^2$. Hence we do not propose the use of the parallel projection in the case $n \geq 5$. For $n=3$ the co-ordinate chart is conveniently computed in terms of the barycentric co-ordinate system as

$$\phi_{j-1,j} := (u_j, v_j) := (b_{j+1}, b_{j-1}) \tag{5.4}$$

5.2 The Diffeomorphism $\Phi_{j-1,j}$

Conditions (4.20), which hold for the central projection co-ordinate charts, are not valid for the parallel projection co-ordinate charts. Hence the diffeomorphism $\Phi_{j-1,j}$ cannot be constructed as in (4.22). Now

$$(u_{j+1}, v_{j+1}) = (v_j, 1 - u_j - v_j), \qquad (u_{j-1}, v_{j-1}) = (1 - u_j - v_j, u_j) \tag{5.5}$$

for the parallel projections on the triangle. Hence, in terms of the variables (u_j, v_j), Proposition 2 requires the construction of $\Phi_{j-1,j}$ which matches

$$\psi_j = \alpha(u_j)(u_j, v_j) + \alpha(1 - u_j - v_j)(u_j + v_j, v_j), \tag{5.6}$$

to order $k-1$ on $v_j = 0$, and

$$\sigma^{-1} \circ \psi_{j-1} = \alpha(v_j)(u_j, v_j) + \alpha(1 - u_j - v_j)(u_j, u_j + v_j), \tag{5.7}$$

to order $k-1$ on $u_j = 0$. Now for $0 \leq l, m \leq k-1$,

$$\partial_{l,m}\psi_j|_{(0,0)} = \partial_{l,m}(\sigma^{-1} \circ \psi_{j-1})|_{(0,0)} = \partial_{l,m}\phi_{j-1,j}|_{(0,0)} = \partial_{l,m}(u_j, v_j)|_{(0,0)}. \tag{5.8}$$

which shows that ψ_j and $\sigma^{-1} \circ \psi_{j-1}$ are compatible to order $k-1$ in each of the variables u_j and v_j at $(u_j, v_j) = (0,0)$. We can thus define $\Phi_{j-1,j}$ by the Boolean sum Taylor interpolant

$$\Phi_{j-1,j} := \sum_{l=0}^{k-1} \frac{u_j^l}{l!} \partial_{l,0}(\sigma \circ \psi_{j-1})|_{(0,v_j)} + \sum_{m=0}^{k-1} \frac{v_j^m}{m!} \partial_{0,m}\psi_j|_{(u_j,0)}$$

$$- \sum_{l=0}^{k-1} \sum_{m=0}^{k-1} \frac{u_j^l}{l!} \frac{v_j^m}{m!} \partial_{l,m}\phi_{j-1,j}|_{(0,0)} \tag{5.9}$$

In particular, with α defined by (4.25), then it can be shown that for $k=1$ (tangent plane continuity)

$$\Phi_{j-1,j} := \phi_{j-1,j} := (u_j, v_j), \tag{5.10}$$

and for $k \doteq 2$ (curvature continuity)

$$\Phi_{j-1,j} := (u_j + 9u_j^2 v_j - 8u_j^3 v_j, \, v_j + 9u_j v_j^2 - 8u_j v_j^3). \tag{5.11}$$

The triangular patch is constructed as in (4.2), where the diffeomorphism $\Phi_{j-1,j}$ in (4.1) is defined by the above Boolean sum construction. The weights w_j in (4.2) can be computed by the rational form (4.24). Alternatively, the polynomial weight

$$w_j := b_j^{k+1} \sum_{l=0}^{k} \sum_{m=0}^{k} \frac{(k+l+m)!}{k!\,l!\,m!} b_{j-1}^l b_{j+1}^m \tag{5.12}$$

can be used, since it satisfies properties (4.3) and (4.4) on the triangular domain Ω. In this case the triangular patch $\mathbf{p}$ will be a polynomial form in the barycentric co-ordinates, if the $\mathbf{q}_{j-1,j}$ extensions of the rectangular patch complex are polynomial.

6. Implementation

We consider the implementation of the polygonal patch schemes for the specific case of uniform bicubic B-spline surfaces which contain polygonal holes. Two possible arrangements for the control points of C^2 bicubic B-spline patch complexes $\mathbf{q}_j, j = 0, \ldots, n-1$, about triangular holes are shown in Figs. 6 and 7. The first example gives one patch adjacent to each edge of a triangular hole and involves the use of a central control point of multiplicity four. The second example gives two patches adjacent to each edge. Clearly, any number of patches adjacent to the hole can be obtained by the addition of further control points. Also, n-sided holes can be obtained by generalizations of these arrangements of the control points.

Given that the surrounding patch complex has a C^2 bicubic B-spline representation, then it is natural to construct the rectangular patch extension $\mathbf{q}_{j-1,j}$ as a bicubic B-spline surface. For Fig. 6, the extension will be a single patch, requiring the definition of an additional control point. For Fig. 7, the extension will consist of four bicubic B-spline patches, requiring the definition of four additional control points. Additional control points will be required for each extension $\mathbf{q}_{j-1,j}, j = 0, \ldots, n-1$, and can be considered as degrees of freedom which allow some control over the shape of the final polygonal patch. Also, the diffeomorphism $\Phi_{j-1,j}$, which is used in the reparameterization of $\mathbf{q}_{j-1,j}$, involves a degree of freedom (a_j, b_j) to

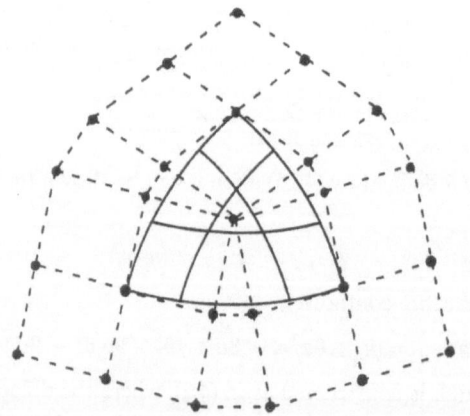

Figure 6. Six bicubic B-spline patches about a triangular hole

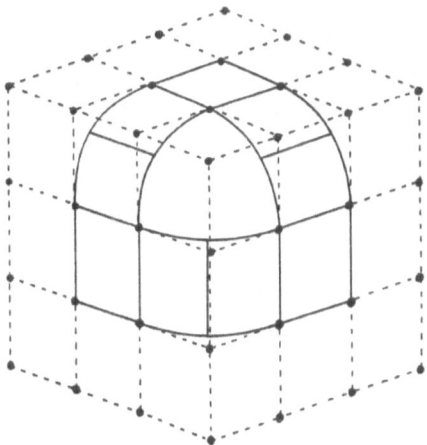

Figure 7. Nine bicubic B-spline patches about a triangular hole

manipulate the shape of the polygonal surface. However, experimental results indicate that this degree of freedom may introduce undesirable "hump" effects around the central region of the surface. Hence, in practice, we set $(a_j, b_j) = (0, 0)$, $j = 0, \ldots, n - 1$.

In order to plot the polygonal patch given by (4.1) and (4.2), it is convenient to reparameterize it as a complex of rectangular patch mappings. This is achieved by quadrilateral subdivision of the polygonal domain about its centre, introducing additional bilinear maps from the unit square onto the quadrilaterals. The problem of calculating normals and curvatures of the patch on the polygonal domain is solved here by adopting a "procedural" method, that is, the chain and product rules

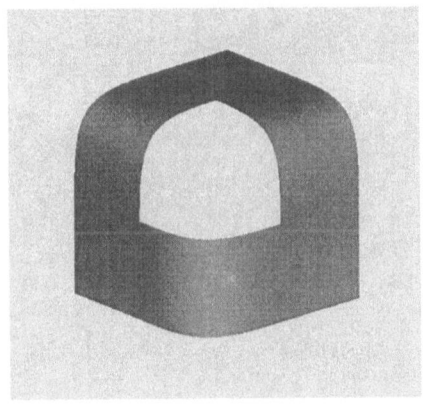 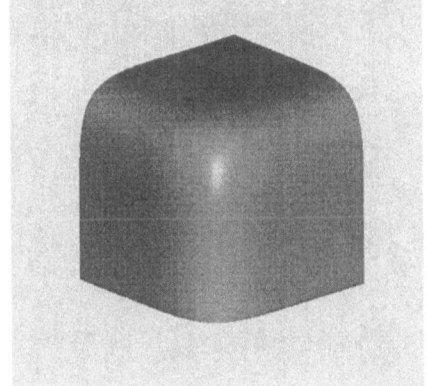

a   b

Figure 8

required in the differentiation of the patch are programmed as procedures, rather than performing these tasks by hand. Alternatively, a numerical differentiation technique applied to the rectangular patch mappings could be used and is probably the simplest technique to apply in practice.

We conclude by giving some model examples, where the control points are arranged to give two bicubic B-spline patches adjacent to each edge. Figure 8 shows the filling of the triangular hole for the case of Fig. 7, displayed using a standard lighting model based on the use of unit normals. Figure 9(a) shows the same surface plotted with a Gaussian curvature map on the surface, displayed together with the lighting

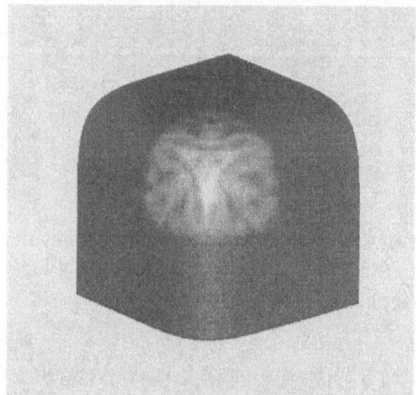 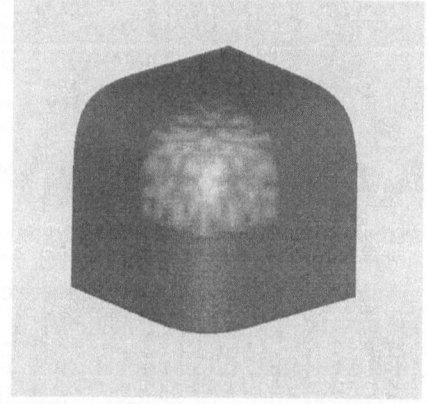

a   b

Figure 9

 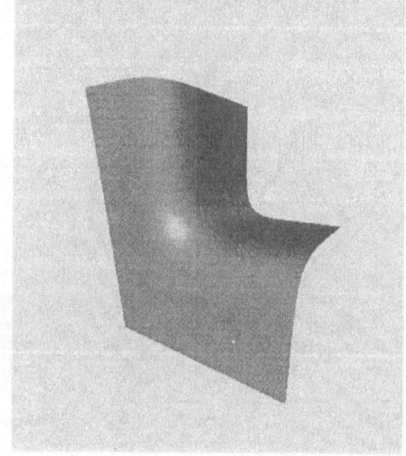

a   b

Figure 10

model. Figure 9(b) shows a C^1 fill of the triangular hole as a comparison. The curvature mapping technique exhibits the lack of curvature continuity across the boundary of the triangular patch in Fig. 9(b). Figures 10 and 11 show a similar display for the case of a pentagonal patch. The figures indicate some oscillation of Gaussian curvature over the polygonal patches, compared with the constant zero Gaussian curvature of the surrounding patch network for the model problems. This, we think, reflects the sensitivity of Gaussian curvature maps, rather than being due to the mathematical complexity of the blended patch definition. For example, Fig. 12(a) shows a perturbation of the surrounding bicubic patch network for the

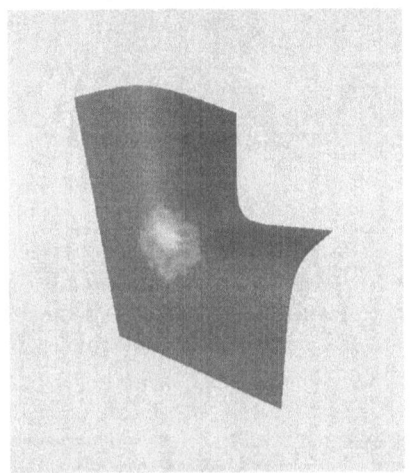 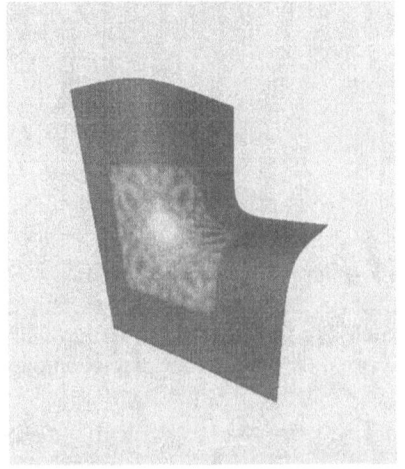

a   b

Figure 11

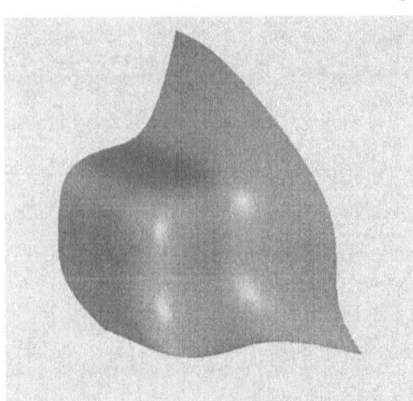 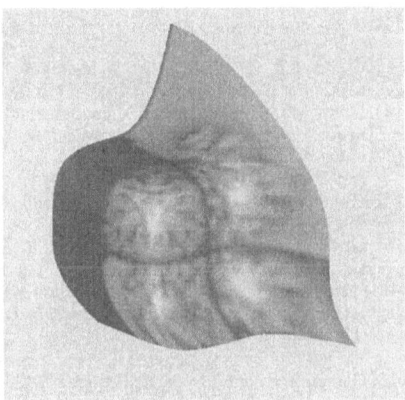

a   b

Figure 12

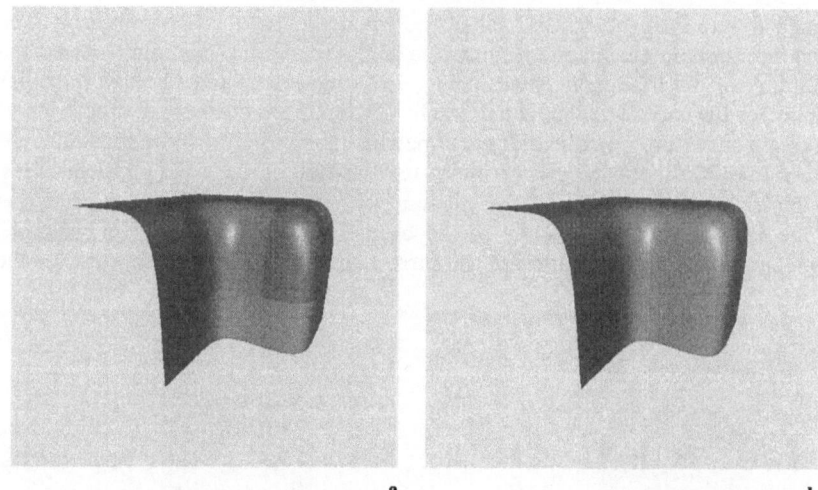

a   b

Figure 13

triangular patch model problem. Here, Fig. 12(b) shows that the Gaussian curvature maps on the bicubic patches exhibit the same sensitivity as that for the triangular patch. The final Fig. 13 shows the combination of a triangular and pentagonal patch within a C^2 rectangular patch complex.

Acknowledgements

This work was supported by the ACME Directorate of the Science and Engineering Research Council, with the grant GR/E 25092. We are also pleased to acknowledge the help of P. K. Yuen, who wrote the software for the curvature lighting model.

References

[1] Charrot, P., Gregory, J. A.: A pentagonal surface patch for computer aided geometric design. Computer Aided Geometric Design *1*, 87–94 (1984).
[2] Gregory, J. A., Hahn, J. M.: Polygonal patches of high order continuity, Technical Report TR/01/87, Mathematics & Statistics Department, Brunel University, 1987a.
[3] Gregory, J. A., Hahn, J. M.: Geometric continuity and convex combination patches. Computer Aided Geometric Design *4*, 79–89 (1987b).
[4] Gregory, J. A., Hahn, J. M.: A C^2 polygonal surface patch. Computer Aided Geometric Design *6*, 69–75 (1989).
[5] Gregory, J. A., Lau, V. K. H., Zhou, J.: Smooth parametric surfaces and *n*-sided patches. In: Dahmen, W., Gasca, M., Micchelli, C. A. (eds.) Kluwer Academic Publishers 1990. Computation of curves and surfaces. (NATO ASI Series C, 307)
[6] Hahn, J. M.: Geometric continuous patch complexes. Computer Aided Geometric Design *6*, 55–67 (1989).
[7] Varady, T.: Overlap patches: a new scheme for interpolating curve networks with *n*-sided regions. Computer Aided Geometric Design *8*, 7–27 (1991).

John A. Gregory
Department of Mathematics and Statistics
Brunel University
Uxbridge, Middlesex UB8 3PH, U.K.

Computing Suppl. 8, 133–138 (1993)

Variational Design of Smooth Rational Bezier-Surfaces

H. Hagen, Kaiserslautern and **G. P. Bonneau,** Paris

Abstract. "NURBS" are currently seen as the most promising curve- and surface-form in CAD/CAM-applications. Rational Bezier-surfaces are special non-uniform rational B-Splines. In this paper we describe a calculus of variation approach to design the weights of a rational surface in a way to achieve a smooth surface in the sense of an energy integral.

Key words: NURBS, rational Bezier-surfaces, variational design, smoothing algorithms

1. Introduction

Computer Aided Geometric Design has emerged from the needs of free form surfaces in CAD/CAM technologies and it has become a major topic in computer science with direct applications for all engineering sciences in the last few years.

A central problem of geometric modelling is the construction of "technically smooth" surface representations (i.e. the data should be directly usable in NC processing). One of the most promising curve- and surface modelling methods is the NURBS-technique (NURBS- non uniform rational B-splines). The fundamental idea of the rational Bezier- and B-Spline algorithms is to evaluate and manipulate the curves and surfaces by (small) number of control points and weights.

The purpose of this paper is to present an algorithm to assign to the weights appropriate values to achieve technical smooth surfaces. The standard fairness criterion for surfaces in engineering is to minimize the strain energy of flexure and torsion in a thin rectangular elastic plate of small deflection. The functional $\int_S (\kappa_1^2 + \kappa_2^2)\, ds$ (κ_1 and κ_2 are the principal curvatures of the surface S) is a mathematical model for the energy stored in such a plate. We use a quadrature formula of this integral as a criterion. A calculus of variation approach based upon this criterion with respect to the weights of the rational surfaces, leads to a linear system of equations. The unique solution of this system gives appropriate weights for smooth surface design.

2. Rational Bezier-Surfaces

We assume that the reader is familiar with the concepts of nonrational Bezier- and B-Spline curves (see [Farin, 90] and [Hoschek-Lasser, 89]). In this chapter we give the fundamental concepts of rational curves and surfaces, following [Farin, 90].

A rational Bezier curve of degree n in E^3 is the projection of an nth degree Bezier curve in E^4 onto the hyperplane $w = 1$.

$$X(t) = \frac{\omega_0 b_0 B_0^n(t) + \cdots + \omega_n b_n B_n^n(t)}{\omega_0 B_0^n(t) + \cdots + \omega_n B_n^n(t)} \tag{2.1}$$

$X(t), b_i \in E^3$; $\omega_i > 0$ are called (scalar) weights.

A rational Bezier-surface $X(u, v)$ is defined by

$$X(u, v) = \frac{\sum_{i=0}^n \sum_{j=0}^m \omega_{ij} b_{ij} B_i^n(u) B_j^m(v)}{\sum_{i=0}^n \sum_{j=0}^m \omega_{ij} B_i^n(u) B_j^m(v)}, \qquad \omega_{ij} > 0 \tag{2.2}$$

as the projection of a 4D-tensor product Bezier-surface. But it is a common misconception to call (2.2) a tensor product surface itself.

3. Variational Principles in Surface Design

Free form objects are an essential part of powerful CAD systems. A major topic is the generation of smooth curves and surfaces which can be immediately supplied to the NC-process. The fundamental idea of our method is the use of modelling tools which minimize a certain functional that can be interpreted in the sense of physics or/and geometry.

In the case of surfaces a thin elastic plate of small deflection can serve as a model for a fair shape. Such a plate tends to take a position of least strain energy of flexure and torsion. The energy stored in this plate is proportional to the integral

$$G := \int_S (\kappa_1^2 + \kappa_2^2)\, ds \tag{3.1}$$

κ_1 and κ_2 are the principal curvatures of the surface S.

In our NURBS-setting:

$$\int_S (\kappa_1^2 + \kappa_2^2)\, ds = \int_{u_1}^{u_2} \int_{v_1}^{v_2} \frac{(g_{11}h_{22} - 2g_{12}h_{12} + g_{22}h_{11})^2 - 2gh}{g^2} \sqrt{g}\, du\, dv \tag{3.2}$$

where

$$h_{11} := \langle X_{uu}, N \rangle \qquad g_{11} := \langle X_u, X_u \rangle$$

$$h_{22} := \langle X_{vv}, N \rangle \qquad g_{22} := \langle X_v, X_v \rangle$$

$$h_{12} := \langle X_{uv}, N \rangle \qquad g_{12} := \langle X_u, X_v \rangle$$

$$g := g_{11}g_{22} - g_{12}^2$$

$$h := h_{11}h_{22} - h_{12}^2$$

$$N := \frac{[X_u, X_v]}{\|[X_u, Xv]\|}.$$

$\langle \ , \ \rangle$ is the scalar product of E^3 and $[\ , \ ]$ is the vector product of E^3)

For general NURBS, the integral (3.2) is a transcendental function of the weigths. To get a fractional function, we use the quadrature formula

$$\int_{u_1}^{u_2} \int_{v_1}^{v_2} f(u,v)\,du\,dv$$

$$\rightarrow \frac{(u_2 - u_1)(v_2 - v_1)}{4}(f(u_1,v_1) + f(u_2,v_1) + f(u_1,v_2) + f(u_2,v_2)) \tag{3.3}$$

Using the quadrature formula (3.3) for a rational bicubic Bezier-surface, we get

$$\int_S (\kappa_1^2 + \kappa_2^2)\,ds \rightarrow$$

$$\frac{1}{2\omega_{10}^3\omega_{01}^3(\langle b_{10} - b_{00}, b_{10} - b_{00}\rangle\langle b_{01} - b_{00}, b_{01} - b_{00}\rangle - \langle b_{10} - b_{00}, b_{01} - b_{00}\rangle^2)^{3/2}}$$

$$\times\, [2\omega_{10}^4\omega_{02}^2\langle b_{10} - b_{00}, b_{10} - b_{00}\rangle^2\langle b_{02} - b_{00}, N_{00}\rangle^2$$

$$+\, 9\omega_{10}^2\omega_{01}^2\omega_{11}^2\langle b_{10} - b_{00}, b_{01} - b_{00}\rangle^2\langle b_{11} - b_{00}, N_{00}\rangle^2$$

$$+\, 2\omega_{01}^4\omega_{20}^2\langle b_{01} - b_{00}, b_{01} - b_{00}\rangle^2\langle b_{20} - b_{00}, N_{00}\rangle^2$$

$$-\, 12\omega_{10}^3\omega_{01}\omega_{11}\omega_{02}\langle b_{10} - b_{00}, b_{10} - b_{00}\rangle\langle b_{10} - b_{00}, b_{01} - b_{00}\rangle\langle b_{11} - b_{00}, N_{00}\rangle\langle b_{02} - b_{00}, N_{00}\rangle$$

$$-\, 12\omega_{10}\omega_{01}^3\omega_{20}\omega_{11}\langle b_{10} - b_{00}, b_{01} - b_{00}\rangle\langle b_{01} - b_{00}, b_{01} - b_{00}\rangle\langle b_{20} - b_{00}, N_{00}\rangle\langle b_{11} - b_{00}, N_{00}\rangle$$

$$+\, 9\omega_{10}^2\omega_{01}^2\omega_{11}^2\langle b_{10} - b_{00}, b_{10} - b_{00}\rangle\langle b_{01} - b_{00}, b_{01} - b_{00}\rangle\langle b_{11} - b_{00}, N_{00}\rangle^2$$

$$+\, 4\omega_{10}^2\omega_{01}^2\omega_{20}\omega_{02}\langle b_{10} - b_{00}, b_{01} - b_{00}\rangle^2\langle b_{20} - b_{00}, N_{00}\rangle\langle b_{02} - b_{00}, N_{00}\rangle]$$

$$+\, \frac{1}{2\omega_{20}^2\omega_{31}^3(\langle b_{20} - b_{30}, b_{20} - b_{30}\rangle\langle b_{31} - b_{03}, b_{31} - b_{03}\rangle - \langle b_{20} - b_{30}, b_{31} - b_{30}\rangle^2)^{3/2}}$$

$$\times\, [2\omega_{20}^4\omega_{32}^4\langle b_{20} - b_{30}, b_{20} - b_{30}\rangle^2\langle b_{32} - b_{30}, N_{30}\rangle^2$$

$$+\, 9\omega_{20}^2\omega_{31}^2\omega_{21}^2\langle b_{20} - b_{30}, b_{31} - b_{30}\rangle^2\langle b_{21} - b_{30}, N_{30}\rangle^2$$

$$+\, 2\omega_{31}^4\omega_{10}^2\langle b_{31} - b_{30}, b_{31} - b_{30}\rangle^2\langle b_{10} - b_{30}, N_{30}\rangle^2$$

$$-\, 12\omega_{20}^3\omega_{31}\omega_{21}\omega_{32}\langle b_{20} - b_{30}, b_{20} - b_{30}\rangle\langle b_{20} - b_{30}, b_{31} - b_{30}\rangle\langle b_{21} - b_{30}, N_{30}\rangle\langle b_{32} - b_{30}, N_{30}\rangle$$

$$-\, 12\omega_{20}\omega_{31}^3\omega_{10}\omega_{21}\langle b_{20} - b_{30}, b_{31} - b_{30}\rangle\langle b_{31} - b_{30}, b_{31} - b_{30}\rangle\langle b_{10} - b_{30}, N_{30}\rangle\langle b_{21} - b_{30}, N_{30}\rangle$$

$$+\, 9\omega_{20}^2\omega_{31}^2\omega_{21}^2\langle b_{20} - b_{30}, b_{20} - b_{30}\rangle\langle b_{31} - b_{30}, b_{31} - b_{30}\rangle\langle b_{21} - b_{30}, N_{30}\rangle^2$$

$$+\, 4\omega_{20}^2\omega_{31}^2\omega_{10}\omega_{32}\langle b_{20} - b_{30}, b_{31} - b_{30}\rangle^2\langle b_{10} - b_{30}, N_{30}\rangle\langle b_{32} - b_{30}, N_{30}\rangle]$$

$$+\, \frac{1}{2\omega_{13}^3\omega_{02}^3(\langle b_{13} - b_{03}, b_{13} - b_{03}\rangle\langle b_{02} - b_{03}, b_{02} - b_{03}\rangle - \langle b_{13} - b_{03}, b_{02} - b_{03}\rangle^2)^{3/2}}$$

$$\times\, [2\omega_{13}^4\omega_{01}^2\langle b_{13} - b_{03}, b_{13} - b_{03}\rangle^2\langle b_{01} - b_{03}, N_{03}\rangle^2$$

$$+\, 9\omega_{13}^2\omega_{02}^2\omega_{12}^2\langle b_{13} - b_{03}, b_{02} - b_{03}\rangle^2\langle b_{12} - b_{03}, N_{03}\rangle^2$$

$$+\, 2\omega_{02}^4\omega_{23}^2\langle b_{02} - b_{03}, b_{02} - b_{03}\rangle^2\langle b_{23} - b_{03}, N_{03}\rangle^2$$

$$-\, 12\omega_{13}^3\omega_{02}\omega_{12}\omega_{01}\langle b_{13} - b_{03}, b_{13} - b_{03}\rangle\langle b_{13} - b_{03}, b_{02} - b_{03}\rangle\langle b_{12} - b_{03}, N_{03}\rangle\langle b_{01} - b_{03}, N_{03}\rangle$$

$$-\, 12\omega_{13}\omega_{02}^3\omega_{23}\omega_{12}\langle b_{13} - b_{03}, b_{02} - b_{03}\rangle\langle b_{02} - b_{03}, b_{02} - b_{03}\rangle\langle b_{23} - b_{03}, N_{03}\rangle\langle b_{12} - b_{03}, N_{03}\rangle$$

$$+ 9\omega_{13}^2\omega_{02}^2\omega_{12}^2\langle b_{13}-b_{03}, b_{13}-b_{03}\rangle\langle b_{02}-b_{03}, b_{02}-b_{03}\rangle\langle b_{12}-b_{03}, N_{03}\rangle^2$$

$$+ 4\omega_{13}^2\omega_{02}^2\omega_{23}\omega_{01}\langle b_{13}-b_{03}, b_{02}-b_{03}\rangle^2\langle b_{23}-b_{03}, N_{03}\rangle\langle b_{01}-b_{03}, N_{03}\rangle]$$

$$+ \frac{1}{2\omega_{23}^3\omega_{32}^3(\langle b_{23}-b_{33}, b_{23}-b_{33}\rangle\langle b_{32}-b_{33}, b_{32}-b_{33}\rangle - \langle b_{32}-b_{33}, b_{23}-b_{33}\rangle^2)^{3/2}}$$

$$\times [2\omega_{23}^4\omega_{31}^2\langle b_{23}-b_{33}, b_{23}-b_{33}\rangle^2\langle b_{31}-b_{33}, N_{33}\rangle^2$$

$$+ 9\omega_{23}^2\omega_{32}^2\omega_{22}^2\langle b_{32}-b_{33}, b_{23}-b_{33}\rangle^2\langle b_{22}-b_{33}, N_{33}\rangle^2$$

$$+ 2\omega_{32}^4\omega_{13}^2\langle b_{32}-b_{33}, b_{32}-b_{33}\rangle^2\langle b_{13}-b_{33}, N_{33}\rangle^2$$

$$- 12\omega_{23}^3\omega_{32}\omega_{22}\omega_{31}\langle b_{23}-b_{33}, b_{23}-b_{33}\rangle\langle b_{32}-b_{33}, b_{23}-b_{33}\rangle\langle b_{22}-b_{33}, N_{33}\rangle\langle b_{31}-b_{33}, N_{33}\rangle$$

$$- 12\omega_{23}\omega_{32}^3\omega_{13}\omega_{22}\langle b_{32}-b_{33}, b_{23}-b_{33}\rangle\langle b_{32}-b_{33}, b_{32}-b_{33}\rangle\langle b_{13}-b_{33}, N_{33}\rangle\langle b_{22}-b_{33}, N_{33}\rangle$$

$$+ 9\omega_{23}^2\omega_{32}^2\omega_{22}^2\langle b_{23}-b_{33}, b_{23}-b_{33}\rangle\langle b_{32}-b_{33}, b_{32}-b_{33}\rangle\langle b_{22}-b_{33}, N_{33}\rangle^2$$

$$+ 4\omega_{23}^2\omega_{32}^2\omega_{13}\omega_{31}\langle b_{32}-b_{33}, b_{23}-b_{33}\rangle^2\langle b_{13}-b_{33}, N_{33}\rangle\langle b_{31}-b_{33}, N_{33}\rangle]$$

This polynom of degree two in the weights ω_{11}, ω_{12}, ω_{21}, ω_{22}, is now used for a calculus of variation approach, with ω_{11}, ω_{12}, ω_{21}, ω_{22}, as variation parameters. This leads to a linear system of equations.

Assuming that none of the triples $\{b_{10}-b_{00}, b_{01}-b_{00}, b_{11}-b_{00}\}$, $\{b_{20}-b_{30}, b_{31}-b_{30}, b_{21}-b_{30}\}$, $\{b_{02}-b_{03}, b_{13}-b_{03}, b_{12}-b_{03}\}$, $\{b_{23}-b_{33}, b_{32}-b_{33}, b_{22}-b_{33}\}$ are coplanar (meaning that the twists of the non rational Bezier-surface defined by the control points (b_{ij}) don't vanish in the corners), this linear system has the unique solution:

$$\omega_{11} = \frac{\omega_{10}^2\omega_{20}\langle b_{10}-b_{00}, b_{10}-b_{00}\rangle\langle b_{20}-b_{00}, N_{00}\rangle + \omega_{01}^2\omega_{02}\langle b_{01}-b_{00}, b_{01}-b_{00}\rangle\langle b_{02}-b_{00}, N_{00}\rangle}{\omega_{10}\omega_{01}(\langle b_{10}-b_{00}, b_{10}-b_{00}\rangle\langle b_{01}-b_{00}, b_{01}-b_{00}\rangle + \langle b_{10}-b_{00}, b_{01}-b_{00}\rangle^2)}$$

$$\times \frac{\langle b_{10}-b_{00}, b_{01}-b_{00}\rangle}{\langle b_{11}-b_{00}, N_{00}\rangle}$$

$$\omega_{12} = \frac{\omega_{13}^2\omega_{23}\langle b_{13}-b_{03}, b_{13}-b_{03}\rangle\langle b_{23}-b_{03}, N_{03}\rangle + \omega_{02}^2\omega_{01}\langle b_{02}-b_{03}, b_{02}-b_{03}\rangle\langle b_{01}-b_{03}, N_{03}\rangle}{\omega_{13}\omega_{02}(\langle b_{13}-b_{03}, b_{13}-b_{03}\rangle\langle b_{02}-b_{03}, b_{02}-b_{03}\rangle + \langle b_{13}-b_{03}, b_{02}-b_{03}\rangle^2)}$$

$$\times \frac{\langle b_{13}-b_{03}, b_{02}-b_{03}\rangle}{\langle b_{12}-b_{03}, N_{03}\rangle}$$

$$\omega_{21} = \frac{\omega_{20}^2\omega_{10}\langle b_{20}-b_{30}, b_{20}-b_{30}\rangle\langle b_{10}-b_{30}, N_{30}\rangle + \omega_{31}^2\omega_{32}\langle b_{31}-b_{30}, b_{31}-b_{30}\rangle\langle b_{32}-b_{30}, N_{30}\rangle}{\omega_{20}\omega_{31}(\langle b_{20}-b_{30}, b_{20}-b_{30}\rangle\langle b_{31}-b_{30}, b_{31}-b_{30}\rangle + \langle b_{20}-b_{30}, b_{31}-b_{30}\rangle^2)}$$

$$\times \frac{\langle b_{20}-b_{30}, b_{31}-b_{30}\rangle}{\langle b_{21}-b_{30}, N_{30}\rangle}$$

$$\omega_{22} = \frac{\omega_{23}^2\omega_{13}\langle b_{23}-b_{33}, b_{23}-b_{33}\rangle\langle b_{13}-b_{33}, N_{33}\rangle + \omega_{32}^2\omega_{31}\langle b_{32}-b_{33}, b_{32}-b_{33}\rangle\langle b_{31}-b_{33}, N_{33}\rangle}{\omega_{23}\omega_{32}(\langle b_{23}-b_{33}, b_{23}-b_{33}\rangle\langle b_{32}-b_{33}, b_{32}-b_{33}\rangle + \langle b_{32}-b_{33}, b_{23}-b_{33}\rangle^2)}$$

$$\times \frac{\langle b_{32}-b_{33}, b_{23}-b_{33}\rangle}{\langle b_{22}-b_{33}, N_{33}\rangle}$$

This weight coefficients are considered optimal in the sense (3.3) → min.

4. Applications

The next picture shows a bicubic rational Bezier-surface with weights equal to one and with our optimal weights.

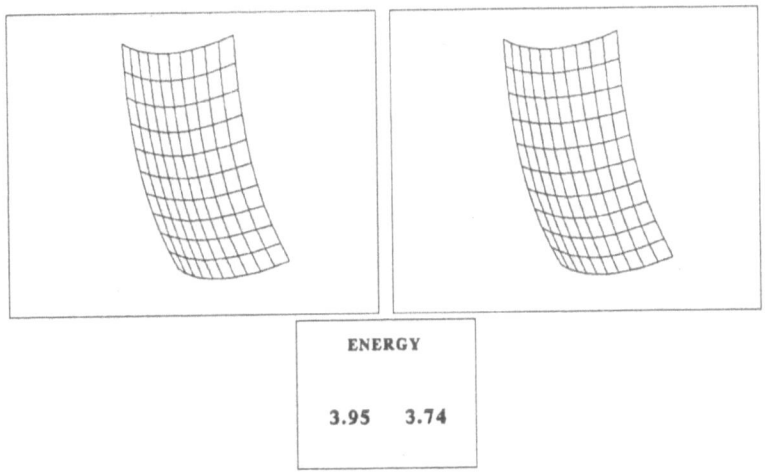

ENERGY

3.95   3.74

Figure 1. Turbine blade

The following picture shows the distribution of the energy on the same surface, with weights equal to one and with our optimal weights.

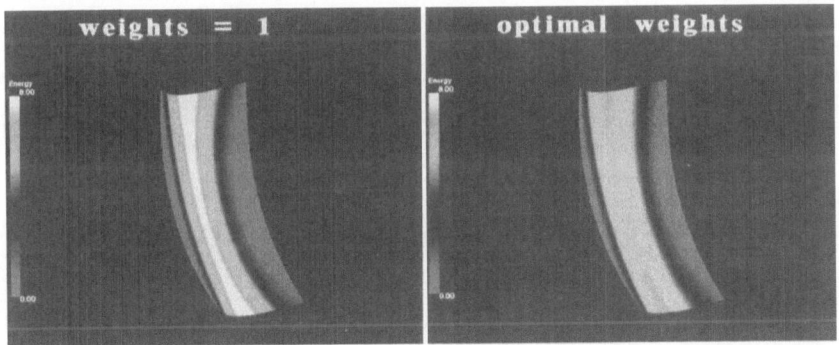

Figure 2

5. Remarks

1. The use of the quadrature formula (3.3) as a criterion ensure a local minimization of the bending energy in each corner of the patch. Some works are currently done to achieve a more global minimization of this bending energy.

2. Other methods are based on a global minimization of an integral criterion ([Hagen-Santarelli]). These methods are working very well for non-rational curves or surfaces, but can't be applied to rational curves or surfaces.

3. For more clarity, the results are presented for a bicubic rational Bezier patch. The same method still leads to a linear system for rational Bezier patches of higher degrees.

References

[1] Farin, G.: Curves and surfaces for computer aided geometric design, 2nd edn. Boston: Academic Press 1990.
[2] Hagen, H., Bonneau, G.P.: Variational design of smooth rational bezier-curves. CAGD 8, 393–399 (1991).
[3] Hagen, H., Santarelli, P.: (1991) Variational design of smooth bezier-curves. (to be published).
[4] Hagen, H., Santarelli, P.: Variational design of smooth bezier-surfaces. In: Hagen (ed.) Topics in surface modelling. SIAM Book, 1991.
[5] Hoschek, J., Lasser, D.: Grundlagen der geometrischen Datenverarbeitung. Stuttgart, Teubner 1989.

Prof. Dr. H. Hagen
Universität Kaiserslautern
Fachbereich Informatik
Postfach 3049
D-W-6750 Kaiserslautern
Federal Republic of Germany

Computing Suppl. 8, 139–153 (1993)

Computing
© Springer-Verlag 1993

Curvature Approximation for Triangulated Surfaces

B. Hamann, Mississippi State University

Abstract. Given a set of points and normals on a surface and a triangulation associated with them a simple scheme for approximating the principal curvatures at these points is developed. The approximation is based on the fact that a surface can locally be represented as the graph of a bivariate function. Quadratic polynomials are used for this local approximation. The principal curvatures at a point on the graph of such a quadratic polynomial is used as the approximation of the principal curvatures at an original surface point.

Key words: Approximation, curvature, Gauss-Weingarten map, platelet, surface, triangulation.

1. Introduction

Methods for exactly calculating and approximating curvatures are important in geometric modeling for two reasons. In order to judge the quality of a surface one commonly computes curvatures for points on the surface, renders the surface's curvature as a texture map onto the surface and can thereby detect regions with undesired curvature behavior, such as surface regions locally changing from an elliptic to a hyperbolic shape. On the other hand, surface schemes are being developed requiring higher order geometric information as input, e.g., normal vectors and normal curvatures.

Definitions and theorems from classical differential geometry are reviewed as far as they are needed for the discussion. In classical differential geometry a surface is understood as a mapping from $\mathbb{R}^2$ to $\mathbb{R}^3$,

$$\mathbf{x}(\mathbf{u}) = (x(u,v), y(u,v), z(u,v))^T \subset \mathbb{R}^3, \qquad \mathbf{u} \in D \subset \mathbb{R}^2. \tag{1}$$

The standard formulae are then used to derive techniques for approximating normal curvatures when a two-dimensional triangulation of a finite point set with associated outward unit normal vectors is given in three-dimensional space. Consequently, curvature estimates can be incorporated into existing surface generating schemes allowing curvature input. The quality of the curvature approximation is tested for triangulated surfaces obtained from a known parametric surface of the form $\mathbf{x}(\mathbf{u}) = (u, v, f(u,v))^T$.

Introductions to differential geometry are [Brauner '81], [do Carmo '76], [Lipschutz '80], [Strubecker '55, '58, '59], and [Struik '61]. Differential geometry is treated more analytically in [O'Neill '69]. One of the most comprehensive works on this subject is [Spivak '70]. Another reference in this field is [Farin '92]. Estimating Gaussian curvature from a discrete, triangulated point set is described in [Calladine '86]. Related triangle-based approximation and interpolation methods are discussed in [Akima '84], [Hagen & Pottmann '89], [Lawson '84], [Nielson & Franke '84], and [Renka & Cline '84]. Modelling triangulations arising in the context of contouring trivariate functions is treated in [Hamann '92].

2. Essential Terms of Differential Geometry

Some basic definitions of differential geometry are reviewed.

Definition 1. A **regular parametric two-dimensional surface** of class C^m ($m \geq 1$) is the point set S in real three-dimensional space $\mathbb{R}^3$ defined by the mapping

$$\mathbf{x} = \mathbf{x}(\mathbf{u}) = (x(u,v), y(u,v), z(u,v))^T \tag{2}$$

of an open set $U \subset \mathbb{R}^2$ into $\mathbb{R}^3$ such that all partial derivatives of x, y, and z of order m or less are continuous in U, and $\mathbf{x}_u \times \mathbf{x}_v \neq (0,0,0)^T$ for all $(u,v) \in U$.

Definition 2. The **tangent plane** at a point $\mathbf{x}_0 = \mathbf{x}(\mathbf{u}_0)$ on a regular parametric two-dimensional surface in three-dimensional space is defined as the set of all points y in $\mathbb{R}^3$ satisfying the equation

$$y = \mathbf{x}_0 + a\mathbf{x}_u(\mathbf{u}_0) + b\mathbf{x}_v(\mathbf{u}_0), \qquad a, b \in \mathbb{R}. \tag{3}$$

Definition 3. The **outward unit normal vector** $\mathbf{n}_0 = \mathbf{n}(\mathbf{u}_0)$ of a regular parametric surface at a point $\mathbf{x}_0$ is given by

$$\mathbf{n}_0 = \frac{\mathbf{x}_u(\mathbf{u}_0) \times \mathbf{x}_v(\mathbf{u}_0)}{\|\mathbf{x}_u(\mathbf{u}_0) \times \mathbf{x}_v(\mathbf{u}_0)\|} = \frac{\mathbf{x}_u \times \mathbf{x}_v}{\|\mathbf{x}_u \times \mathbf{x}_v\|}, \tag{4}$$

where "$\| \ \|$" indicates the Euclidean norm.

Definition 4. Let $\mathbf{x}(\mathbf{u})$ be a regular parametric surface of class m, $m \geq 2$, and $\mathbf{c}(t) = \mathbf{c}(u(t), v(t))$ be a (regular) curve of class 2 on the surface through the point $\mathbf{x}_0 = \mathbf{x}(\mathbf{u}_0)$. The **normal curvature vector** to $\mathbf{c}(t)$ at $\mathbf{x}_0$ is the projection of the curvature vector $\mathbf{k} = \dot{\mathbf{t}}/\|\dot{\mathbf{t}}\|$, $\mathbf{t} = \dot{\mathbf{c}}/\|\dot{\mathbf{c}}\|$, onto the unit surface normal vector $\mathbf{n}_0$,

$$\mathbf{k}_n = (\mathbf{k} \cdot \mathbf{n}_0)\mathbf{n}_0. \tag{5}$$

The proportionality factor $\mathbf{k} \cdot \mathbf{n}_0$ is called the **normal curvature**, denoted by κ_n.

Definition 5. The second degree polynomial

$$I(du, dv) = \mathbf{x}_u \cdot \mathbf{x}_u \, du^2 + 2\mathbf{x}_u \cdot \mathbf{x}_v \, du \, dv + \mathbf{x}_v \cdot \mathbf{x}_v \, dv^2$$
$$= E \, du^2 + 2F \, du \, dv + G \, dv^2, \tag{6}$$

where $du, dv \in \mathbb{R}$, is called the **first fundamental form** of a regular parametric surface $\mathbf{x}(\mathbf{u})$. The coefficients E, F, and G are called the **first fundamental coefficients**.

Definition 6. Assuming that the regular parametric surface $\mathbf{x(u)}$ is at least of order 2, the second degree polynomial

$$II(du, dv) = -\mathbf{x}_u \cdot \mathbf{n}_u \, du^2 - (\mathbf{x}_u \cdot \mathbf{n}_v + \mathbf{x}_v \cdot \mathbf{n}_u) \, du \, dv - \mathbf{x}_v \cdot \mathbf{n}_v \, dv^2$$

$$= \mathbf{x}_{uu} \cdot \mathbf{n} \, du^2 + 2\mathbf{x}_{uv} \cdot \mathbf{n} \, du \, dv + \mathbf{x}_{vv} \cdot \mathbf{n} \, dv^2$$

$$= L \, du^2 + 2M \, du \, dv + N \, dv^2, \tag{7}$$

where $du, dv \in \mathbb{R}$, is called the **second fundamental form** of $\mathbf{x(u)}$. The coefficients L, M, and N are called the **second fundamental coefficients**.

Definition 7. The two (real) eigenvalues κ_1 and κ_2 of the matrix

$$-A = -\begin{pmatrix} a_{1,1} & a_{1,2} \\ a_{2,1} & a_{2,2} \end{pmatrix} = \begin{pmatrix} L & M \\ M & N \end{pmatrix}\begin{pmatrix} E & F \\ F & G \end{pmatrix}^{-1}, \tag{8}$$

where

$$a_{1,1} = \frac{MF - LG}{EG - F^2}, \qquad a_{1,2} = \frac{LF - ME}{EG - F^2},$$

$$a_{2,1} = \frac{NF - MG}{EG - F^2}, \qquad a_{2,2} = \frac{MF - NE}{EG - F^2},$$

of a regular surface of class of at least 2 at a point $\mathbf{x}_0$ are called **principal curvatures** of the regular parametric surface at $\mathbf{x}_0$. The associated eigenvectors determine the **principal curvature directions**. Therefore, the principal curvatures are the (real) roots of the characteristic polynomial of $-A$, the quadratic polynomial

$$\kappa^2 + (a_{1,1} + a_{2,2})\kappa + a_{1,1}a_{2,2} - a_{1,2}a_{2,1}. \tag{9}$$

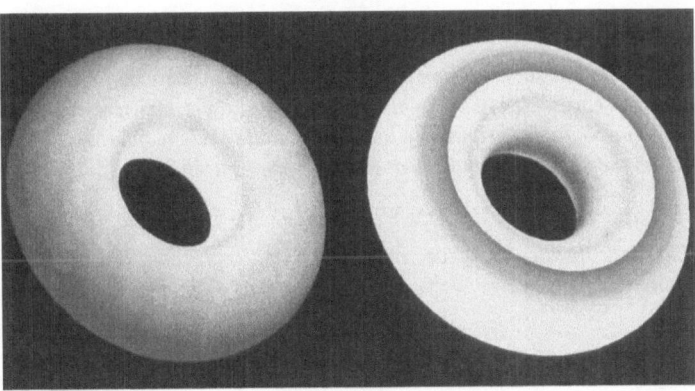

Figure 1. Texture map of mean and Gaussian curvature onto a torus, $((2 + \cos u)\cos v, (2 + \cos u)\sin v, \sin u)^T$, $u, v \in [0, 2\pi]$; green/yellow representing negative curvature values, magenta/blue representing positive curvature values

Definition 8. The average H of the two principal curvatures κ_1 and κ_2 is called the **mean curvature**, the product K is called the **Gaussian curvature** of the regular parametric surface $x(u)$ at x_0,

$$H = \tfrac{1}{2}(\kappa_1 + \kappa_2), \qquad K = \kappa_1 \kappa_2. \tag{10}$$

3. Curvature Approximation for Triangulated two-Dimensional Surfaces

The graph of an explicit bivariate function $f(x, y)$ can be viewed as a special parametric surface with the parametrization $x(u, v) = u$, $y(u, v) = v$, and $z(u, v) = f(u, v)$,

$$\mathbf{x(u)} = (u, v, f(u, v))^T, \qquad (u, v) \in D \subset \mathbb{R}^2, \tag{11}$$

The following formulae will be needed later on. Therefore, some basic facts are summarized next. For this particular surface, the unit normal vector is given by

$$\mathbf{n(u)} = \frac{\mathbf{x}_u \times \mathbf{x}_v}{\|\mathbf{x}_u \times \mathbf{x}_v\|} = \frac{(-f_u, -f_v, 1)^T}{\sqrt{1 + f_u^2 + f_v^2}}, \tag{12}$$

and the first and second fundamental coefficients are

$$E = 1 + f_u^2, \qquad F = f_u f_v \qquad G = 1 + f_v^2,$$

$$L = \frac{f_{uu}}{\sqrt{1 + f_u^2 + f_v^2}}, \quad M = \frac{f_{uv}}{\sqrt{1 + f_u^2 + f_v^2}}, \quad \text{and} \quad N = \frac{f_{vv}}{\sqrt{1 + f_u^2 + f_v^2}}. \tag{13}$$

The Gauss-Weingarten map is

$$-A = -\begin{pmatrix} a_{1,1} & a_{1,2} \\ a_{2,1} & a_{2,2} \end{pmatrix} = \frac{1}{l}\begin{pmatrix} f_{uu} & f_{uv} \\ f_{uv} & f_{vv} \end{pmatrix}\begin{pmatrix} 1 + f_u^2 & f_u f_v \\ f_u f_v & 1 + f_v^2 \end{pmatrix}^{-1}, \tag{14}$$

where $l = \sqrt{1 + f_u^2 + f_v^2}$.

Theorem 1. *Each regular parametric two-dimensional surface $x(u)$ of class m, $m \geq 2$, can locally be represented in the explicit form $z = z(x, y)$ which is at least C^2. Choosing a surface point x_0 as the origin of a local coordinate system and the z-axis in the same direction as the surface normal n_0 at x_0, z can be written as*

$$z(x, y) = \tfrac{1}{2}(c_{2,0}x^2 + 2c_{1,1}xy + c_{0,2}y^2) + \cdots, \tag{15}$$

*Choosing appropriate basis vectors yields the representation of the **osculating paraboloid** at x_0, given by*

$$z(x, y) = \tfrac{1}{2}(c_{2,0}^* x^2 + c_{0,2}^* y^2),$$

such that the two principal curvatures at x_0 coincide with the coefficients of this paraboloid, $\kappa_1 = c_{2,0}^*$ and $\kappa_2 = c_{0,2}^*$.

Proof. See [Strubecker '58, '59] or [Struik '61].

The principal curvature approximation method to be introduced is based on bivariate polynomials. It is essential to prove a certain property of such functions before describing the approximation technique. Given an origin in the plane, the graph of a bivariate polynomial f consisting of all the points in the set $\{(x, y, f(x, y))^T \mid x, y \in \mathbb{R}\}$ is independent of the choice of the orientation of the two unit vectors determining an orthonormal coordinate system for the plane. This fact implies that the principal curvatures of the graph, a two-dimensional surface, are independent of the two unit vectors as well.

Lemma 1. *The equation*

$$\sum_{k=0}^{i} (-1)^k \binom{i}{k} (x \cos^2 \alpha + y \sin \alpha \cos \alpha)^{i-k} (-x \sin^2 \alpha + y \sin \alpha \cos \alpha)^k = x^i \quad (16)$$

holds for all $x, y, \alpha \in \mathbb{R}$ and $i \geq 0$.

Proof. It is easy to show that Eq. (16) is valid for $i = 0$:

$$1 = x^0.$$

The induction hypothesis is made that Eq. (16) is true for $i - 1$. Thereby one proves that

$$\sum_{k=0}^{i} (-1)^k \binom{i}{k} (x \cos^2 \alpha + y \sin \alpha \cos \alpha)^{i-k} (-x \sin^2 \alpha + y \sin \alpha \cos \alpha)^k$$

$$= ((x \cos^2 \alpha + y \sin \alpha \cos \alpha) - (-x \sin^2 \alpha + y \sin \alpha \cos \alpha))$$

$$\sum_{k=0}^{i-1} (-1)^k \binom{i-1}{k} (x \cos^2 \alpha + y \sin \alpha \cos \alpha)^{i-1-k} (-x \sin^2 \alpha + y \sin \alpha \cos \alpha)^k$$

$$= x(\cos^2 \alpha + \sin^2 \alpha) x^{i-1} = x x^{i-1} = x^i. \qquad \square$$

Lemma 2. *The equation*

$$\sum_{l=0}^{j} \binom{j}{l} (x \sin \alpha \cos \alpha + y \sin^2 \alpha)^{j-l} (-x \sin \alpha \cos \alpha + y \cos^2 \alpha)^l = y^j \quad (17)$$

holds for all $x, y, \alpha \in \mathbb{R}$ and $j \geq 0$.

Proof. Follows the proof of lemma 1.

Theorem 2. *Let f be the bivariate polynomial*

$$f(x, y) = \sum_{\substack{i+j \leq n \\ i,j \geq 0}} c_{i,j} x^i y^j, \quad (18)$$

where a point in the plane has coordinates x and y with respect to a coordinate system given by an origin o and two orthonormal basis vectors d_1 and d_2; rotating d_1 and d_2 around the origin o changes the representation of the bivariate polynomial, but not its graph.

Proof. Let d_1 and d_2 be two unit vectors determining a first orthonormal coordinate system together with the origin o, and let $\bar{d}_1$ and $\bar{d}_2$ be a second pair of unit vectors obtained by rotating d_1 and d_2 by an angle α around o. A point in the plane may have coordinates $(x, y)^T$ with respect to the first coordinate system and coordinates

$$\begin{pmatrix} \bar{x} \\ \bar{y} \end{pmatrix} = \begin{pmatrix} \cos \alpha & \sin \alpha \\ -\sin \alpha & \cos \alpha \end{pmatrix} \begin{pmatrix} x \\ y \end{pmatrix} \tag{19}$$

with respect to the second coordinate system. Assuming (18) is the representation of the polynomial f with respect to the first coordinate system, f can be rewritten using the inverse map of (19):

$$f(x = \bar{x}\cos\alpha - \bar{y}\sin\alpha, y = \dot{x}\sin\alpha + \bar{y}\cos\alpha)$$

$$= \sum_{\substack{i+j \le n \\ i,j \ge 0}} c_{i,j}(\bar{x}\cos\alpha - \bar{y}\sin\alpha)^i(\bar{x}\sin\alpha + \bar{y}\cos\alpha)^j. \tag{20}$$

Evaluating f at the point $(\bar{x}, \bar{y})^T = (x\cos\alpha + y\sin\alpha, -x\sin\alpha + y\cos\alpha)^T$, considering the binomial theorem, lemma 1, and lemma 2, one derives the equation

$$f(\bar{x} = x\cos\alpha + y\sin\alpha, \bar{y} = -x\sin\alpha + y\cos\alpha)$$

$$= \sum_{\substack{i+j \le n \\ i,j \ge 0}} c_{i,j}(\cos\alpha(x\cos\alpha + y\sin\alpha) - \sin\alpha(-x\sin\alpha + y\cos\alpha))^i$$

$$(\sin\alpha(x\cos\alpha + y\sin\alpha) + \cos\alpha(-x\sin\alpha + y\cos\alpha))^j$$

$$= \sum_{\substack{i+j \le n \\ i,j \ge 0}} c_{i,j}\left(\sum_{k=0}^{i}(-1)^k\binom{i}{k}(\cos\alpha(x\cos\alpha + y\sin\alpha))^{i-k}(\sin\alpha(-x\sin\alpha + y\cos\alpha))^k\right.$$

$$\left.\sum_{l=0}^{j}\binom{j}{l}(\sin\alpha(x\cos\alpha + y\sin\alpha))^{j-l}(\cos\alpha(-x\sin\alpha + y\cos\alpha))^l\right)$$

$$= \sum_{\substack{i+j \le n \\ i,j \ge 0}} c_{i,j}\left(\sum_{k=0}^{i}(-1)^k\binom{i}{k}(x\cos^2\alpha + y\sin\alpha\cos\alpha)^{i-k}(-x\sin^2\alpha + y\sin\alpha\cos\alpha)^k\right.$$

$$\left.\sum_{l=0}^{j}\binom{j}{l}(x\sin\alpha\cos\alpha + y\sin^2\alpha)^{j-l}(-x\sin\alpha\cos\alpha + y\cos^2\alpha)^l\right)$$

$$= \sum_{\substack{i+j \le n \\ i,j \ge 0}} c_{i,j}x^iy^j = f(x, y). \qquad \square$$

The curvature approximation method is based on a localization of a two-dimensional triangulation. The local neighborhood around a point x_i is its platelet.

Definition 9. Given a two-dimensional triangulation in two- or three-dimensional space, the **platelet** $\mathcal{P}_i$ associated with a point x_i in the triangulation is the set of all triangles (determined by the index-triples (j_1, j_2, j_3) specifying their vertices) sharing x_i as a common vertex,

$$\mathcal{P}_i = \bigcup \{(j_1, j_2, j_3) | i = j_1 \vee i = j_2 \vee i = j_3\}. \tag{21}$$

The vertices constituting $\mathcal{P}_i$ are referred to as **platelet points**.

In order to approximate the principal curvatures at a point x_i in a two-dimensional triangulation a bivariate polynomial is constructed for a certain neighborhood around this point. Considering the facts that a two-dimensional surface can locally be represented explicitly (theorem 1) and that the graph of a bivariate polynomial is independent of the orientation of the two unit vectors determining an orthonormal coordinate system for the plane (theorem 2), the following sequence of computations is proposed.

(i)   Determine the platelet points associated with x_i.
(ii)   Compute the plane P passing through x_i and having n_i (exact or approximated normal at x_i) as its normal.
(iii) Define an orthonormal coordinate system in P with x_i as its origin and two arbitrary unit vectors in P.
(iv)   Compute the distances of all platelet points from the plane P.
(v)   Project all platelet points onto the plane, P and represent their projections with respect to the local coordinate system in P.
(vi)   Interpret the projections in P as abscissae values and the distances of the original platelet points from P as ordinate values.
(vii) Construct a bivariate polynomial f approximating these ordinate values.
(viii) Compute the principal curvatures of f's graph at x_i.

Let $\{y_j = (x_j, y_j, z_j)^T | j = 0 \ldots n_i\}$ be the set of all platelet points associated with the point x_i such that $y_0 = x_i$, and let $n = (n^x, n^y, n^z)^T$ be the outward unit normal vector at y_0. The implicit equation for the plane P is then given by

$$
\begin{aligned}
n \cdot (x - y_0) &= n^x(x - x_0) + n^y(y - y_0) + n^z(z - z_0) \\
&= n^x x + n^y y + n^z z - (n^x x_0 + n^y y_0 + n^z z_0) \\
&= Ax + By + Cz + D = 0.
\end{aligned}
\tag{22}
$$

Depending on the outward unit normal vector n one chooses a vector a perpendicular to n $(a \cdot n = 0)$ among the possibilities

$$
a = \begin{cases}
\dfrac{1}{n^x}(-(n^y + n^z), n^x, n^x)^T, & n^x \neq 0, \\[2mm]
\dfrac{1}{n^y}(n^y, -(n^x + n^z), n^y)^T, & n^y \neq 0, \\[2mm]
\dfrac{1}{n^z}(n^z, n^z, -(n^x + n^y))^T, & n^z \neq 0,
\end{cases}
$$

in order to obtain the first unit basis vector b_1,

$$
b_1 = \frac{a}{\|a\|}, \qquad \|a\| = \sqrt{(a \cdot a)}.
$$

The second unit basis vector b_2 is defined as the cross product of n and b_1,

$$
b_2 = n \times b_1.
$$

The perpendicular signed distances $d_j, j = 0 \ldots n_i$, of all platelet points y_j from the plane P are

$$d_j = \text{dist}(\mathbf{y}_j, P) = \frac{Ax_j + By_j + Cz_j + D}{\sqrt{A^2 + B^2 + C^2}} = Ax_j + By_j + Cz_j + D. \qquad (23)$$

Projecting all platelet points $\mathbf{y}_j$ onto P yields the points $\mathbf{y}_j^P$,

$$\mathbf{y}_j^P = \mathbf{y}_j - d_j\mathbf{n}. \qquad (24)$$

Considering $\mathbf{y}_0$ as the origin and $\mathbf{b}_1$ and $\mathbf{b}_2$ as the two unit basis vectors of a local two-dimensional orthonormal coordinate system for the plane P, each point $\mathbf{y}_j^P$ in P can be expressed in terms of that coordinate system. Therefore, one computes the difference vectors

$$\mathbf{d}_j = \mathbf{y}_j^P - \mathbf{y}_0, \qquad j = 0 \ldots n_i,$$

and expresses them as linear combinations of the two unit basis vectors $\mathbf{b}_1$ and $\mathbf{b}_2$ in P. Each difference vector $\mathbf{d}_j$ can be represented in the form

$$\mathbf{d}_j = (\mathbf{d}_j \cdot \mathbf{b}_1)\mathbf{b}_1 + (\mathbf{d}_j \cdot \mathbf{b}_2)\mathbf{b}_2, \qquad (25)$$

defining the local coordinates u_j and v_j of the point $\mathbf{y}_j^P$ in terms of the local coordinate system:

$$(u_j, v_j)^T = (\mathbf{d}_j \cdot \mathbf{b}_1, \mathbf{d}_j \cdot \mathbf{b}_2)^T. \qquad (26)$$

Interpreting the local coordinates u_j and v_j as abscissae values and the signed distances d_j as ordinate values (in direction of the normal $\mathbf{n}$), a polynomial $f(u, v)$ of degree two (see theorem 1) is constructed approximating these ordinate values. Forcing the polynomial f to satisfy $f(0,0) = f_u(0,0) = f_v(0,0) = 0$, the constraints

$$f(u_j, v_j) = \tfrac{1}{2}(c_{2,0}u_j^2 + 2c_{1,1}u_jv_j + c_{0,2}v_j^2) = d_j, \qquad j = 1 \ldots n_i,$$

remain. Written in matrix representation these constraints are

$$\begin{bmatrix} u_1^2 & 2u_1v_1 & v_1^2 \\ \vdots & \vdots & \vdots \\ u_{n_i}^2 & 2u_{n_i}v_{n_i} & v_{n_i}^2 \end{bmatrix} \begin{bmatrix} c_{2,0} \\ c_{1,1} \\ c_{0,2} \end{bmatrix} = U\mathbf{c} = \mathbf{d} = \begin{bmatrix} d_1 \\ \vdots \\ d_{n_i} \end{bmatrix}. \qquad (27)$$

This overdetermined system of linear equations is solved using a least squares approach (see [Davis '75]). The resulting normal equations are

$$U^T U\mathbf{c} = U^T\mathbf{d}. \qquad (28)$$

Provided the determinant of $U^T U$ does not vanish this system can immediately be solved using Cramer's rule. In the case that the determinant of $U^T U$ vanishes (e.g., when $\mathbf{x}_i$ is a point on the boundary of the triangulation) one considers additional points connected to $\mathbf{x}_i$'s platelet by an edge in the triangulation.

Theorem 3. *The principal curvatures κ_1 and κ_2 of the graph $(u, v, f(u, v))^T \subset \mathbb{R}^3$, u, $v \in \mathbb{R}$, of the bivariate polynomial*

$$f(u,v) = \tfrac{1}{2}(c_{2,0}u^2 + 2c_{1,1}uv + c_{0,2}v^2) \qquad (29)$$

at the point $(0, 0, f(0, 0))^T$ are given by the two real roots of the quadratic equation

$$\kappa^2 - (c_{2,0} + c_{0,2})\kappa + c_{2,0}c_{0,2} - c_{1,1}^2 = 0. \tag{30}$$

Proof. According to definition 7 and Eq. (14), the principal curvatures of f's graph are the eigenvalues of the matrix

$$-A = \frac{1}{l}\begin{pmatrix} f_{uu} & f_{uv} \\ f_{uv} & f_{vv} \end{pmatrix}\begin{pmatrix} 1 + f_u^2 & f_u f_v \\ f_u f_v & 1 + f_v^2 \end{pmatrix}^{-1},$$

where $l = \sqrt{1 + f_u^2 + f_v^2}$. Evaluating $-A$ for $u = v = 0$, one obtains the matrix

$$-A = \begin{pmatrix} c_{2,0} & c_{1,1} \\ c_{1,1} & c_{0,2} \end{pmatrix},$$

having the characteristic polynomial in (30). □

Solving the normal Eq. (28) and determining the roots of the characteristic polynomial in (30), one finally obtains the desired approximations for the principal curvatures at the point x_i.

The above construction is illustrated in Fig. 2. Shown are the platelet points around the point x_i, the tangent plane P, its local orthonormal coordinate system (origin x_i and basis vectors b_1 and b_2), and the projections of the platelet points (y_j^P) onto P.

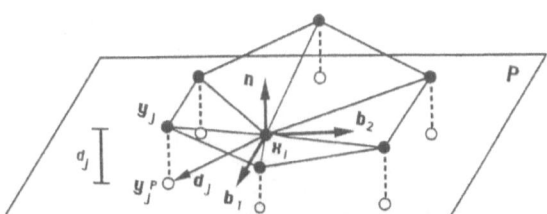

Figure 2. Construction of a bivariate polynomial for platelet points in a two-dimensional triangulation

4. Test Results

The presented technique for principal curvature approximation is tested for graphs of several bivariate functions. The exact principal curvatures κ_1^{ex} and κ_2^{ex} are compared with the approximated principal curvatures κ_1^{app} and κ_2^{app}; the exact mean curvature $H^{ex} = \frac{1}{2}(\kappa_1^{ex} + \kappa_2^{ex})$ is compared with the average of the approximated principal curvatures $H^{app} = \frac{1}{2}(\kappa_1^{app} + \kappa_2^{app})$ and the exact Gaussian curvature $K^{ex} = \kappa_1^{ex}\kappa_2^{ex}$ with the product of the approximated principal curvatures $K^{app} = \kappa_1^{app}\kappa_2^{app}$.

All bivariate test functions $f(x, y)$ are defined over $[-1, 1] \times [-1, 1]$ and evaluated on a $51 \cdot 51$—grid with equidistant spacing,

$$(x_i, y_j)^T = \left(-1 + \frac{i}{25}, -1 + \frac{j}{25}\right)^T, \qquad i,j = 0\ldots 50,$$

determining a finite set of three-dimensional points on their graphs,

$$\{(x_i, y_j, f(x_i, y_j))^T | i,j = 0\ldots 50\}.$$

The triangulation of a function's graph is obtained by splitting each quadrilateral specified by its index quadruple

$$((i,j), (i+1,j), (i+1,j+1), (i,j+1))$$

into the two triangles $T_{i,j}^1$ and $T_{i,j}^2$ identified by their index triples,

$$T_{i,j}^1 = ((i,j), (i+1,j), (i+1,j+1)) \quad \text{and} \quad T_{i,j}^2 = ((i,j), (i+1,j+1), (i,j+1)).$$

The root-mean-square error (RMS error) is a common error measure and is computed for each test example and curvature type. The RMS error is defined as

$$\sqrt{\frac{1}{n}\sum_{i=0}^{n-1}(f_i^{ex} - f_i^{app})^2}, \tag{31}$$

where n is the total number of exact (or approximated) values $f_i^{ex}(f_i^{app})$. Here, n equals $51 \cdot 51$; depending on the curvature type approximated f_i^{ex} can represent the exact values for $\kappa_1^{ex}, \kappa_2^{ex}, H^{ex}$ or K^{ex}, and f_i^{app} can represent the approximated values for $\kappa_1^{app}, \kappa_2^{app}, H^{app}$ or K^{app}, respectively. Table 1 summarizes the test results for the approximation of the principal curvatures, the mean, and the Gaussian curvature.

Table 1. RMS errors of curvature approximation for graphs of bivariate functions

Function	κ_1	κ_2	H	K
1. Cylinder: $\sqrt{2-x^2}$.	.000291	.000035	.000132	.000025
2. Sphere: $\sqrt{4-(x^2+y^2)}$.	.000159	.000046	.000080	.000080
3. Paraboloid: $.4(x^2+y^2)$.	.003073	.001342	.001358	.001684
4. Hyperboloid: $.4(x^2-y^2)$.	.002058	.002058	.001057	.001767
5. Monkey saddle: $.2(x^3-3xy^2)$.	.004483	.004483	.001591	.007247
6. Cubic polynomial: $.15(x^3+2x^2y-xy+2y^2)$.	.002258	.003598	.001665	.002242
7. Exponential function: $e^{-1/2(x^2+y^2)}$.	.001757	.005546	.002722	.002602
8. Trigonometric function: $.1(\cos(\pi x) + \cos(\pi y))$.	.002998	.002821	.001013	.003541

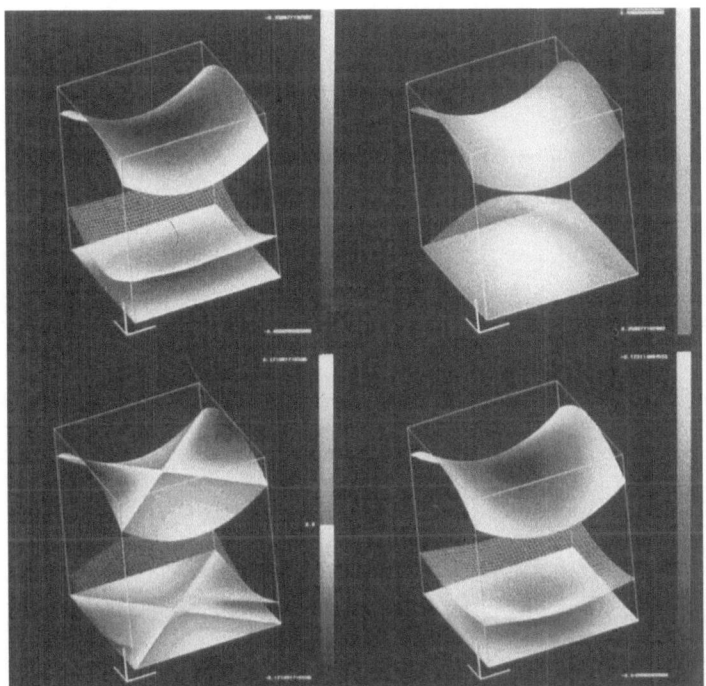

Figure 3. Exact curvatures κ_1^{ex}, κ_2^{ex}, H^{ex}, and K^{ex} on the graph of $f(x, y) = .4(x^2 - y^2)$, $x, y \in [-1, 1]$

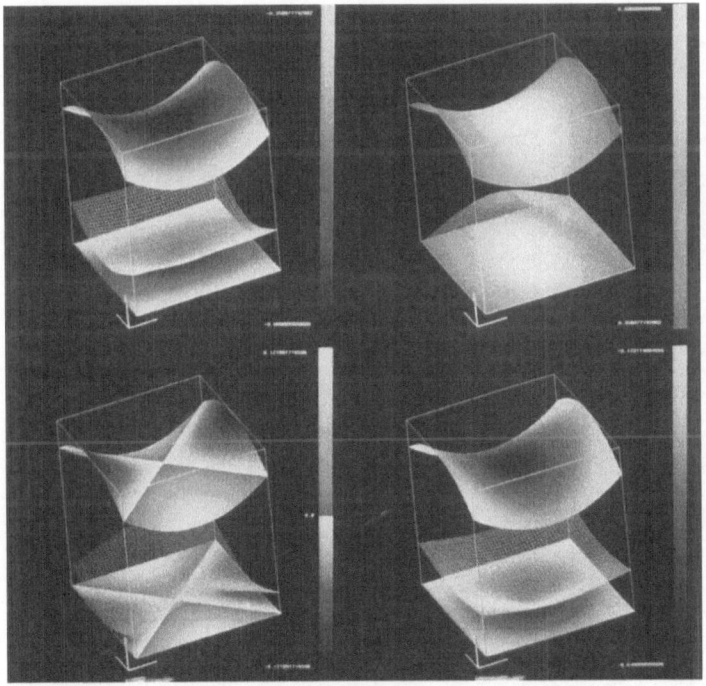

Figure 4. Approximated curvatures κ_1^{app}, κ_2^{app}, H^{app}, and K^{app} on the graph of $f(x, y) = .4(x^2 - y^2)$, $x, y \in [-1, 1]$

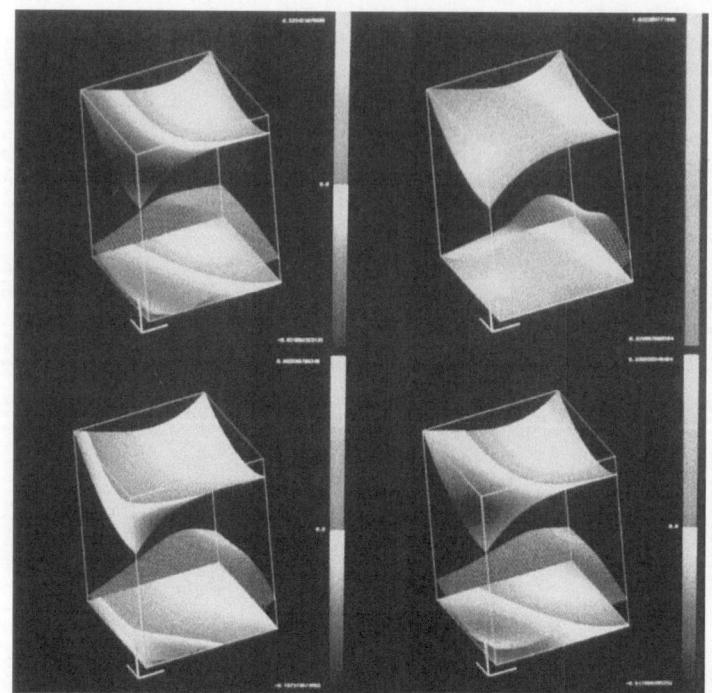

Figure 5. Exact curvatures κ_1^{ex}, κ_2^{ex}, H^{ex}, and K^{ex} on the graph of $f(x, y) = .15(x^3 + 2x^2y - xy + 2y^2)$, $x, y \in [-1, 1]$

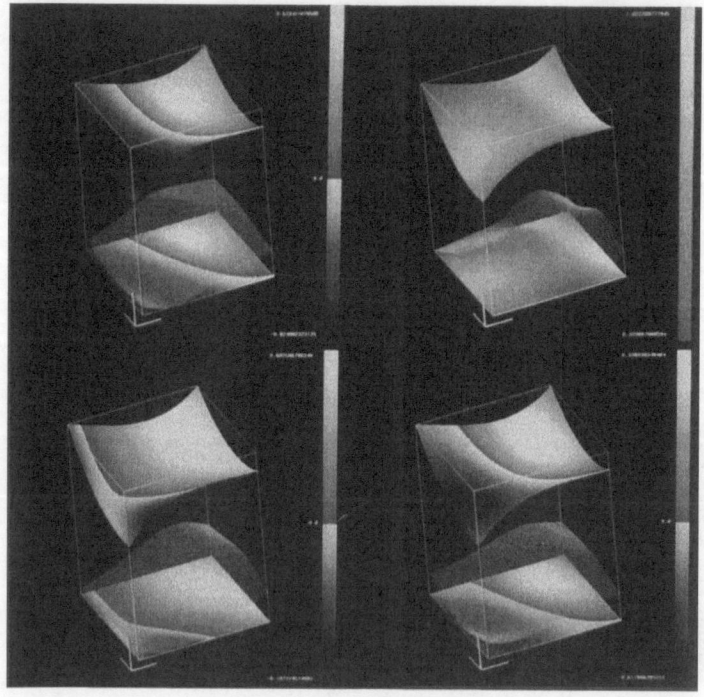

Figure 6. Approximated curvatures κ_1^{app}, κ_2^{app}, H^{app}, and K^{app} on the graph of $f(x, y) = .15(x^3 + 2x^2y - xy + 2y^2)$, $x, y \in [-1, 1]$

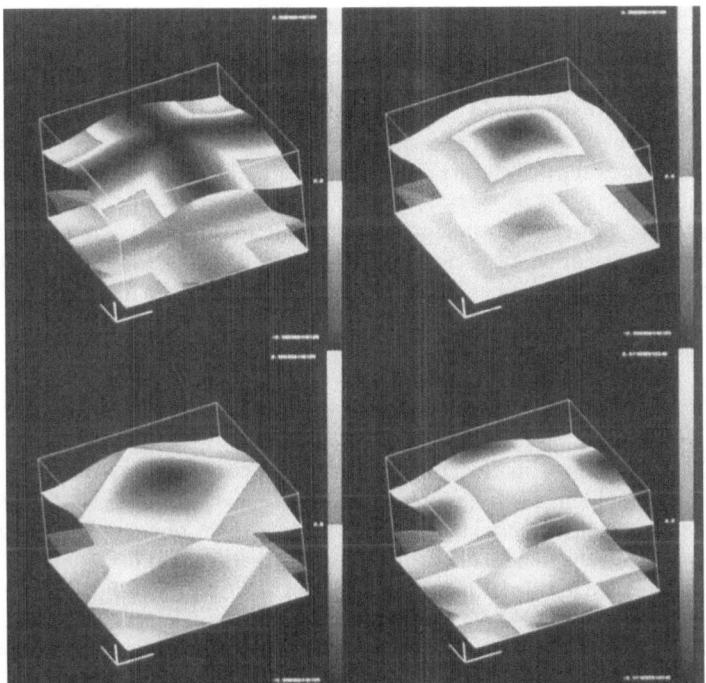

Figure 7. Exact curvatures κ_1^{ex}, κ_2^{ex}, H^{ex}, and K^{ex} on the graph of $f(x,y) = .1(\cos(\pi x) + \cos(\pi y))$, x, $y \in [-1, 1]$

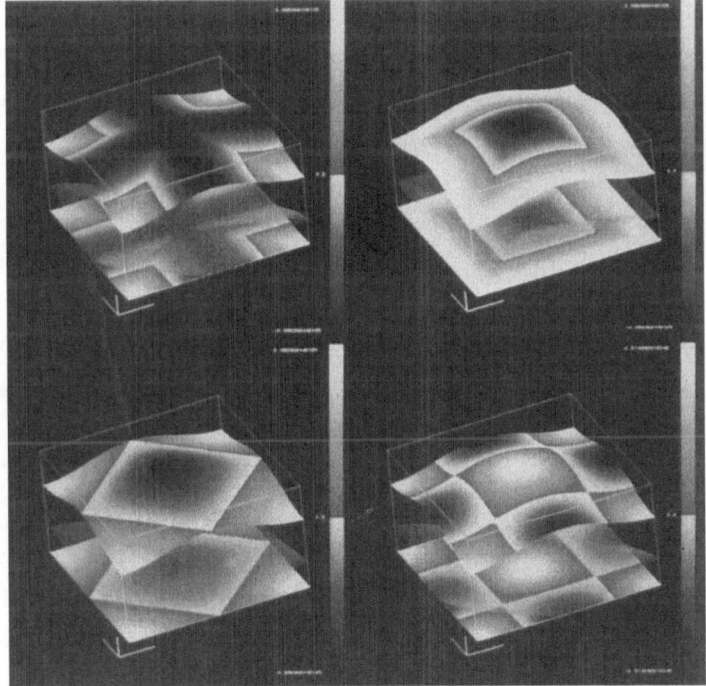

Figure 8. Approximated curvatures κ_1^{app}, κ_2^{app}, H^{app}, and K^{app} on the graph of $f(x,y) = .1(\cos(\pi x) + \cos(\pi y))$, $x, y \in [-1, 1]$

In the figures, the four particular curvatures used in Table 1 are mapped as textures onto the hyperboloid (function 5), the graph of the cubic polynomial (function 7) and the graph of the trigonometric function (function 9). Pairs of consecutive figures show the exact (upper figure) and the approximated curvatures (lower figure). The principal curvature κ_1 is visualized in the upper-left, κ_2 in the upper-right, the mean curvature H in the lower-left and the Gaussian curvature K in the lower-right corner of each figure. The Figs. 3 and 4 show the exact and approximated curvature values for function 5, the Figs. 5 and 6 for function 7, and the Figs. 7 and 8 for function 9.

5. Conclusions

A technique for approximating the two principal curvatures at the vertices in a two-dimensional surface triangulation has been developed. The test examples chosen are all graphs of bivariate functions leading to an obvious error measure. Nevertheless, the scheme should perform well for general surface triangulations, since all surfaces can locally be viewed as graphs of bivariate functions. At this point, it has not been investigated how to adjust the scheme to platelets which can not be described in terms of a function. One could use an implicit surface approximation whenever necessary.

Acknowledgements

The work presented was supported by the Department of Energy under contract DE-FG02-87ER25041 and by the National Science Foundation under contract DDM 8807747 to Arizona State University. I wish to thank all members of the Computer Aided Geometric Design research group in the Computer Science Department at Arizona State University, especially Gregory M. Nielson, for their helpful suggestions. I am particularly thankful for the invitation to the Dagstuhl conference and want to express my gratitude to the organizers.

References

[1] Akima, H.: On estimating partial derivatives for bivariate interpolation of scattered data. The Rocky Mountain Journal of Mathematics *14*(1), 41–52 (1984).
[2] Brauner, H.: Differentialgeometrie (Differential Geometry, in German). Braunschweig: Vieweg 1981.
[3] Calladine, C. R.: Gaussian curvature and shell structures. In: Gregory, J. (ed.) Mathematics of surfaces, pp. 179–196. Oxford: Clarendon Press 1986.
[4] Davis, P. J.: Interpolation and approximation. New York: Dover Publications 1975.
[5] do Carmo, M. P.: Differential geometry of curves and surfaces. New Jersey: Prentice Hall 1976.
[6] Farin, G.: Curves and surfaces for computer aided geometric design, 3rd edn. San Diego: Academic Press 1992.
[7] Hagen, H., Pottmann, H.: Curvature continuous triangular interpolants. In: Lyche, T., Schumaker, L. L. (eds.) Mathematical methods in computer aided geometric design, pp. 373–384. New York: Academic Press 1989.
[8] Hamann, B.: Modeling contours of trivariate data. Mathematical modelling and numerical analysis (Modélisation Mathématique et Analysis Numérique) *26*(1), 51–75 (1992).
[9] Lawson, C. L.: C^1 surface interpolation for scattered data on a sphere. The Rocky Mountain Journal of Mathematics *14*(1), 177–202 (1984)

[10] Lipschutz, M. M.: Differential geometry. Schaum's outline series. New York: McGraw-Hill 1980.
[11] Nielson, G. M., Franke, R.: A method for construction of surfaces under tension. The Rocky Mountain Journal of Mathematics *14*(1), 203–221 (1984).
[12] O'Neill, B.: Elementary differential geometry, 3rd edn. New York: Academic Press 1969.
[13] Renka, R. J., Cline, A. K.: A triangle-based C^1 interpolation method. The Rocky Mountain Journal of Mathematics *14*(1), 223–237 (1984).
[14] Spivak, M.: Comprehensive introduction to differential geometry. Vol. 1–5. Publish or Perish. Massachusetts: Waltham 1970.
[15] Strubecker, K.: Differentialgeometrie I (Differential Geometry I, in German). Berlin: De Gruyter 1955.
[16] Strubecker, K.: Differentialgeometrie II (Differential Geometry II, in German). Berlin: De Gruyter 1958.
[17] Strubecker, K.: Differentialgeometrie III (Differential Geometry III, in German). Berlin: De Gruyter 1959.
[18] Struik, D. J.: Lectures on classical differential geometry. New York: Dover Publications 1961.

Bernd Hamann
Department of Computer Science
Mississippi State University
P.O. Drawer CS
Mississippi State, MS 39762-5623, U.S.A.
Electronic mail (internet):
hamann@cs.msstate.edu

Engineering Research Center for
Computational Field Simulation
Mississippi State University
P.O. Box 6176
Mississippi State, MS 39762, U.S.A.
Electronic mail (internet): hamann@erc.msstate.edu

Computing Suppl. 8, 155–172 (1993)

Composition of Tensor Product Bézier Representations

D. Lasser, Kaiserslautern

Abstract. Trimming of surfaces and volumes, curve and surface modeling via Bézier's idea of destortion, segmentation, reparametrization, geometric continuity are examples of applications of functional composition. This paper shows how to compose polynomial and rational tensor product Bézier representations. The problem of composing Bézier splines and B-spline representations will also be addressed in this paper.

Key words: Bézier, tensor product, composition, trimming, free-form deformation.

I. Introduction

The purpose of this paper is to show how the composition of two Bézier representations can again be represented as a tensor product Bézier representation (of higher degree). Figure 1 illustrates the idea of composition in the case of a planar Bézier curve $\mathbf{K}(t)$: $\mathbb{R} \to \mathbb{R}^2$ and a tensor product Bézier surface $\mathbf{F}(u, v)$: $\mathbb{R}^2 \to \mathbb{R}^d$, $d = 2, 3$.

Some applications of composition have been pointed out by deRose [14], who is discussing the composition of Bézier simplex forms. Some simple examples of composition are the evaluation, subdivision and polynomial/rational reparametrization of polynomial/rational Bézier representations. The later one might be of importance in context of GC^r-continuity.

A more interesting application is given by the idea of curve and surface modeling in the sense of free-form deformations (FFDs), first described by Bézier [1], and in the following subject of [15], [6] and [3]. The FFD idea is to embed an object in a deformable medium and, then manipulate the object by deforming the medium that surrounds it (see also [4], [8]).

A second, very important application concerns the subject of trimmed surfaces and trimmed volumes which are of high interest in solid modeling and surface design. Different aspects of trimmed subjects have been discussed in [2], [16], [13], [7] and [11].

The results of this paper can be used for example to exactly represent FFDs as well as trim curves and trim surfaces in coordinate space, in the sense that given

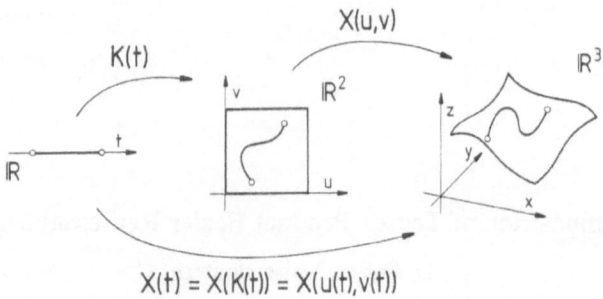

$$X(t) = X(K(t)) = X(u(t), v(t))$$

Figure 1. Composition $F(K(t))$ of a planar Bézier curve $K(t)$ and a tensor product Bézier surface $F(u, v)$

Theorems and Corollaries can be used to directly and exactly compute the control points for deformed subjects and trim curves and surfaces, respectively. Both implementations are presently under development and are being tested against an approximative method and also against a combination of the exact with an approximative method (see [9, 10]). Section II reviews definitions of tensor product Bézier representations and introduces the notation of this paper. In Sections III and IV explicit representations of the composition of polynomial and of rational tensor product representations are given. Section V is concerned with spline representations, but no in-depth treatment is given there.

II. Bézier Representations

A **Bézier curve** of degree l in u is defined by

$$X(u) = \sum_{i=0}^{l} b_i B_i^l(u), \qquad u \in [0, 1],$$

where $b_i \in \mathbb{R}^d$, $d \in \mathbb{N}$, and

$$B_i^l(u) = \binom{l}{i} u^i (1 - u)^{l-i}$$

are the (ordinary) **Bernstein polynomials** of degree l in u. The coefficients b_i are called **Bézier points**. They form in their natural ordering, given by their subscripts, the vertices of the Bézier polygon.

All properties of Bézier curves are a direct consequence of properties of the Bernstein polynomials. We list the ones which are of importance for the following calculations:

Recursion formula:

$$B_I^N(t) = (1 - t)B_I^{N-1}(t) + tB_{I-1}^{N-1}(t)$$

Partition of unity:

$$\sum_{I=0}^{N} B_I^N(t) = 1$$

Product formula:

$$\prod_{k=1}^{\alpha} B_{I_k}^{N_k}(t) = \frac{\sum_{k=1}^{\alpha} \binom{N_k}{I_k}}{\binom{|\mathbf{N}|}{|\mathbf{I}|}} B_{|\mathbf{I}|}^{|\mathbf{N}|}(t)$$

where $\mathbf{I} = (I_1, \ldots, I_\alpha), |\mathbf{I}| = I_1 + \cdots + I_\alpha$, and $\mathbf{N} = (N_1, \ldots, N_\alpha), |\mathbf{N}| = N_1 + \cdots + N_\alpha$.

The Bézier description of a curve is a very powerful tool because the expansion in terms of Bernstein polynomials yields, firstly, a numerically very stable behavior of all the curve algorithms. And, secondly, a geometric relationship between a curve and its defining Bézier points.

A **tensor product Bézier surface**—briefly **TPB-surface**—of degree (l, m) is defined by

$$\mathbf{X}(u, v) = \sum_{i=0}^{l} \sum_{j=0}^{m} \mathbf{b}_{i,j} B_i^l(u) B_j^m(v), \qquad u, v \in [0, 1],$$

and a **tensor product Bézier volume**—briefly **TPB-volume**—of degree (l, m, n) is defined by

$$\mathbf{X}(u, v, w) = \sum_{i=0}^{l} \sum_{j=0}^{m} \sum_{k=0}^{n} \mathbf{b}_{i,j,k} B_i^l(u) B_j^m(v) B_k^n(w), \qquad u, v, w \in [0, 1].$$

By reason of the tensor product definition the properties of Bézier surfaces and volumes are similar to the ones for curves and can easily be deduced from properties of the underlying Bézier curve scheme.

Also, as a consequence of the tensor product definition algorithms in u, in v and in w commute, and the result is independent of the order.

A **rational Bézier curve** of degree l in u is defined by

$$\mathbf{X}(u) = \frac{\sum_{i=0}^{l} \beta_i \mathbf{b}_i B_i^l(u)}{\sum_{i=0}^{l} \beta_i B_i^l(u)}, \qquad u \in [0, 1],$$

with **weights** $\beta_i \in \mathbb{R}$, and **rational TPB-surfaces** and **TPB-volumes** analogously. If we demand positive weights, we have all the properties and algorithms for rational Bézier curves, surfaces and volumes which we have for non-rational representations.

For an extensive coverage of properties of Bernstein polynomials and Bézier representations see e.g. [4], [8].

III. Composition of Polynomial TPB-Representations

III.1 Composition of Bézier Curves and TPB-Surfaces

Theorem 1 is fundamental for the composition $\mathbf{F}(t) = \mathbf{F}(\mathbf{K}(t)) = \mathbf{F}(u(t), v(t))$ of Bézier curves $\mathbf{K}(t)$ and TPB-surfaces $\mathbf{F}(u, v)$:

Theorem 1. Bézier curves and TPB-surfaces
Let $\mathbf{K}(t)\colon \mathbb{R} \to \mathbb{R}^2$ be a planar polynomial Bézier curve of degree N,

$$\mathbf{K}(t) = \sum_{I=0}^{N} \mathbf{k}_I B_I^N(t), \qquad t \in [0, 1],$$

where $\mathbf{K}(t) = (u(t), v(t))$, and Bézier points $\mathbf{k}_I = (u_I, v_I)$. And let $\mathbf{F}(u, v)\colon \mathbb{R}^2 \to \mathbb{R}^d$, $d = 2, 3$, be a polynomial TPB-surface of degree (l, m),

$$\mathbf{F}(u, v) = \sum_{i=0}^{l} \sum_{j=0}^{m} \mathbf{b}_{i,j} B_i^l(u) B_j^m(v), \qquad u, v \in [0, 1],$$

where $\mathbf{F}(u, v) = (x(u, v), y(u, v), z(u, v))$, and with Bézier points $\mathbf{b}_{i,j} = (x_{i,j}, y_{i,j}, z_{i,j})$. If $d = 2$ (2D-solid): $z_{i,j} = 0$, for all i, j.

For each $r = \alpha + \beta$ where $\alpha \in \{0, \ldots, l\}$; $\beta \in \{0, \ldots, m\}$, i.e. $r \in \{0, \ldots, l + m\}$, we have

$$\mathbf{F}(t) = \mathbf{F}(\mathbf{K}(t)) = \sum_{R=0}^{rN} \mathbf{B}_R B_R^{rN}(t), \tag{1}$$

where

$$\mathbf{B}_R = \sum_{|\mathbf{I}|=R} C_R^{\alpha,\beta}(N, \mathbf{I}) \mathbf{F}^{\alpha,\beta} \tag{2}$$

with

$$\mathbf{F}^{\alpha,\beta} = \sum_{i=0}^{l-\alpha} \sum_{j=0}^{m-\beta} \mathbf{b}_{i,j}^{i+\alpha, j+\beta} B_i^{l-\alpha}(u(t)) B_j^{m-\beta}(v(t)), \tag{3}$$

where $\mathbf{b}_{i,j}^{i+\alpha,j+\beta} = \mathbf{b}_{i,j}^{i+\alpha,j+\beta}(u_{\mathbf{I}^u}^{\alpha}, v_{\mathbf{I}^v}^{\beta})$, and with constants

$$C_R^{\alpha,\beta}(N, \mathbf{I}) = \frac{\displaystyle\prod_{Q^u=1}^{\alpha} \binom{N}{I_{Q^u}^u} \prod_{Q^v=1}^{\beta} \binom{N}{I_{Q^v}^v}}{\displaystyle\binom{rN}{R}}.$$

$\sum_{|\mathbf{I}|=R}$ has the meaning of summation over all $\mathbf{I} = (\mathbf{I}^u, \mathbf{I}^v)$ where $\mathbf{I}^u = (I_1^u, \ldots, I_\alpha^u)$, $\mathbf{I}^v = (I_1^v, \ldots, I_\beta^v)$ and where $0 \le I_1^u, \ldots, I_\alpha^u \le N$ and $0 \le I_1^v, \ldots, I_\beta^v \le N$ and $|\mathbf{I}| = |\mathbf{I}^u| + |\mathbf{I}^v| = I_1^u + \cdots + I_\alpha^u + I_1^v + \cdots + I_\beta^v = R$.

The $\mathbf{b}_{i,j}^{i+\alpha,j+\beta}(u_{\mathbf{I}^u}^{\alpha}, v_{\mathbf{I}^v}^{\beta})$ are defined recursively by de Casteljau's construction, i.e. for the u parameter direction by

$$\mathbf{b}_{i,j}^{i+\alpha,j+\beta}(u_{\mathbf{I}^u}^{\alpha}, v_{\mathbf{I}^v}^{\beta}) = (1 - u_{I_\alpha^u}) \mathbf{b}_{i,j}^{i+\alpha-1,j+\beta}(u_{\mathbf{I}^u}^{\alpha-1}, v_{\mathbf{I}^v}^{\beta}) + u_{I_\alpha^u} \mathbf{b}_{i+1,j}^{i+\alpha,j+\beta}(u_{\mathbf{I}^u}^{\alpha-1}, v_{\mathbf{I}^v}^{\beta}),$$

and for the v parameter direction by

$$\mathbf{b}_{i,j}^{i+\alpha,j+\beta}(u_{\mathbf{I}^u}^{\alpha}, v_{\mathbf{I}^v}^{\beta}) = (1 - v_{I_\beta})\mathbf{b}_{i,j}^{i+\alpha,j+\beta-1}(u_{\mathbf{I}^u}^{\alpha}, v_{\mathbf{I}^v}^{\beta-1}) + v_{I_\beta}\mathbf{b}_{i,j+1}^{i+\alpha,j+\beta}(u_{\mathbf{I}^u}^{\alpha}, v_{\mathbf{I}^v}^{\beta-1}),$$

where $\mathbf{b}_{i,j}^{i,j} = \mathbf{b}_{i,j}$.

According to (1), $\mathbf{F}(\mathbf{K}(t))$ is polynomial and can be represented as Bézier curve of degree rN. Bézier points of this representation are given as, (2), convex combinations of auxiliary points $\mathbf{F}^{\alpha,\beta}$ which are calculated, (3), for parameter values $(u_{\mathbf{I}^u}^{\alpha}, v_{\mathbf{I}^v}^{\beta})$ via the blossoming principle (see e.g. [12]). This means: The argument $(u_{\mathbf{I}^u}^{\alpha}, v_{\mathbf{I}^v}^{\beta})$ has the meaning that $\mathbf{b}_{i,j}^{i+\alpha,j+\beta}$ has to be calculated by performing α de Casteljau constructions in u direction for the u parameter values given by the indices $\mathbf{I}^u = (I_1^u, \ldots, I_\alpha^u)$, i.e. for the parameter values $u_{I_1^u}, \ldots, u_{I_\alpha^u}$ and β de Casteljau constructions in v direction for the v parameter values given by the indices $\mathbf{I}^v = (I_1^v, \ldots, I_\beta^v)$, i.e. for the parameter values $v_{I_1^v}, \ldots, v_{I_\beta^v}$. Calculations for different parameter values commute, and the order of performed calculations does not affect the final result.

To further illustrate the notation, we give the example of $l = m = 3$, $N = 3$ and $\alpha = 3$, $\beta = 2$. In this case $\mathbf{I}^u$ and $\mathbf{I}^v$ can be given by

$$\mathbf{I}^u(0,0,2), \quad \mathbf{I}^v = (0,1) \rightarrow \mathbf{b}_{i,j}^{i+3,j+2}(u_{\mathbf{I}^u}^{3}, v_{\mathbf{I}^v}^{2}) = \mathbf{b}_{i,j}^{i+3,j+2}(u_0, u_0, u_2, v_0, v_1)$$

$$\mathbf{I}^u(0,1,1), \quad \mathbf{I}^v = (0,3) \rightarrow \mathbf{b}_{i,j}^{i+3,j+2}(u_{\mathbf{I}^u}^{3}, v_{\mathbf{I}^v}^{2}) = \mathbf{b}_{i,j}^{i+3,j+2}(u_0, u_1, u_1, v_0, v_3)$$

for example, etc.

Next we prove the statement of Theorem 1.

Proof of Theorem 1. By induction on $r = \alpha + \beta$, analogously to [14].

Base case. $r = 0$, i.e. $\alpha = \beta = 0$. This is trivially true, because for $r = 0$, Theorem 1 yields $\mathbf{F}(u, v)$.

Inductive hypothesis. We assume,

$$\mathbf{F}(t) = \mathbf{F}(\mathbf{K}(t)) = \sum_{R=0}^{rN} \mathbf{B}_R B_R^{rN}(t)$$

is valid. On the one side for $r = (\alpha - 1) + \beta$, i.e.

$$\mathbf{B}_R = \sum_{|\mathbf{I}|=R} C_R^{\alpha-1,\beta}(N, \mathbf{I}) \mathbf{F}^{\alpha-1,\beta}(u_{\mathbf{I}^u}^{\alpha-1}, v_{\mathbf{I}^v}^{\beta})$$

and

$$\mathbf{F}^{\alpha-1,\beta} = \sum_{i=0}^{l-\alpha+1} \sum_{j=0}^{m-\beta} \mathbf{b}_{i,j}^{i+\alpha-1,j+\beta}(u_{\mathbf{I}^u}^{\alpha-1}, v_{\mathbf{I}^v}^{\beta}) B_i^{l-\alpha+1}(u(t)) B_j^{m-\beta}(v(t)),$$

where $\mathbf{I}^u = (I_1^u, \ldots, I_{\alpha-1}^u)$ and $\mathbf{I}^v = (I_1^v, \ldots, I_\beta^v)$; on the other side for $r = \alpha + (\beta - 1)$, i.e.

$$\mathbf{B}_R = \sum_{|\mathbf{I}|=R} C_R^{\alpha,\beta-1}(N, \mathbf{I}) \mathbf{F}^{\alpha,\beta-1}(u_{\mathbf{I}^u}^{\alpha}, v_{\mathbf{I}^v}^{\beta-1})$$

and

$$\mathbf{F}^{\alpha,\beta-1} = \sum_{i=0}^{l-\alpha} \sum_{j=0}^{m-\beta+1} \mathbf{b}_{i,j}^{i+\alpha,j+\beta-1}(u_{\mathbf{I}^u}^{\alpha}, v_{\mathbf{I}^v}^{\beta-1}) B_i^{l-\alpha}(u(t)) B_j^{m-\beta+1}(v(t)),$$

where $\mathbf{I}^u = (I_1^u, \ldots, I_\alpha^u)$ and $\mathbf{I}^v = (I_1^v, \ldots, I_{\beta-1}^v)$.

Inductive proof. Both cases are done similarly. We only prove the first case: Applying the recursive definition of the Bernstein polynomials to $B_i^{l-\alpha+1}(u(t))$ gives

$$B_i^{l-\alpha+1}(u(t)) = (1 - u(t))B_i^{l-\alpha}(u(t)) + u(t)B_{i-1}^{l-\alpha}(u(t)).$$

With that and an index transformation, the sum over i is

$$\sum_{i=0}^{l-\alpha} [(1 - u(t))\mathbf{b}_{i,j}^{i+\alpha-1,j+\beta}(u_{\mathbf{I}^u}^{\alpha-1}, v_{\mathbf{I}^v}^\beta) + u(t)\mathbf{b}_{i+1,j}^{i+\alpha,j+\beta}(u_{\mathbf{I}^u}^{\alpha-1}, v_{\mathbf{I}^v}^\beta)]B_i^{l-\alpha}(u(t)).$$

Using the Bernstein representation of $u(t)$ and the partition of unity property of the Bernstein polynomials, the term in squared brackets can be written as

$$\sum_{I_\alpha^u=0}^{N} \{(1 - u_{I_\alpha^u})\mathbf{b}_{i,j}^{i+\alpha-1,j+\beta}(u_{\mathbf{I}^u}^{\alpha-1}, v_{\mathbf{I}^v}^\beta) + u_{I_\alpha^u}\mathbf{b}_{i+1,j}^{i+\alpha,j+\beta}(u_{\mathbf{I}^u}^{\alpha-1}, v_{\mathbf{I}^v}^\beta)\}B_{I_\alpha^u}^N(t).$$

Substituting

$$\mathbf{b}_{i,j}^{i+\alpha,j+\beta}(u_{\mathbf{I}^u}^\alpha, v_{\mathbf{I}^v}^\beta) = (1 - u_{I_\alpha^u})\mathbf{b}_{i,j}^{i+\alpha-1,j+\beta}(u_{\mathbf{I}^u}^{\alpha-1}, v_{\mathbf{I}^v}^\beta) + u_{I_\alpha^u}\mathbf{b}_{i+1,j}^{i+\alpha,j+\beta}(u_{\mathbf{I}^u}^{\alpha-1}, v_{\mathbf{I}^v}^\beta),$$

where $\mathbf{I}^u = (I_1^u, \ldots, I_{\alpha-1}^u, I_\alpha^u)$, on the left side, while $\mathbf{I}^u = (I_1^u, \ldots, I_{\alpha-1}^u)$, on the right side, results in

$$\sum_{i=0}^{l-\alpha} \left[\sum_{I_\alpha^u=0}^{N} \mathbf{b}_{i,j}^{i+\alpha,j+\beta}(u_{\mathbf{I}^u}^\alpha, v_{\mathbf{I}^v}^\beta)B_{I_\alpha^u}^N(t) \right] B_i^{l-\alpha}(u(t)).$$

Therefore, the expression for $\mathbf{B}_R$ becomes

$$\mathbf{B}_R = \sum_{|\mathbf{I}|=R} C_R^{\alpha-1,\beta}(N, \mathbf{I}) \sum_{I_\alpha^u=0}^{N} \mathbf{F}^{\alpha,\beta}(u_{\mathbf{I}^u}^\alpha, v_{\mathbf{I}^v}^\beta)B_{I_\alpha^u}^N(t)$$

with

$$\mathbf{F}^{\alpha,\beta} = \sum_{i=0}^{l-\alpha} \sum_{j=0}^{m-\beta} \mathbf{b}_{i,j}^{i+\alpha,j+\beta}(u_{\mathbf{I}^u}^\alpha, v_{\mathbf{I}^v}^\beta)B_i^{l-\alpha}(u(t))B_j^{m-\beta}(v(t)).$$

Regrouping the terms of the sums $\sum_{|\mathbf{I}|=R}$ and $\sum_{I_\alpha^u=0}^{N}$ into one sum and considering the product formula for Bernstein polynomials for the product of $B_R^N(t)$ and $B_{I_\alpha^u}^N(t)$ completes the proof. $\square$

The restriction $d = 2, 3$ of Theorem 1 is not necessary and could be relaxed, because it is not needed for the proof of Theorem 1, but the cases $d = 2, 3$ are the ones of interest for most practical applications. Now, setting $\alpha = l$ and $\beta = m$ results in:

Corollary 1. Bézier curves and TPB-surfaces
Let $\mathbf{F}(u, v)$ and $\mathbf{K}(t)$ be given as in Theorem 1. For $\mathbf{F}(t) = \mathbf{F}(\mathbf{K}(t)) = \mathbf{F}(u(t), v(t))$ we have

$$\mathbf{F}(t) = \sum_{R=0}^{rN} \mathbf{B}_R B_R^{rN}(t),$$

where $r = l + m$ and with Bézier points

$$\mathbf{B}_R = \sum_{|\mathbf{I}|=R} C_R^{l,m}(N, \mathbf{I})\mathbf{b}_{0,0}^{l,m}(u_{\mathbf{I}^u}^l, v_{\mathbf{I}^v}^m).$$

Remark 1. **Bézier curves and surfaces**

- **Triangle Bézier surface of degree n**
  Parameter lines and lines of parameter space in general position map to Bézier curves of degree n.
- **TPB-surface of degree (l, m)**
  Parameter lines map to Bézier curves of degree l or m, respectively.
  Lines of parameter space in general position map to Bézier curves of degree $l + m$.

Example 1. **Straight Line**
Let $\mathbf{F}(u, v)$ be a biquadratic tensor product surface, i.e. $l = m = 2 \to r = 4$. And let $\mathbf{K}(t)$ be a straight line (in general position), i.e. $N = 1 \to rN = 4$. Bézier points $\mathbf{B}_R$ of the quartic Bézier curve $\mathbf{F}(t)$ are convex combinations of points $\mathbf{b}_{0;0}^{2;2}(u_{\bar{i}u}^2, v_{\bar{i}v}^2)$:

$$\mathbf{B}_0 = \mathbf{b}_{0;0}^{2;2}(u_0, u_0, v_0, v_0)$$

$$\mathbf{B}_1 = \tfrac{1}{4}[\mathbf{b}_{0;0}^{2;2}(u_0, u_0, v_0, v_1) + \mathbf{b}_{0;0}^{2;2}(u_0, u_0, v_1, v_0)$$
$$+ \mathbf{b}_{0;0}^{2;2}(u_0, u_1, v_0, v_0) + \mathbf{b}_{0;0}^{2;2}(u_1, u_0, v_0, v_0)]$$

$$\mathbf{B}_2 = \tfrac{1}{6}[\mathbf{b}_{0;0}^{2;2}(u_0, u_0, v_1, v_1) + \mathbf{b}_{0;0}^{2;2}(u_0, u_1, v_0, v_1) + \mathbf{b}_{0;0}^{2;2}(u_0, u_1, v_1, v_0)$$
$$+ \mathbf{b}_{0;0}^{2;2}(u_1, u_0, v_0, v_1) + \mathbf{b}_{0;0}^{2;2}(u_1, u_0, v_1, v_0) + \mathbf{b}_{0;0}^{2;2}(u_1, u_1, v_0, v_0)]$$

$$\mathbf{B}_3 = \tfrac{1}{4}[\mathbf{b}_{0;0}^{2;2}(u_0, u_1, v_1, v_1) + \mathbf{b}_{0;0}^{2;2}(u_1, u_0, v_1, v_1)$$
$$+ \mathbf{b}_{0;0}^{2;2}(u_1, u_1, v_0, v_1) + \mathbf{b}_{0;0}^{2;2}(u_1, u_1, v_1, v_0)]$$

$$\mathbf{B}_4 = \mathbf{b}_{0;0}^{2;2}(u_1, u_1, v_1, v_1)$$

The auxiliary points $\mathbf{b}_{0;0}^{2;2}(u_{\bar{i}u}^2, v_{\bar{i}v}^2)$ are a result of the merging of several de Casteljau algorithms, i.e. they are given by the polar form (blossom) values of $\mathbf{F}(u, v)$ for parameter values $u \in \{u_0, u_1\}$ and $v \in \{v_0, v_1\}$.

Because de Casteljau constructions commute, and the polar form of $\mathbf{F}(u, v)$ is symmetric in the argument, we have

$$\mathbf{b}_{0;0}^{2;2}(u_0, u_1, v_0, v_0) = \mathbf{b}_{0;0}^{2;2}(u_0, v_0, u_1, v_0) = \mathbf{b}_{0;0}^{2;2}(u_1, v_0, u_0, v_0),$$

and so on, and therefore,

$$\mathbf{B}_0 = \mathbf{b}_{0;0}^{2;2}(u_0, u_0, v_0, v_0)$$

$$\mathbf{B}_1 = \tfrac{1}{2}[\mathbf{b}_{0;0}^{2;2}(u_0, u_0, v_0, v_1) + \mathbf{b}_{0;0}^{2;2}(u_0, u_1, v_0, v_0)]$$

$$\mathbf{B}_2 = \tfrac{1}{6}[\mathbf{b}_{0;0}^{2;2}(u_0, u_0, v_1, v_1) + 4\mathbf{b}_{0;0}^{2;2}(u_0, u_1, v_0, v_1) + \mathbf{b}_{0;0}^{2;2}(u_1, u_1, v_0, v_0)]$$

$$\mathbf{B}_3 = \tfrac{1}{2}[\mathbf{b}_{0;0}^{2;2}(u_0, u_1, v_1, v_1) + \mathbf{b}_{0;0}^{2;2}(u_1, u_1, v_0, v_1)]$$

$$\mathbf{B}_4 = \mathbf{b}_{0;0}^{2;2}(u_1, u_1, v_1, v_1)$$

Figure 2a illustrates $\mathbf{K}(t) = (u(t), v(t))$, having Bézier points $\mathbf{k}_I = (u_I, v_I)$, embedded in the domain of $\mathbf{F}(u, v)$. Auxiliary points defined by the intersection of parameter lines and given by (u_0, v_1) and (u_1, v_0) are marked, too.

Figure 2b gives the Bézier net of the biquadratic tensor product Bézier surface.

D. Lasser

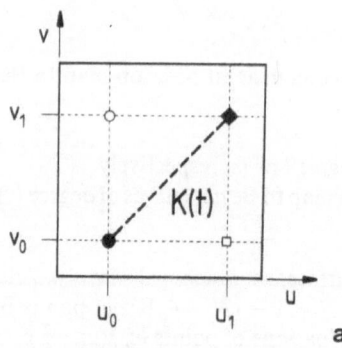

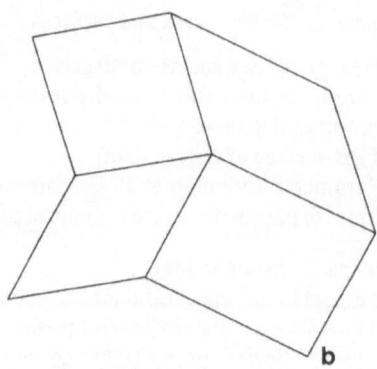

Figure 2a. $K(t)$ and domain of $F(u, v)$

Figure 2b. Bézier net of $F(u, v)$

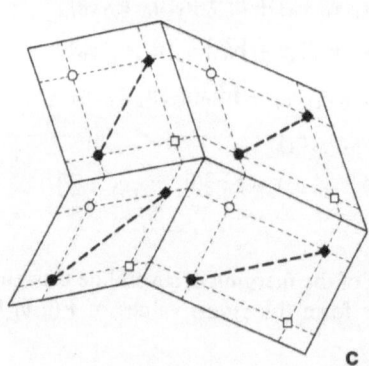

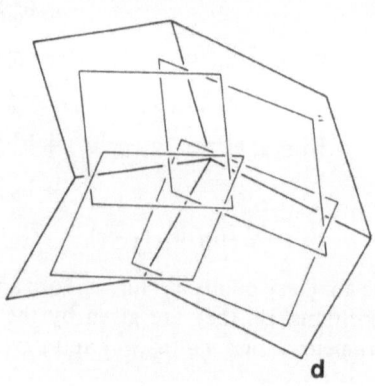

Figure 2c. Auxiliary points $b_{i,j}^{i+1,j+1}(u_{Iu}^1, v_{Iv}^1)$

Figure 2d. New bilinear Bézier nets

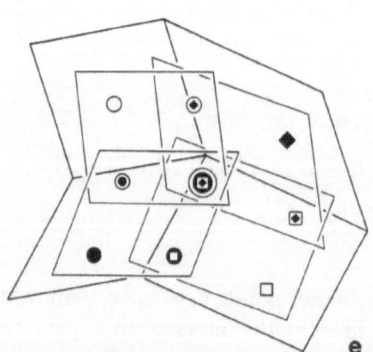

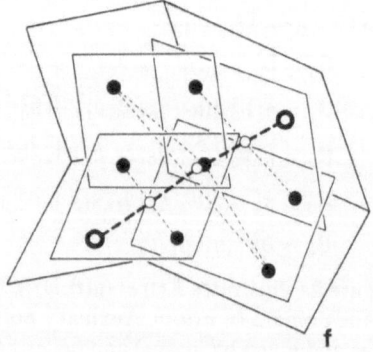

Figure 2e. Auxiliary points $b_{0;0}^{2;2}(u_{Iu}^2, v_{Iv}^2)$

Figure 2f. Bézier points B_R

Figure 2c shows the result of one de Casteljau step in u-direction and one de Casteljau step in v-direction, for all possible (i.e. four) pairs of parameter values. Thus, Figure 2c shows the image of the parameter space situation under the affine map of $\mathbf{F}(u, v)$ for each of the quadrilaterals of the Bézier net.

Figure 2d constructs four new bilinear Bézier nets using the auxiliary points $\mathbf{b}_{i,j}^{i+1,j+1}(u_{\bar{1}u}^1, v_{\bar{1}v}^1)$ of Fig. 2c.

Figure 2e repeats the procedure for each of the four bilinear Bézier nets, resulting in 16 auxiliary points $\mathbf{b}_{0,0}^{2,2}(u_{\bar{1}u}^2, v_{\bar{1}v}^2)$. Only nine of them are distinct (see the remark above). This is due to the connection with the polar form of $\mathbf{F}(u, v)$.

Figure 2f depicts the construction of Bézier points $\mathbf{B}_R$ as convex combinations of auxiliary points $\mathbf{b}_{0,0}^{2,2}(u_{\bar{1}u}^2, v_{\bar{1}v}^2)$.

Auxiliary points $\mathbf{b}_{0,0}^{2,2}(u_{\bar{1}u}^2, v_{\bar{1}v}^2)$ are at the same time Bézier points of the surface subsegment of $\mathbf{F}(u, v)$ which is defined for $u \in [u_0, u_1]$ and $v \in [v_0, v_1]$ (compare with the subdivision procedure for Bézier triangles [5], see also [14]).

III.2 Composition of TPB-Surfaces and TPB-Volumes

The foundation for the composition $\mathbf{V}(\mu, v) = \mathbf{V}(\mathbf{F}(\mu, v)) = \mathbf{V}(u(\mu, v), v(\mu, v), w(\mu, v))$ of TPB-surfaces $\mathbf{F}(\mu, v)$ and TPB-volumes $\mathbf{V}(u, v, w)$ is given by Theorem 2:

Theorem 2. TPB-surfaces and TPB-volumes

Let $\mathbf{F}(\mu, v): \mathbb{R}^2 \to \mathbb{R}^3$ be a polynomial TPB-surface of degree (L, M),

$$\mathbf{F}(\mu, v) = \sum_{I=0}^{L} \sum_{J=0}^{M} \mathbf{f}_{I,J} B_I^L(\mu) B_J^M(v), \qquad \mu, v \in [0, 1],$$

where $\mathbf{F}(\mu, v) = (u(\mu, v), v(\mu, v), w(\mu, v))$, and Bézier points $\mathbf{f}_{I,J} = (u_{I,J}, v_{I,J}, w_{I,J})$. And let $\mathbf{V}(u, v, w): \mathbb{R}^3 \to \mathbb{R}^3$ be a polynomial TPB-volume of degree (l, m, n),

$$\mathbf{V}(u, v, w) = \sum_{i=0}^{l} \sum_{j=0}^{m} \sum_{k=0}^{n} \mathbf{b}_{i,j,k} B_i^l(u) B_j^m(v) B_k^n(w), \qquad u, v, w \in [0, 1],$$

where $\mathbf{V}(u, v, w) = (x(u, v, w), y(u, v, w), z(u, v, w))$, Bézier points $\mathbf{b}_{i,j,k} = (x_{i,j,k}, y_{i,j,k}, z_{i,j,k})$.

For each $r = \alpha + \beta + \gamma$ where $\alpha \in \{0, \ldots, l\}$ and $\beta \in \{0, \ldots, m\}$ and $\gamma \in \{0, \ldots, n\}$, i.e. $r \in \{0, \ldots, l + m + n\}$, we have

$$\mathbf{V}(\mu, v) = \mathbf{V}(\mathbf{F}(\mu, v)) = \sum_{R=0}^{rL} \sum_{S=0}^{rM} \mathbf{B}_{R,S} B_R^{rL}(\mu) B_S^{rM}(v), \tag{4}$$

where

$$\mathbf{B}_{RS} = \sum_{|\mathbf{I}|=R} \sum_{|\mathbf{J}|=S} C_R^{\alpha,\beta,\gamma}(L, \mathbf{I}) C_S^{\alpha,\beta,\gamma}(M, \mathbf{J}) \mathbf{V}^{\alpha,\beta,\gamma} \tag{5}$$

with

$$\mathbf{V}^{\alpha,\beta,\gamma} = \sum_{i=0}^{l-\alpha} \sum_{j=0}^{m-\beta} \sum_{k=0}^{n-\gamma} \mathbf{b}_{i,j,k}^{i+\alpha,j+\beta,k+\gamma} B_i^{l-\alpha}(u(\mu, v)) B_j^{m-\beta}(v(\mu, v)) B_k^{n-\gamma}(w(\mu, v)), \tag{6}$$

where $\mathbf{b}_{i,j,k}^{i+\alpha,j+\beta,k+\gamma} = \mathbf{b}_{i,j,k}^{i+\alpha,j+\beta,k+\gamma}(u_{\mathbf{I}^u,\mathbf{J}^u}^{\alpha}, v_{\mathbf{I}^v,\mathbf{J}^v}^{\beta}, w_{\mathbf{I}^w,\mathbf{J}^w}^{\gamma}),$ *and with constants*

$$C_R^{\alpha,\beta,\gamma}(L,\mathbf{I}) = \frac{\prod\limits_{Q^u=1}^{\alpha}\binom{L}{I_{Q^u}^u}\prod\limits_{Q^v=1}^{\beta}\binom{L}{I_{Q^v}^v}\prod\limits_{Q^w=1}^{\gamma}\binom{L}{I_{Q^w}^w}}{\binom{rL}{R}}$$

and $C_S^{\alpha,\beta,\gamma}(M,\mathbf{J})$ *similarly.*

$\sum_{|\mathbf{I}|=R}$ *has the same meaning as for Theorem 1, but with* $\mathbf{I} = (\mathbf{I}^u,\mathbf{I}^v,\mathbf{I}^w),$ *and* $\sum_{|\mathbf{J}|=S}$ *analogously.* $\mathbf{b}_{i,j,k}^{i+\alpha,j+\beta,k+\gamma}$ *is defined recursively by de Casteljau's construction.*

According to (4), $\mathbf{V}(\mathbf{F}(\mu,v))$ *is polynomial and can be represented as TPB-surface of degree* (rL,rM). *Bézier points of this representation are given as, (5), convex combinations of auxiliary points* $\mathbf{V}^{\alpha,\beta,\gamma}$ *which are calculated, (6), for parameter values* $(u_{\mathbf{I}^u,\mathbf{J}^u}^{\alpha}, v_{\mathbf{I}^v,\mathbf{J}^v}^{\beta}, w_{\mathbf{I}^w,\mathbf{J}^w}^{\gamma})$ *via the blossoming principle (cf. Section 1) applying de Casteljau's algorithm.*

Proof of Theorem 2. Essentially like the proof for Theorem 1. The difference is that higher dimensions are involved, and to see how this will be put down into the calculations. Because of the similarity to the proof of Theorem 1, only the proof of the inductive statement for the *u* parameter direction is drawn out here. Thus, we assume,

$$\mathbf{V}(\mu,v) = \mathbf{V}(\mathbf{F}(\mu,v)) = \sum_{R=0}^{rL}\sum_{S=0}^{rM}\mathbf{B}_{R,S}B_R^{rL}(\mu)B_S^{rM}(v),$$

with $r = (\alpha - 1) + \beta + \gamma$ is valid, i.e.

$$\mathbf{B}_{R,S} = \sum_{|\mathbf{I}|=R}\sum_{|\mathbf{J}|=S}C_R^{\alpha-1,\beta,\gamma}(L,\mathbf{I})C_S^{\alpha-1,\beta,\gamma}(M,\mathbf{J})\mathbf{V}^{\alpha-1,\beta,\gamma}$$

with

$$\mathbf{V}^{\alpha-1,\beta,\gamma} = \sum_{i=0}^{l-\alpha+1}\sum_{j=0}^{m-\beta}\mathbf{b}_{i,j,k}^{i+\alpha-1,j+\beta,k+\gamma}B_i^{l-\alpha+1}(u(\mu,v))B_j^{m-\beta}(v(\mu,v))B_k^{n-\gamma}(w(\mu,v)),$$

where $\mathbf{b}_{i,j,k}^{i+\alpha-1,j+\beta,k+\gamma} = \mathbf{b}_{i,j,k}^{i+\alpha-1,j+\beta,k+\gamma}(u_{\mathbf{I}^u,\mathbf{J}^u}^{\alpha-1}, v_{\mathbf{I}^v,\mathbf{J}^v}^{\beta}, w_{\mathbf{I}^w,\mathbf{J}^w}^{\gamma}),$ and $\mathbf{I} = (\mathbf{I}^u,\mathbf{I}^v,\mathbf{I}^w)$ with $\mathbf{I}^u = (I_1^u,\ldots,I_{\alpha-1}^u),$ $\mathbf{I}^v = (I_1^v,\ldots,I_\beta^v),$ $\mathbf{I}^w = (I_1^w,\ldots,I_\gamma^w),$ and $\mathbf{J} = (\mathbf{J}^u,\mathbf{J}^v,\mathbf{J}^w)$ with $\mathbf{J}^u = (J_1^u,\ldots,J_{\alpha-1}^u),$ $\mathbf{J}^v = (J_1^v,\ldots,J_\beta^v),$ $\mathbf{J}^w = (J_1^w,\ldots,J_\gamma^w).$

As in the proof of Theorem 1, we get for the sum over *i*

$$\sum_{i=0}^{l-\alpha}[(1 - u(\mu,v))\mathbf{b}_{i,j,k}^{i+\alpha-1,j+\beta,k+\gamma} + u(\mu,v)\mathbf{b}_{i+1,j,k}^{i+\alpha,j+\beta,k+\gamma}]B_i^{l-\alpha}(u(\mu,v)),$$

where the argument of b_{***}^{***} is $(u_{\mathbf{I}^u,\mathbf{J}^u}^{\alpha-1}, v_{\mathbf{I}^v,\mathbf{J}^v}^{\beta}, w_{\mathbf{I}^w,\mathbf{J}^w}^{\gamma}).$

Now, the term in squared brackets can be written as

$$\sum_{I_\alpha^u=0}^{L}\sum_{J_\alpha^u=0}^{M}\{(1 - u_{I_\alpha^u,J_\alpha^u})\mathbf{b}_{i,j,k}^{i+\alpha-1,j+\beta,k+\gamma} + u_{I_\alpha^u,J_\alpha^u}\mathbf{b}_{i+1,j,k}^{i+\alpha,j+\beta,k+\gamma}\}B_{I_\alpha^u}^{L}(\mu)B_{J_\alpha^u}^{M}(v).$$

Using the substitution

$$\mathbf{b}_{i,j,k}^{i+\alpha,j+\beta,k+\gamma} = (1 - u_{I_\alpha^u,J_\alpha^u})\mathbf{b}_{i,j,k}^{i+\alpha-1,j+\beta,k+\gamma} + u_{I_\alpha^u,J_\alpha^u}\mathbf{b}_{i+1,j,k}^{i+\alpha,j+\beta,k+\gamma},$$

where the argument is $(u_{I_\alpha^u,J_\alpha^u}^\alpha, v_{I_\alpha^v,J_\alpha^v}^\beta, w_{I_\alpha^w,J_\alpha^w}^\gamma)$ with $\mathbf{I}^u = (I_1^u, \ldots, I_{\alpha-1}^u, I_\alpha^u)$, on the left side, while it is $(u_{I_\alpha^u,J_\alpha^u}^{\alpha-1}, v_{I_\alpha^v,J_\alpha^v}^\beta, w_{I_\alpha^w,J_\alpha^w}^\gamma)$ with $\mathbf{I}^u = (I_1^u, \ldots, I_{\alpha-1}^u)$, on the right side, yields

$$\sum_{i=0}^{l-\alpha}\left[\sum_{I_\alpha^u=0}^{L}\sum_{J_\alpha^u=0}^{M}\mathbf{b}_{i,j,k}^{i+\alpha,j+\beta,k+\gamma}(u_{I_\alpha^u,J_\alpha^u}^\alpha, v_{I_\alpha^v,J_\alpha^v}^\beta, w_{I_\alpha^w,J_\alpha^w}^\gamma)B_{I_\alpha^u}^N(t)\right]B_i^{l-\alpha}(u(\mu,v)).$$

Therefore, the expression for $\mathbf{B}_{R,S}$ becomes

$$\mathbf{B}_{R,S} = \sum_{|\mathbf{I}|=R}\sum_{|\mathbf{J}|=S}C_R^{\alpha-1,\beta,\gamma}(L,\mathbf{I})C_S^{\alpha-1,\beta,\gamma}(M,\mathbf{J})\sum_{I_\alpha^u=0}^{L}\sum_{J_\alpha^u=0}^{M}\mathbf{V}^{\alpha,\beta,\gamma}B_{I_\alpha^u}^L(\mu)B_{J_\alpha^u}^M(v)$$

with $\mathbf{V}^{\alpha,\beta,\gamma} = \mathbf{V}^{\alpha,\beta,\gamma}(u_{I_\alpha^u,J_\alpha^u}^\alpha, v_{I_\alpha^v,J_\alpha^v}^\beta, w_{I_\alpha^w,J_\alpha^w}^\gamma)$, now given as in (6).

Regrouping the terms of the sums $\sum_{|\mathbf{I}|=R}$ and $\sum_{I_\alpha^u=0}^L$ into one sum, as well as the terms of the sums $\sum_{|\mathbf{J}|=S}$ and $\sum_{J_\alpha^u=0}^M$ into another sum, and considering the product formula for Bernstein polynomials for the product of $\mathbf{B}_R^L(\mu)$ and $B_{I_\alpha^u}^L(\mu)$, and for the product of $B_S^{rM}(v)$ and $B_{J_\alpha^u}^M(v)$, this completes the proof. $\qquad\Box$

Setting $\alpha = l$, $\beta = m$ and $\gamma = n$ results in:

Corollary 2. Bézier surfaces and TPB-volumes
Let $\mathbf{V}(u,v,w)$ and $\mathbf{F}(\mu,v)$ be given as in Theorem 2. For $\mathbf{V}(\mu,v) = \mathbf{V}(\mathbf{F}(\mu,v)) = \mathbf{V}(u(\mu,v),v(\mu,v),w(\mu,v))$ we have

$$\mathbf{V}(\mu,v) = \sum_{R=0}^{rL}\sum_{S=0}^{rM}\mathbf{B}_{R,S}B_R^{rL}(\mu)B_S^{rM}(v),$$

where $r = l + m + n$ and with Bézier points

$$\mathbf{B}_{R,S} = \sum_{|\mathbf{I}|=R}\sum_{|\mathbf{J}|=S}C_R^{l,m,n}(L,\mathbf{I})C_S^{l,m,n}(M,\mathbf{J})\mathbf{b}_{0,0,0}^{l,m,n}(u_{I_\alpha^u,J_\alpha^u}^l, v_{I_\alpha^v,J_\alpha^v}^m, w_{I_\alpha^w,J_\alpha^w}^n).$$

Remark 2. **Bézier surfaces and volumes**

- **Tetrahedra Bézier volume of degree n**
  Parameter planes and planes of parameter space in general position map to triangle Bézier surfaces of degree n.
- **TPB-volume of degree (l, m, n)**
  Parameter planes map to tensor product Bézier surfaces of degree (l,m), (l,n) or (m,n), respectively.
  Planes of parameter space in general position map to tensor product Bézier surfaces of degree $(l + m + n, l + m + n)$.

Now, a generalization to the case of composing tensor products of arbitrary dimensions is straight forward.

IV. Composition of Rational TPB-Representations

IV.1 Composition of Rational TPB-Curves and TPB-Surfaces

Theorem 3 forms the foundation for the composition of rational curves $\mathbf{K}(t)$ and rational TPB-surfaces $\mathbf{F}(u, v)$:

Theorem 3. Rational Bézier curves and TPB-surfaces

Let $\mathbf{K}(t)$: $\mathbb{R} \to \mathbb{R}^2$ be a planar rational Bézier curve of degree N,

$$\mathbf{K}(t) = \frac{\sum_{I=0}^{N} \beta_I \mathbf{k}_I B_I^N(t)}{\sum_{I=0}^{N} \beta_I B_I^N(t)}, \qquad t \in [0, 1],$$

where $\mathbf{K}(t) = (u(t), v(t))$, Bézier points $\mathbf{k}_I = (u_I, v_I)$, and weights $\beta_I \in \mathbb{R}$. And let $\mathbf{F}(u, v)$: $\mathbb{R}^2 \to \mathbb{R}^d$, $d = 2, 3$, be a rational TPB-surface of degree (l, m),

$$\mathbf{F}(u, v) = \frac{\sum_{i=0}^{l} \sum_{j=0}^{m} \omega_{i,j} \mathbf{b}_{i,j} B_i^l(u) B_j^m(v)}{\sum_{i=0}^{l} \sum_{j=0}^{m} \omega_{i,j} B_i^l(u) B_j^m(v)}, \qquad u, v \in [0, 1],$$

where $\mathbf{F}(u, v) = (x(u, v), y(u, v), z(u, v))$, Bézier points $\mathbf{b}_{i,j} = (x_{i,j}, y_{i,j}, z_{i,j})$, and weights $\omega_{i,j} \in \mathbb{R}$. If $d = 2$ (2D-Solid): $z_{i,j} = 0$, for all i, j.

For each $r = \alpha + \beta$ where $\alpha \in \{0, \dots, l\}$, $\beta \in \{0, \dots, m\}$, i.e. $r \in \{0, \dots, l + m\}$, we have

$$\mathbf{F}(t) = \mathbf{F}(\mathbf{K}(t)) = \frac{\sum_{R=0}^{rN} \Omega_R \mathbf{B}_R B_R^{rN}(t)}{\sum_{R=0}^{rN} \Omega_R B_R^{rN}(t)}, \tag{7}$$

where

$$\Omega_R \mathbf{B}_R = \sum_{|\mathbf{I}|=R} B_R^{\alpha, \beta}(N, \mathbf{I}) \mathbf{F}^{\alpha, \beta}$$

$$\Omega_R = \sum_{|\mathbf{I}|=R} B_R^{\alpha, \beta}(N, \mathbf{I}) G^{\alpha, \beta} \tag{8}$$

with

$$\mathbf{F}^{\alpha, \beta} = \sum_{i=0}^{l-\alpha} \sum_{j=0}^{m-\beta} \omega_{i,j}^{i+\alpha, j+\beta} \mathbf{b}_{i,j}^{i+\alpha, j+\beta} B_i^{l-\alpha}(u(t)) B_j^{m-\beta}(v(t))$$

$$G^{\alpha, \beta} = \sum_{i=0}^{l-\alpha} \sum_{j=0}^{m-\beta} \omega_{i,j}^{i+\alpha, j+\beta} B_i^{l-\alpha}(u(t)) B_j^{m-\beta}(v(t)), \tag{9}$$

where $\mathbf{b}_{i,j}^{i+\alpha, j+\beta} = \mathbf{b}_{i,j}^{i+\alpha, j+\beta}(u_{\mathbf{I}^u}^\alpha, v_{\mathbf{I}^v}^\beta)$, $\omega_{i,j}^{i+\alpha, j+\beta} = \omega_{i,j}^{i+\alpha, j+\beta}(u_{\mathbf{I}^u}^\alpha, v_{\mathbf{I}^v}^\beta)$, and with constants

$$B_R^{\alpha,\beta}(N,I) = \frac{\prod_{Q^u=1}^{\alpha} \beta_{I_Q^u} \binom{N}{I_{Q^u}^u} \prod_{Q^v=1}^{\beta} \beta_{I_{Q^v}^v} \binom{N}{I_{Q^v}^v}}{\binom{rN}{R}}.$$

$\sum_{|I|=R}$ and $I = (I^u, I^v)$ *have the same meaning as for Theorem 1.* $\omega_{i,j}^{i+\alpha,j+\beta} \mathbf{b}_{i,j}^{i+\alpha,j+\beta}$ *and* $\omega_{i,j}^{i+\alpha,j+\beta}$ *are defined recursively by de Casteljau's construction, analogously to Theorem 1.*

Again, the restriction $d = 2, 3$ is not needed in the proof and could be relaxed to $d \in \mathbb{N}$.

Proof of Theorem 3. Essentially like the foregoing proofs. The difference is that rational representations are now involved. Therefore, only the proof of the statement for the u parameter direction is drawn out briefly: We assume, (7) is valid, with

$$\Omega_R \mathbf{B}_R = \sum_{|I|=R} B_R^{\alpha-1,\beta}(N,I) \mathbf{F}^{\alpha-1,\beta}(u_{I^u}^{\alpha-1}, v_{I^v}^{\beta})$$

$$\Omega_R = \sum_{|I|=R} B_R^{\alpha-1,\beta}(N,I) G^{\alpha-1,\beta}(u_{I^u}^{\alpha-1}, v_{I^v}^{\beta})$$

and

$$\mathbf{F}^{\alpha-1,\beta}(u_{I^u}^{\alpha-1}, v_{I^v}^{\beta}) = \sum_{i=0}^{l-\alpha+1} \sum_{j=0}^{m-\beta} \omega_{i,j}^{i+\alpha-1,j+\beta} \mathbf{b}_{i,j}^{i+\alpha-1,j+\beta} B_i^{l-\alpha+1}(u(t)) B_j^{m-\beta}(v(t))$$

$$G^{\alpha-1,\beta}(u_{I^u}^{\alpha-1}, v_{I^v}^{\beta}) = \sum_{i=0}^{l-\alpha+1} \sum_{j=0}^{m-\beta} \omega_{i,j}^{i+\alpha-1,j+\beta} B_i^{l-\alpha+1}(u(t)) B_j^{m-\beta}(v(t)),$$

where $I = (I^u, I^v)$ and $I^u = (I_1^u, \ldots, I_{\alpha-1}^u)$, $I^v = (I_1^v, \ldots, I_\beta^v)$. Then, as above, we first get

$$\sum_{i=0}^{l-\alpha} [(1 - u(t))\omega_{i,j}^{i+\alpha-1,j+\beta} \mathbf{b}_{i,j}^{i+\alpha-1,j+\beta} + u(t)\omega_{i+1,j}^{i+\alpha,j+\beta} \mathbf{b}_{i+1,j}^{i+\alpha,j+\beta}] B_i^{l-\alpha}(u(t))$$

$$\sum_{i=0}^{l-\alpha} [(1 - u(t))\omega_{i,j}^{i+\alpha-1,j+\beta} + u(t)\omega_{i+1,j}^{i+\alpha,j+\beta}] B_i^{l-\alpha}(u(t)),$$

and in the following

$$\sum_{i=0}^{l-\alpha} \left[\sum_{I_\alpha^u=0}^{N} \omega_{i,j}^{i+\alpha,j+\beta} \mathbf{b}_{i,j}^{i+\alpha,j+\beta} B_{I_\alpha^u}^N(t) \right] B_i^{l-\alpha}(u(t)) \bigg/ \left(\sum_{I=0}^{N} \beta_I B_I^N(t) \right)^\alpha$$

$$\sum_{i=0}^{l-\alpha} \left[\sum_{I_\alpha^u=0}^{N} \omega_{i,j}^{i+\alpha,j+\beta} B_{I_\alpha^u}^N(t) \right] B_i^{l-\alpha}(u(t)) \bigg/ \left(\sum_{I=0}^{N} \beta_I B_I^N(t) \right)^\alpha,$$

where we have used the substitutions

$$\omega_{i,j}^{i+\alpha,j+\beta} \mathbf{b}_{i,j}^{i+\alpha,j+\beta} = (1 - u_{I^u})\omega_{i,j}^{i+\alpha-1,j+\beta} \mathbf{b}_{i,j}^{i+\alpha-1,j+\beta} + u_{I^u}\omega_{i+1,j}^{i+\alpha,j+\beta} \mathbf{b}_{i+1,j}^{i+\alpha,j+\beta}$$

$$\omega_{i,j}^{i+\alpha,j+\beta} = (1 - u_{I^u})\omega_{i,j}^{i+\alpha-1,j+\beta} + u_{I^u}\omega_{i+1,j}^{i+\alpha,j+\beta}.$$

Regrouping the terms of the sums in numerator and denominator results in

$$\sum_{|I|=R} B_R^{\alpha-1,\beta}(N,I) \sum_{I_\alpha^u=0}^{N} \mathbf{F}^{\alpha,\beta}(u_{I^u}^\alpha, v_{I^v}^\beta) B_{I_\alpha^u}^N(t) \bigg/ \left(\sum_{I=0}^{N} \beta_I B_I^N(t) \right)^\alpha$$

$$\sum_{|\mathbf{I}|=R} B_R^{\alpha-1,\beta}(N,\mathbf{I}) \sum_{I_\alpha^u=0}^{N} G^{\alpha,\beta}(u_{\mathbf{I}^u}^\alpha, v_{\mathbf{I}^v}^\beta) B_{I_\alpha^u}^N(t) \Big/ \left(\sum_{I=0}^{N} \beta_I B_I^N(t) \right)^\alpha$$

with $\mathbf{F}^{\alpha,\beta} = \mathbf{F}^{\alpha,\beta}(u_{\mathbf{I}^u}^\alpha, v_{\mathbf{I}^v}^\beta)$ and $G^{\alpha,\beta} = G^{\alpha,\beta}(u_{\mathbf{I}^u}^\alpha, v_{\mathbf{I}^v}^\beta)$, now given like in (9).

Regrouping the terms of the sums $\sum_{|\mathbf{I}|=R}$ and $\sum_{I_\alpha^u=0}^{N}$ into one sum, and considering the product formula for Bernstein polynomials for the product of $B_R^{rN}(t)$ and $B_{I_\alpha^u}^N(t)$ completes the proof, because the term $(\sum_{I=0}^{N} \beta_I B_I^N(t))^\alpha$ cancels out when forming the ratio of $\mathbf{F}(\mathbf{K}(t))$. $\square$

Setting $\alpha = l$ and $\beta = m$ results in:

Corollary 3. Rational Bézier curves and TPB-surfaces
Let $\mathbf{F}(u,v)$ and $\mathbf{K}(t)$ be given as in Theorem 3. For $\mathbf{F}(t) = \mathbf{F}(\mathbf{K}(t)) = \mathbf{F}(u(t), v(t))$ we have

$$\mathbf{F}(t) = \frac{\displaystyle\sum_{R=0}^{rN} \Omega_R \mathbf{B}_R B_R^{rN}(t)}{\displaystyle\sum_{R=0}^{rN} \Omega_R B_R^{rN}(t)},$$

where $r = l + m$ and weights are given by

$$\Omega_R = \sum_{|\mathbf{I}|=R} B_R^{l,m}(N,\mathbf{I}) \omega_{0,0}^{l,m}(u_{\mathbf{I}^u}^l, v_{\mathbf{I}^v}^m),$$

and Bézier points are given by $\mathbf{B}_R = \dfrac{\Omega_R \mathbf{B}_R}{\Omega_R}$, where

$$\Omega_R \mathbf{B}_R = \sum_{|\mathbf{I}|=R} B_R^{l,m}(N,\mathbf{I}) \omega_{0,0}^{l,m}(u_{\mathbf{I}^u}^l, v_{\mathbf{I}^v}^m) \mathbf{b}_{0,0}^{l,m}(u_{\mathbf{I}^u}^l, v_{\mathbf{I}^v}^m).$$

Remark 3. **Rational Bézier curves and surfaces**

- **Rational triangle Bézier surface of degree n**
  Parameter lines and lines of parameter space in general position map to rational Bézier curves of degree n.
- **Rational TPB-surface of degree (l, m)**
  Parameter lines map to rational Bézier curves of degree l or m, respectively.
  Lines of parameter space in general position map to rational Bézier curves of degree $l + m$.

Note, that three important special cases are included in Theorem 3 and Corollary 3:

- If $\mathbf{K}(t)$ is polynomial (i.e. $\beta_I = 1$, for all I) and $\mathbf{F}(u,v)$ is polynomial (i.e. $\omega_{i,j} = 1$, for all i,j):
  The statement of Theorem 1 results.
- If $\mathbf{K}(t)$ is polynomial (i.e. $\beta_I = 1$, for all I) and $\mathbf{F}(u,v)$ is rational:
  Constants $B_R^{\alpha,\beta}(N,\mathbf{I})$ reduce to constants $C_R^{\alpha,\beta}(N,\mathbf{I})$ of Theorem 1, but $\mathbf{F}(\mathbf{K}(t))$ is (still) rational.
- If $\mathbf{K}(t)$ is rational and $\mathbf{F}(u,v)$ is polynomial (i.e. $\omega_{i,j} = 1$, for all i,j):
  $G^{\alpha,\beta} = 1$, for all α, β, $\omega_{0,0}^{l,m} = 1$, and $\mathbf{F}^{\alpha,\beta}$ is given as in Theorem 1, but $\mathbf{F}(\mathbf{K}(t))$ is (still) rational.

IV.2 Composition of Rational TPB-Surfaces and TPB-Volumes

Theorem 4. Rational TPB-surfaces and TPB-volumes

Let $\mathbf{F}(\mu, v)\colon \mathbb{R}^2 \to \mathbb{R}^3$ be a rational TPB-surface of degree (L, M),

$$
\mathbf{F}(\mu, v) = \frac{\displaystyle\sum_{I=0}^{L}\sum_{J=0}^{M}\beta_{I,J}\mathbf{f}_{I,J}B_I^L(\mu)B_J^M(v)}{\displaystyle\sum_{I=0}^{L}\sum_{J=0}^{M}\beta_{I,J}B_I^L(\mu)B_J^M(v)}, \qquad \mu, v \in [0, 1],
$$

where $\mathbf{F}(\mu, v) = (u(\mu, v), v(\mu, v), w(\mu, v))$, Bézier points $\mathbf{f}_{I,J} = (u_{I,J}, v_{I,J}, w_{I,J})$, and weights $\beta_{I,J} \in \mathbb{R}$.

Let $\mathbf{V}(u, v, w)\colon \mathbb{R}^3 \to \mathbb{R}^3$ be a rational TPB-volume of degree (l, m, n),

$$
\mathbf{V}(u, v, w) = \frac{\displaystyle\sum_{i=0}^{l}\sum_{j=0}^{m}\sum_{k=0}^{n}\omega_{i,j,k}\mathbf{b}_{i,j,k}B_i^l(u)B_j^m(v)B_k^n(w)}{\displaystyle\sum_{i=0}^{l}\sum_{j=0}^{m}\sum_{k=0}^{n}\omega_{i,j,k}B_i^l(u)B_j^m(v)B_k^n(w)}, \qquad u, v, w \in [0, 1],
$$

where $\mathbf{V}(u, v, w) = (x(u, v, w), y(u, v, w), z(u, v, w))$, Bézier points $\mathbf{b}_{i,j,k} = (x_{i,j,k}, y_{i,j,k}, z_{i,j,k})$ and weights $\omega_{i,j,k} \in \mathbb{R}$.

For each $r = \alpha + \beta + \gamma$ where $\alpha \in \{0, \ldots, l\}$ and $\beta \in \{0, \ldots, m\}$ and $\gamma \in \{0, \ldots, n\}$, i.e. $r \in \{0, \ldots, l + m + n\}$, we have

$$
\mathbf{V}(\mu, v) = \mathbf{V}(\mathbf{F}(\mu, v)) = \frac{\displaystyle\sum_{R=0}^{rL}\sum_{S=0}^{rM}\Omega_{R,S}\mathbf{B}_{R,S}B_R^{rL}(\mu)B_S^{rM}(v)}{\displaystyle\sum_{R=0}^{rL}\sum_{S=0}^{rM}\Omega_{R,S}B_R^{rL}(\mu)B_S^{rM}(v)}, \tag{10}
$$

where

$$
\Omega_{R,S}\mathbf{B}_{R,S} = \sum_{|\mathbf{I}|=R}\sum_{|\mathbf{J}|=S}B_R^{\alpha,\beta,\gamma}(L, \mathbf{I})B_S^{\alpha,\beta,\gamma}(M, \mathbf{J})\mathbf{V}^{\alpha,\beta,\gamma}
$$

$$
\Omega_{R,S} = \sum_{|\mathbf{I}|=R}\sum_{|\mathbf{J}|=S}B_R^{\alpha,\beta,\gamma}(L, \mathbf{I})B_S^{\alpha,\beta,\gamma}(M, \mathbf{J})G^{\alpha,\beta,\gamma}
\tag{11}
$$

with

$$
\mathbf{V}^{\alpha,\beta,\gamma} = \sum_{i=0}^{l-\alpha}\sum_{j=0}^{m-\beta}\sum_{k=0}^{n-\gamma}\omega_{i,j,k}^{i+\alpha,j+\beta,k+\gamma}\mathbf{b}_{i,j,k}^{i+\alpha,j+\beta,k+\gamma}B_i^{l-\alpha}(u(\mu, v))B_j^{m-\beta}(v(\mu, v))B_k^{n-\gamma}(w(\mu, v))
$$

$$
G^{\alpha,\beta,\gamma} = \sum_{i=0}^{l-\alpha}\sum_{j=0}^{m-\beta}\sum_{k=0}^{n-\gamma}\omega_{i,j,k}^{i+\alpha,j+\beta,k+\gamma}B_i^{l-\alpha}(u(\mu, v))B_j^{m-\beta}(v(\mu, v))B_k^{n-\gamma}(w(\mu, v)),
$$

and with constants

$$
B_R^{\alpha,\beta,\gamma}(L, \mathbf{I}) = \frac{\displaystyle\prod_{Q^u=1}^{\alpha}\beta_{I_{Q^u}^u}\binom{L}{I_{Q^u}^u}\prod_{Q^v=1}^{\beta}\beta_{I_{Q^v}^v}\binom{L}{I_{Q^v}^v}\prod_{Q^w=1}^{\gamma}\beta_{I_{Q^w}^w}\binom{L}{I_{Q^w}^w}}{\dbinom{rL}{R}}
$$

and $B_S^{\alpha,\beta,\gamma}(M, \mathbf{J})$ similarly.

Proof of Theorem 4. Essentially like the foregoing proofs.

Setting $\alpha = l$, $\beta = m$ and $\gamma = n$ results in:

Corollary 4. Rational Bézier surfaces and TPB-volumes
Let $\mathbf{V}(u, v, w)$ *and* $\mathbf{F}(\mu, v)$ *be given as in Theorem 4. For* $\mathbf{V}(\mu, v) = \mathbf{V}(\mathbf{F}(\mu, v)) = \mathbf{V}(u(\mu, v), v(\mu, v), w(\mu, v))$ *we have*

$$\mathbf{V}(\mu, v) = \frac{\sum\limits_{R=0}^{rL} \sum\limits_{S=0}^{rM} \Omega_{R,S} \mathbf{B}_{R,S} B_R^{rL}(\mu) B_S^{rM}(v)}{\sum\limits_{R=0}^{rL} \sum\limits_{S=0}^{rM} \Omega_{R,S} B_R^{rL}(\mu) B_S^{rM}(v)},$$

where $r = l + m + n$ *and weights are given by*

$$\Omega_{R,S} = \sum_{|\mathbf{I}|=R} \sum_{|\mathbf{J}|=S} B_R^{l,m,n}(L, \mathbf{I}) B_S^{l,m,n}(M, \mathbf{J}) \omega_{0,0,0}^{l,m,n},$$

and Bézier points are given by $\mathbf{B}_{R,S} = \dfrac{\Omega_{R,S} \mathbf{B}_{R,S}}{\Omega_{R,S}}$, *where*

$$\Omega_{R,S} \mathbf{B}_{R,S} = \sum_{|\mathbf{I}|=R} \sum_{|\mathbf{J}|=S} B_R^{l,m,n}(L, \mathbf{I}) B_S^{l,m,n}(M, \mathbf{J}) \omega_{0,0,0}^{l,m,n} \mathbf{b}_{0,0,0}^{l,m,n}.$$

Remark 4. **Rational Bézier surfaces and volumes**

● **Rational tetrahedra Bézier volume of degree n**
  Parameter planes and planes of parameter space in general position map to rational triangle Bézier surfaces of degree n.
● **Rational TPB-volume of degree (l, m, n)**
  Parameter planes map to rational TPB-surfaces of degree (l, m), (l, n) or (m, n), respectively.
  Planes of parameter space in general position map to rational TPB-surfaces of degree $(l + m + n, l + m + n)$.

Theorem 4 and Corollary 4 also include the three special cases of $\mathbf{V}(u, v, w)$ and $\mathbf{F}(\mu, v)$ both being polynomial, of $\mathbf{V}(u, v, w)$ being polynomial and $\mathbf{F}(u, v)$ being rational, and of $\mathbf{V}(u, v, w)$ being rational and $\mathbf{F}(\mu, v)$ being polynomial (cf. remark *rational curves on surfaces* given above in Section IV.1).

An extension to arbitrary dimensions can be done similarly.

V. Spline Representations

V.1 Bézier Spline Representations

It is possible to build up complex Bézier splines from a number of Bézier segments. The conditions for C^r-continuity of adjacent segments can be found in [4] and in [8]. As an example we formulate:

Theorem 5. Bézier spline curves and surfaces

Let $K(t)$ be a C^a-continuous planar polynomial Bézier spline curve of degree N. And let $F(u, v)$ be a C^b-continuous polynomial Bézier surface of degree (l, m):

$F(K(t))$ *is a Bézier subspline curve of degree $N(l + m)$ and is of smoothness C^e with $e = \min\{a, b\}$.*

Proof of Theorem 5. $F(K(t))$ can be calculated using Theorem 1 for each Bézier curve segment. Thus, segments are of degree $N(l + m)$.

C^e-continuity results by applying the chain rule to $F(K(t))$. ☐

V.2 B-Spline Representations

Using the results of the foregoing sections we are able to prove the following Theorem 6:

Theorem 6. B-spline curves and surfaces

Let $K(t)$ be a C^a-continuous planar polynomial B-spline curve of order $N + 1$, i.e. of degree N. And let $F(u, v)$ be a C^b-continuous polynomial B-spline surface of order $(l + 1, m + 1)$, i.e. of degree (l, m):

$F(K(t))$ *is a B-spline curve of order $N(l + m) + 1$, i.e. of degree $N(l + m)$. $F(K(t))$ is C^e-continuous with $e = \min\{a, b\}$. Knots of $F(K(t))$ have multiplicity $\mu = N(l + m) - e$.*

Proof of Theorem 6. The first and second statement are a direct consequence of Theorem 5 and the possibility of representing B-spline curves and surfaces as Bézier spline curves and surfaces. For the third statement, note, that a B-spline curve of order k is C^{k-2}-continuous and that knots of multiplicity μ result in $C^{k-\mu-1}$-continuous curves (see e.g. [4], [8]). Now, $F(K(t))$ is C^e-continuous. Therefore, $k - \mu - 1 = e$, i.e. $\mu = k - e - 1$. It is $k = N(l + m) + 1$. Thus, $\mu = N(l + m) - e$. ☐

References

[1] Bézier, P.: General distortion of an esemble of biparametric surfaces. Computer-Aided Design *10*, 116–120 (1978).

[2] Casale, M. S.: Free-form solid modeling with trimmed surface patches. IEEE Computer Graphics & Applications *7*, 33–43 (1987).

[3] Coquillart, S.: Extended free-form deformation: a sculpturing tool for 3D geometric modeling. ACM Computer Graphics *24*, 386–391 (1990).

[4] Farin, G.: Curves and surfaces for computer aided geometric design. A practical guide, 2. edn. San Diego: Academic Press 1990.

[5] Goldman, R. N.: Subdivision algorithms for Bézier triangles. Computer-Aided Design *15*, 159–166 (1983).

[6] Griessmair, J., Purgathofer, W.: Deformation of solids with trivariate B-splines. In: Hansmann, W., Hopgood, F. R. A., Strasser, W. (eds.) Eurographics '89, pp. 137–148. Amsterdam: North-Holland 1989.

[7] Hoschek, J., Schneider, F. J.: Spline conversion for trimmed rational Bézier- and B-spline surfaces. Computer-Aided Design *22*, 580–590 (1990).

[8] Hoschek, J., Lasser, D.: Grundlagen der Geometrischen Datenverarbeitung, 2. Aufl. Stuttgart: Teubner 1992.

[9] Lasser, D., Bonneau, G. P.: Trimmed Bézier surfaces. Interner Bericht. Informatik, Universität Kaiserslautern 1992.

[10] Lasser, D., Bonneau, G. P.: Free-form deformation of Bézier representations. Interner Bericht. Informatik, Universität Kaiserslautern 1992.

[11] Nishita, T., Sederberg, Th. W., Kakimoto, M.: Ray tracing trimmed rational surface patches. ACM Computer Graphics 24, 337–345 (1990).

[12] Ramshaw, L.: Blossoming: a connect-the-dots approach to splines. Research Report 19, Digital Systems Research Center, Palo Alto 1987.

[13] Rockwood, A. P., Heaton, K., Davis, T.: Real time rendering of trimmed surfaces. ACM Computer Graphics 23, 107–116 (1989).

[14] deRose, A. D.: Composing Bézier simplices. ACM Transactions on Graphics 7, 198–221 (1988).

[15] Sederberg, Th. W., Parry, S. R.: Free-form deformation of solid geometric models. ACM 20, 151–160 (1986).

[16] Shantz, M., Chang, S. L.: Rendering trimmed NURBS with adaptive forward differencing. ACM Computer Graphics 22, 189–198 (1988).

D. Lasser
Computer Science
University of Kaiserslautern
D-W-6750 Kaiserslautern
Federal Republic of Germany

Computing Suppl. 8, 173–190 (1993)

Interpolation with Exponential B-Splines in Tension

P. E. Koch, Trondheim and **T. Lyche,** Oslo

Abstract. In [11, 13] we introduced a basis of B-splines for the exponential splines in tension considered by Schweikert already in 1966. For interpolation with these basis functions we give a necessary and sufficient condition for the existence of a unique interpolant. We consider bicubic interpolation of data on rectangular grids using this basis, and give several examples showing the usefulness of this scheme.

Key words: Splines, B-splines, data fitting, shape preservation, interpolation.

1. Introduction

When fitting spline curves and surfaces to functions and data it is useful to have methods available which preserve the shape of the data both locally and globally. Some methods have extra parameters which can be used to control shape. By increasing one or more of these shape parameters the curve is pulled towards an inherent shape, usually a piecewise linear curve, at the same time keeping the smoothness of the method. Examples include methods requiring only geometric continuity [1, 17], methods based on rational splines, see [25] and references therein, and weighted splines [5, 7, 14, 24]. Here we consider the oldest such method, the (exponential) spline in tension, introduced by Schweikert in 1966, [27]. This method has kept its popularity [6, 9, 11, 12, 13, 16, 18, 19, 20, 21, 22, 23, 28, 29], and has even found its way into a general numerical analysis textbook [10].

Until recently there has been no local support basis available for computation with exponential tension splines. This means that when using splines in tension the choice of methods have been somewhat limited. In particular, in most papers on exponential tension splines, only an analog of cubic spline interpolation is discussed. In [11, 13] we introduced local support basis functions for tension splines allowing the tension parameter to vary from interval to interval. With such a basis it becomes fairly simple to carry out least squares approximation of curves and surfaces, and interpolation of tensor product surfaces. This paper discusses the use of tension B-splines for interpolation. We give a number of examples showing the effect of increasing the tension parameters.

The content of this paper is as follows. In Section 2 and 3 we recall the definition of exponential tension B-splines and their main properties. In Section 4 we discuss univariate interpolation allowing the knots and interpolation points to be different in general. We give a necessary and sufficient condition for the existence and uniqueness of an interpolant. We consider tensor product interpolation with data on a rectangular grid in Section 5.

2. B-Splines

Exponential tension splines of degree r are real valued functions which on each nonempty interval (t_j, t_{j+1}) of a partition $\mathbf{t} = (t_j)$ of an interval $[a, b]$, are linear combinations of the functions $1, x, x^2, \ldots, x^{r-2}, e^{\rho_j x}, e^{-\rho_j x}$. We are mainly interested in $r \leq 3$. (For $r > 3$ see [12].) The given positive number ρ_j controls the shape of the spline on (t_j, t_{j+1}). Increasing ρ_j forces the curve to approach a straight line on (t_j, t_{j+1}). If $\rho_j = 0$ then the spline is defined to be a polynomial of degree r on (t_j, t_{j+1}). Thus an exponential tension spline reduces to a polynomial spline if all tension parameters are zero.

It is convenient to have exponential analogs of the polynomial power functions x^r for various integers r. For $x \in \mathbb{R}$, and $\rho, d \geq 0$, corresponding to x^3, x^2 and x^1 we define

$$\gamma^3(x|\rho, d) = \begin{cases} (\sinh \rho x - \rho x)/(\rho^2 \sinh \rho d), & \text{if } \rho > 0, \quad d > 0, \\ x^3/(6d), & \text{if } \rho = 0, \quad d > 0, \\ 0, & \text{if } d = 0, \end{cases} \quad (2.1)$$

$$\gamma^r = \frac{\partial}{\partial x} \gamma^{r+1}, \qquad r = 2, 1. \quad (2.2)$$

Normally we consider $\gamma^r = \gamma^r(x)$ as a function of x with ρ and d fixed parameters. Note that γ^r is similar to x^r in that it has a zero of order $r - 1$ at the origin while the rth derivative is nonzero. Moreover, γ^r is monotone and convex for $x > 0$. The normalization is such that $\gamma^1(d|\rho, d) = 1$ for $d > 0$. With $\mathbf{t}$ and ρ finite or infinite sequences of real numbers we set

$$\mathbf{t} = (t_j),$$

$$\Delta t_j = t_{j+1} - t_j,$$

$$\rho = (\rho_j),$$

$$u_j^r(x) = \gamma^r(x - t_j|\rho_j, \Delta t_j), \quad (2.3)$$

$$v_j^r(x) = \gamma^r(t_{j+1} - x|\rho_j, \Delta t_j),$$

$$\beta_j^r = \gamma^r(\Delta t_j|\rho_j, \Delta t_j).$$

The *knots* $\mathbf{t}$ are supposed to be nondecreasing, i.e., multiple knots are allowed. The sequence ρ of *tension parameters* is assumed to be nonnegative. On each nonempty subinterval (t_j, t_{j+1}) an exponential tension spline is a linear combination of the functions $1, x, x^2, \ldots, x^{r-2}, u_j^r(x), v_j^r(x)$. The precise definition is in terms of linear combinations of local support exponential B-splines. (cf. Section 3.) In the rest of this section we recall the definition of these functions and give some of their main properties.

Define the exponential hat function

$$B_j^1(x) = \begin{cases} u_j^1(x), & \text{if } t_j \le x < t_{j+1}, \\ v_{j+1}^1(x), & \text{if } t_{j+1} \le x < t_{j+2}, \\ 0, & \text{otherwise}. \end{cases} \tag{2.4}$$

The function B_j^1 is called an exponential tension B-spline of degree 1, and is shown in Fig. 2.1 for $(t_j, t_{j+1}, t_{j+2}) = (0, 1, 2)$, $\rho_j = 1$, and $\rho_{j+1} = 5$. As the ρ's increase from zero to infinity the shape of B_j^1 changes from the piecewise linear hat function to a delta pulse of height one.

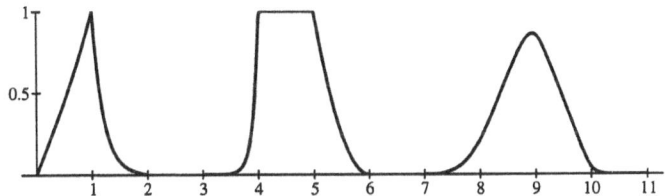

Figure 2.1. Exponential Tension B-splines of degree 1 (left), 2 (middle), and 3 (right)

Higher order B-splines are defined by successive integration,

$$B_j^{r+1} = \Phi_j^r - \Phi_{j+1}^r, \tag{2.5}$$

where

$$\Phi_j^r(x) = \begin{cases} 0, & \text{if } x < t_j, \\ \int_{t_j}^x B_j^r(y)\, dy / \sigma_j^r, & \text{if } t_j \le x < t_{j+r+1}, \\ 1, & \text{otherwise}, \end{cases} \tag{2.6}$$

and

$$\sigma_j^r = \int_{t_j}^{t_{j+r+1}} B_j^r(y)\, dy.$$

We also define $B_j^r \equiv 0$, and $\sigma_j^r = 1$ if $t_{j+r+1} = t_j$.

For convenience we give explicit formulae for the cases $r = 2$ (quadratic) and $r = 3$ (cubic). From (2.4) we find

$$\sigma_j^1 = \int_{t_j}^{t_{j+2}} B_j^1(y)\, dy = \beta_j^2 + \beta_{j+1}^2,$$

$$B_j^2(x) = \begin{cases} u_j^2(x)/\sigma_j^1, & \text{if } t_j \le x < t_{j+1}, \\ 1 - u_{j+1}^2(x)/\sigma_{j+1}^1 - v_{j+1}^2(x)/\sigma_j^1, & \text{if } t_{j+1} \le x < t_{j+2}, \\ v_{j+2}^2(x)/\sigma_{j+1}^1, & \text{if } t_{j+2} \le x < t_{j+3}, \\ 0, & \text{otherwise}. \end{cases} \tag{2.7}$$

A plot of a quadratic B-spline B_j^2 is shown in Fig. 2.1. We have chosen $(t_j, \ldots, t_{j+3}) = (3, 4, 5, 6)$ and tension parameters $(10, 100, 0)$ on the three intervals $(3, 4), (4, 5), (5, 6)$. This function has smoothness C^1. As the ρ's increases from zero to infinity, the function B_j^2 would change shape from a polynomial B-spline to the function which is one on $[4, 5)$ and zero otherwise.

The cubic case requires more work. From the following general result ([13]), explicit expressions for the pieces of a cubic B-spline can be derived. A similar formula for $r = 2$ can be found in [13].

Lemma 2.1. *Given* $t_{i-2} \leq t_{i-1} \leq t_i \leq x < t_{i+1} \leq t_{i+2} \leq t_{i+3}$, *nonnegative tension parameters* $\rho_j, j = i - 2, \ldots, i + 2$, *and B-spline coefficients* $c_{i-3}, c_{i-2}, c_{i-1}, c_i$, *one has*

$$\sum_{j=i-3}^{i} c_j B_j^3(x) = u_i^3(x) \nabla_3^2 c_i + v_i^3(x) \nabla_3^2 c_{i-1} + \frac{(x - t_{i-2}^*) c_{i-1} + (t_{i-1}^* - x) c_{i-2}}{t_{i-1}^* - t_{i-2}^*}, \quad (2.8)$$

where

$$\nabla_3 c_j = (c_j - c_{j-1})/\sigma_j^2,$$

$$\nabla_3^2 c_j = (\nabla_3 c_j - \nabla_3 c_{j-1})/\sigma_j^1.$$

$$\sigma_j^2 = \int_{t_j}^{t_{j+3}} B_j^2(y) \, dy = t_j^* - t_{j-1}^*, \quad (2.9)$$

$$t_j^* = \begin{cases} t_{j+2} + \dfrac{\beta_{j+2}^3 - \beta_{j+1}^3}{\beta_{j+2}^2 + \beta_{j+1}^2}, & \text{if } t_{j+3} > t_{j+1}, \\ t_{j+2}, & \text{otherwise}. \end{cases}$$

A cubic B-spline B_j^3 with knots $(7, 8, 9, 10, 11)$ can be seen in Fig. 2.1. The tension parameters on the 4 subintervals are $(1, 5, 10, 10)$. Note that this B_j^3 has smoothness C^2, but is almost piecewise linear on $(9, 11)$.

From the definition (2.1) of the exponential powers and the explicit expressions (2.4), (2.7), and (2.8), it follows that B_j^r is the usual polynomial B-spline of degree r if all tension parameters on (t_j, t_{j+r+1}) are zero.

From (2.5) we immediately obtain the following differentiation formula for the generalized B-splines

$$DB_j^{r+1} = B_j^r/\sigma_j^r - B_{j+1}^r/\sigma_{j+1}^r.$$

In general these functions have the following properties:

Lemma 2.2. *We have*

(i) $B_j^r(x) = 0$ *if* $x < t_j$ *or* $x > t_{j+r+1}$.
(ii) $B_j^r(x) > 0$ *for* $t_j < x < t_{j+r+1}$.
(iii) $\sum_j B_j^r \equiv 1, r \geq 2$.
(iv) B_j^r *has* $r - m$ *continuous derivatives at a knot of multiplicity* m.
(v) B_j^r *is a linear combination of* $u_i^r(x), v_i^r(x)$, *and a polynomial of degree* $r - 2$. *on each interval* (t_i, t_{i+1}).

Proof: Properties (i), (iii), (iv), and (v) follow easily from the definitions (2.4) and (2.5). The positivity follows from general arguments in [13]. ∎

3. Splines

Given knots **t** tension parameters ρ, and a sequence $\mathbf{c} = (c_j)$ of B-spline coefficients, the function

$$f = \sum_j c_j B_j^r$$

is called an *exponential tension spline of degree r*. Results in this section are taken from [13]. We first state

Lemma 3.1. *The B-splines B_j^r are linearly independent on $\mathbb{R}$ provided $t_{j+r+1} > t_j$ for all integers j.*

We are mainly interested in the cubic case $r = 3$. First we derive from Lemma 2.1 the B-spline coefficients of the function $f(x) = x$.

Lemma 3.2. *With the t_j^* defined in (2.9) we have*

$$x = \sum_j t_j^* B_j^3(x). \tag{3.1}$$

Proof: Taking $c_j = t_j^*$ in (2.8) and using (2.9) we find

$$V_3 c_j = (t_j^* - t_{j-1}^*)/\sigma_j^2 = 1,$$

so that $V_3^2 c_j = 0$ for all integers j. Hence we are left only with the linear part in (2.8). But for $c_j = t_j^*$ the linear part reduces to x. ∎

Discussion: We evaluate the expression for t_j^* in (2.9) for $\rho_{j+1} = \rho_{j+2} = 0$ and find

$$t_j^* = (t_{j+1} + t_{j+2} + t_{j+3})/3, \qquad \text{for} \qquad \rho_{j+1} = \rho_{j+2} = 0. \tag{3.2}$$

Thus (3.1) reduces to a well known B-spline identity in this case. We call the t_j^* *knot averages.* ∎

Let us now consider knot insertion. The proof of the following result can be found in [13].

Lemma 3.3. *Suppose τ is a nondecreasing sequence of knots and let $\mathbf{t} = \tau \cup \{t\}$ be nondecreasing and obtained from τ by inserting one knot at the position t. Let $B_{j,\tau}^3$ and $B_{j,\mathbf{t}}^3$ be the B-splines on τ and $\mathbf{t}$, respectively. If*

$$f = \sum_j c_j B_{j,\tau}^3 = \sum_i d_i B_{i,\mathbf{t}}^3,$$

then

$$d_i = \frac{(t_i^* - \tau_{i-1}^*)c_i + (\tau_i^* - t_i^*)c_{i-1}}{\tau_i^* - \tau_{i-1}^*}, \tag{3.3}$$

where τ_i^* and t_i^* are knot averages given by (2.9) on τ and $\mathbf{t}$, respectively. Moreover, *for all integers i*

$$\tau_{i-1}^* \leq t_i^* \leq \tau_i^*. \tag{3.4}$$

Combining (3.2) and (3.3) we see that (3.3) reduces to the usual formula ([2]) for inserting one knot in a polynomial spline curve. In general we observe that inequality (3.4) implies that we are forming d_i from a *convex combination* of c_i and c_{i-1}. This property also holds for more general classes of splines, see [15].

For polynomial splines it is well known that knot insertion can be used to find the value of a spline $f(x)$ at a point x. A similar approach can be used in the exponential case. Indeed, if we insert x three times in the knot sequence τ to obtain a refined knot sequence t and B-spline coefficients (d_j) on t, then $f(x) = d_i$ where the integer i is such that $\tau_i \leq x < \tau_{i+1}$. This follows from properties (i), (iii) and (iv) of Lemma 2.2.

As an example consider the Bernstein/Bézier case where

$$\tau = \{0,0,0,0,1,1,1,1\} = (\tau_j)_{j=0}^7 \quad \text{and} \quad f(x) = \sum_{j=0}^3 c_j B_{j,\tau}^3(x). \tag{3.5}$$

Let $\rho_3 = \rho > 0$ be the nontrivial tension parameter. From (2.9) we find the knot averages

$$(0, \tau^*, 1 - \tau^*, 1) = (\tau_j^*)_{j=0}^3, \quad \text{with} \quad \tau^* = \frac{\sinh \rho - \rho}{\rho(\cosh \rho - 1)}. \tag{3.6}$$

Suppose we insert a triple knot at a point x in the interval $(0, 1)$. For any real numbers a, b, c let t_{abc}^* be the knot average computed from (2.9) using $t_{j+1} = a$, $t_{j+2} = b$, $t_{j+3} = c$, and $\rho_{j+1} = \rho_{j+2} = \rho$. The calculation of $d_i^3 = f(x)$ is shown in Fig. 3.1. This is yet another example of triangular schemes considered recently, see [8].

In Fig. 3.1 we have $d_j^0 = c_j$, and each entry d_j^{r+1} is computed from the two entries d_j^r and d_{j-1}^r by forming convex combinations with the quantities shown. For example

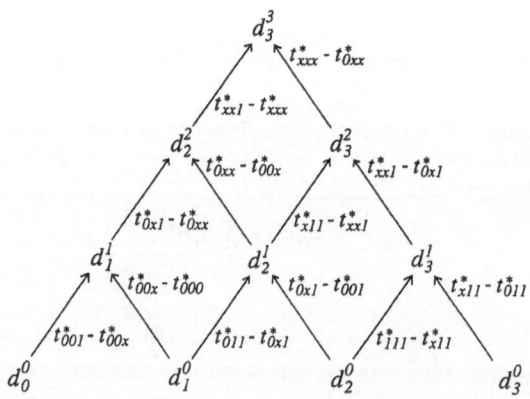

Figure 3.1. A de Casteljau type algorithm for an exponential Bézier curve

$$d_2^2 = \frac{(t_{0xx}^* - t_{00x}^*)d_2^1 + (t_{0x1}^* - t_{0xx}^*)d_1^1}{t_{0x1}^* - t_{00x}^*}$$

For clarity the factors used in the divisions are not shown in Fig. 3.1. However, these factors are uniquely defined from the fact that we are taking convex combinations.

We end this section with a discussion of spline evaluation. Suppose $f = \sum_j c_j B_j^3$ is a known exponential tension spline and we want to evaluate $f(x)$ for some x. We can compute $f(x)$ either using (2.8) or by a triangular scheme similar to the one for Bézier curves in Fig. 3.1. It seems simpler to use (2.8). Some care has to be taken in order to evaluate the exponential powers accurately in floating point arithmetic. Using the formulae as in (2.1) might lead to overflow for large values of the tension parameters and inaccurate evaluation for small values. These problems can easily be avoided. For small values of the tension parameters we approximate γ by a truncated Taylor series, while for large values of the ρ's we rewrite the expressions for the γ's using only exponential functions with nonpositive arguments. We refer to [13] for details. Once we have accurate evaluation of exponential powers we can evaluate the β_j^r in (2.3) and the knot averages t_j^* from (2.9), and form the necessary differences of coefficients in (2.8). When these values have been precomputed we can evaluate $f(x)$ for various values of x fairly efficiently from (2.8).

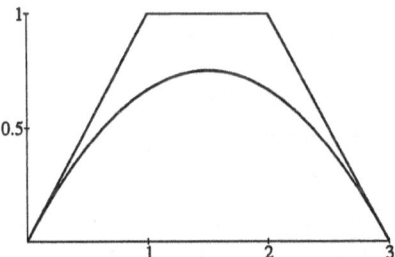

Figure 3.2. A Bézier curve with no tension. The control polygon $c = ((0,0),(1,1),(2,1),(3,0))$ is also shown

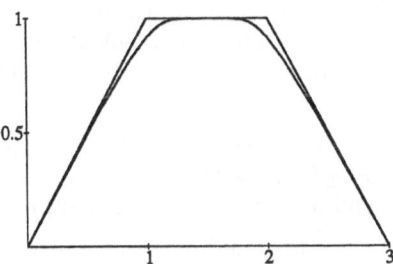

Figure 3.3. An exponential Bézier curve using the same coefficients as in Fig. 3.2, but with tension $\rho = 20$

As an example consider the two curves in Fig. 3.2 and 3.3. In Fig. 3.2 we show a standard polynomial Bézier curve, *i.e.*, zero tension. In Fig. 3.3 we use the same coefficients as in Fig. 3.2, but we increase the tension from zero to 20. Observe how the curve is pulled towards its control polygon.

4. Univariate Interpolation

Given data $(x_i, y_i)_{i=1}^m$ and a knot vector $(t_j)_{j=1}^{m+4}$ we want to find a cubic exponential tension spline

$$f = \sum_{j=1}^m c_j B_j^3$$

such that

$$f(x_i) = y_i, \qquad \text{for} \qquad i = 1, 2, \dots, m. \tag{4.1}$$

Combining the interpolation conditions and the special form of f we obtain a linear system

$$\mathbf{Ac} = \mathbf{y}. \tag{4.2}$$

Here $\mathbf{A}$ is a square $m \times m$ matrix with element $B_j^3(x_i)$ in row i and column j. There exists a unique interpolant f if and only if the matrix $\mathbf{A}$ is nonsingular. The following theorem shows that this matrix is totally positive. Moreover it is nonsingular if and only if the diagonal elements are nonzero. (Recall that a matrix is *totally positive* if all square submatrices have a nonnegative determinant.)

Theorem 4.1. *The matrix* $\mathbf{A}$ *is totally positive. Moreover* $\mathbf{A}$ *is nonsingular if and only if*

$$t_i < x_i < t_{i+4}, \qquad for \qquad i = 1, 2, \dots, m. \tag{4.3}$$

Proof: It is shown in [13] that even with variable tension parameters the theory of Chebyshevian splines applies to the exponential tension splines. The theorem therefore follows from Theorem 9.33 in [26]. ∎

The conditions (4.3) is known in the spline literature as the Schoenberg-Whitney nesting condition. A similar condition is also necessary and sufficient for the existence and uniqueness of a Hermite interpolant, *i.e.*, we specify values and derivatives up to a certain order at each interpolation point. See Theorem 9.33 in [26].

Since $\mathbf{A}$ is totally positive the linear system (4.2) can be solved by Gaussian elimination without pivoting ([3]). Note also that $\mathbf{A}$ is a banded matrix. Each row has a nonzero diagonal and contains at most 4 nonzero neighbouring elements.

As an example consider tension spline interpolation at knots. For a B-spline formulation we choose the knot vector

$$\mathbf{t} = (x_1, x_1, x_1, x_1, x_2, \dots, x_{m-1}, x_m, x_m, x_m, x_m) = (t_j)_{j=1}^{m+6}.$$

We need $m + 2$ interpolation conditions corresponding to the $m + 2$ cubic B-splines which can be defined on t. In addition to the m conditions given by (4.1) we choose 2 boundary conditions. In *complete cubic spline interpolation* these boundary conditions are taken to be derivatives at the ends. For tension splines approaching a piecewise linear shape as the tension parameters tend to infinity, it is natural to choose these derivatives as the slopes of the end secants defined by the data. Thus we seek a spline $f = \sum_{j=1}^{m+2} c_j B_j^3$ such that

$$Df(x_1) = d_1 = (y_2 - y_1)/(x_2 - x_1),$$

$$f(x_i) = y_i, \quad \text{for} \quad i = 1, 2, \ldots, m, \tag{4.4}$$

$$Df(x_m) = d_m = (y_m - y_{m-1})/(x_m - x_{m-1}),$$

From (4.4) we obtain a linear system

$$\mathbf{Ac} = \mathbf{f}, \quad \text{with} \quad \mathbf{c} = (c_1, \ldots, c_{m+2})^T \quad \text{and} \quad \mathbf{f} = (d_1, y_1, \ldots, y_m, d_m)^T.$$

The local support property (i) of Lemma 2.2 implies that the matrix $\mathbf{A}$ is tridiagonal

$$\mathbf{A} = \begin{bmatrix} \alpha_0 & \gamma_0 & & & \\ \delta_1 & \alpha_1 & \gamma_1 & & \\ & \ddots & \ddots & \ddots & \\ & & \delta_m & \alpha_m & \gamma_m \\ & & & \delta_{m+1} & \alpha_{m+1} \end{bmatrix}$$

where

$$\alpha_0 = DB_1^3(x_1), \quad \alpha_{m+1} = DB_{m+2}^3(x_m),$$

$$\gamma_0 = DB_2^3(x_1), \quad \delta_{m+1} = DB_{m+1}^3(x_m),$$

$$\delta_i = B_i^3(x_i), \quad \alpha_i = B_{i+1}^3(x_i), \quad \gamma_i = B_{i+2}^3(x_i).$$

Using Lemma 2.1 one can show that

$$\delta_i = \frac{\beta_{i+3}^3}{\sigma_{i+2}^1 \sigma_{i+1}^2}, \quad \gamma_i = \frac{\beta_{i+2}^3}{\sigma_{i+2}^1 \sigma_{i+2}^2}, \quad \alpha_i = 1 - \delta_i - \gamma_i,$$

where the quantities used are defined in (2.3), (2.7) and (2.9). In particular we have $\delta_1 = \gamma_m = 1$ and $\alpha_1 = \gamma_1 = \delta_m = \alpha_m = 0$. The values of the derivatives of the B-spline can be found from the differentiation formula and from the property that the derivatives of the B-splines at a point sum to zero (cf. property (iii) in Lemma 2.2.) Using this we find that the nonzero derivative values at x_1 and x_m are

$$\alpha_0 = DB_1^3(x_1) = -\gamma_0 = -DB_2^3(x_1) = -\beta_4^2/\beta_4^3,$$

$$\alpha_{m+1} = DB_{m+2}^3(x_m) = -DB_{m+1}^3(x_m) = -\delta_{m+1} = \beta_{m+2}^2/\beta_{m+2}^3.$$

The linear system $\mathbf{Ac} = \mathbf{f}$ is now easily solved by Gaussian elimination. Pivoting is not necessary. For if we eliminate the first and last two equations and unknowns then we obtain a linear system for the remaining $m - 2$ variables and equations which is totally positive by Theorem 4.1.

It has been shown in [19] that as the tension parameters tend to infinity the interpolant approaches the piecewise linear interpolant to the data.

The tension spline converges to the piecewise linear interpolant when the tension parameters tend to infinity. This makes the secant condition (4.4) a reasonable choice, at least for large tension parameters at the end of the range. As an alternative to (4.4) the "natural" conditions

$$D^2 f(x_1) = D^2 f(x_m) = 0,$$

could be used. However the popular *not-a-knot* condition, where we remove the knots at x_2 and x_{m-1} is not recommended for large values of the tension parameters. This is because we then have a conflict at the ends. At the left end for example the spline will at the same time try to be both the straight line between (x_1, y_1) and (x_3, y_3) and the broken line interpolating in addition (x_2, y_2). The resulting curve might not look good. In Fig. 4.1–4.4 we show the effect of using different boundary conditions. The function $f(x) = \sin\left(\dfrac{\pi}{2}x\right)$ is interpolated at the 11 points $x_i = -1 + (i - 1)/5, i = 1, 2, \ldots, 11$. The tension parameters are zero for $x < 0$ and equal

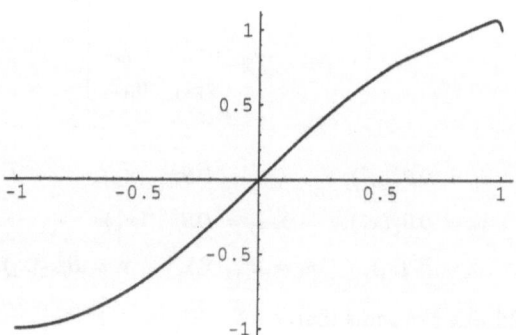

Figure 4.1. Not-a-knot

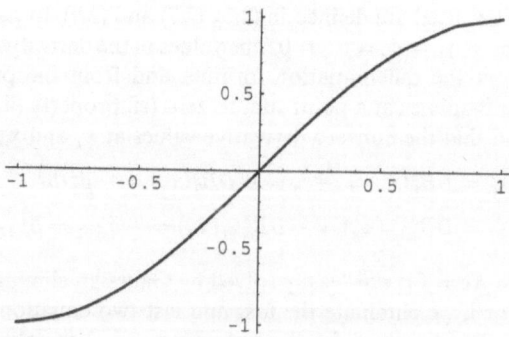

Figure 4.2. Natural

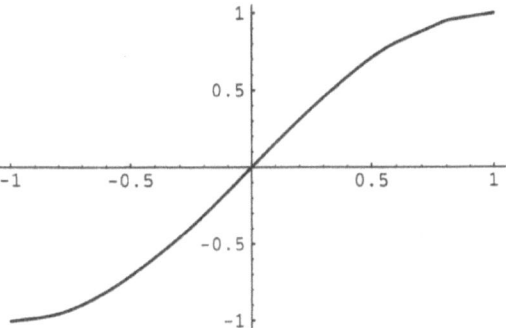

Figure 4.3. Secant

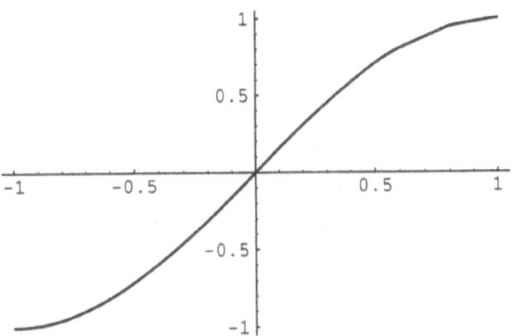

Figure 4.4. Exact 1. derivatives

to $(1, 1, 10, 100, 100)$ on the 5 subintervals to the right of the origin. Figure 4.1 shows that the use of the not-a-knot condition gives a dip at the right hand side of the range. On the open interval (x_{n-2}, x_n) the interpolant will converge to a straght line through (x_{n-2}, y_{n-2}) and (x_{n-1}, y_{n-1}). In order to arrive at (x_n, y_n) the curve dips down almost vertically at x_n. This problem is not encountered with the natural end conditions in Fig. 4.2 and the secant condition (4.4) used in Fig. 4.3. In Fig. 4.4 we have used exact first derivatives at both ends of the range. This gives a better approximation to the original function. When exact derivatives are not known one could use higher order difference approximations instead of (4.4).

5. Bicubic Interpolation

Given data

$$(u_i, v_j, z_{i,j})_{i=1, j=1}^{m, n}$$

where

$$u_1 < u_2 < \cdots < u_m \quad \text{and} \quad v_1 < v_2 < \cdots < v_n.$$

We want to find a surface $f(u, v)$ defined on the rectangle $R = [u_1, u_m] \times [v_1, v_n]$ such that

$$f(u_i, v_j) = z_{i,j}, \qquad i = 1, \ldots, m, \qquad j = 1, \ldots, n. \tag{5.1}$$

The values $z_{i,j}$ are real numbers or points in $\mathbb{R}^3$, corresponding to explicit, or parametric surfaces, respectively. We consider here the case where f is a bicubic exponential tension spline on a rectangular mesh with horizontal and vertical knot lines at the u_i and v_j values. For this we also need nonnegative tension parameters ρ_i on (u_i, u_{i+1}) and σ_j on (v_j, v_{j+1}). We introduce the knot vectors

$$\mathbf{s} = (u_1, u_1, u_1, u_1, u_2, \ldots, u_{m-1}, u_m, u_m, u_m, u_m) = (s_i)_{i=1}^{m+6},$$

$$\mathbf{t} = (v_1, v_1, v_1, v_1, v_2, \ldots, v_{n-1}, v_n, v_n, v_n, v_n) = (t_j)_{j=1}^{n+6},$$

and let $\mathbf{B}_{i,\mathbf{s}}^3(u)$ and $B_{j,\mathbf{t}}^3(v)$ denote the exponential tension B-splines on these knot vectors. The unknown f can then be written in the form

$$f(u, v) = \sum_{p=1}^{m+2} \sum_{q=1}^{n+2} c_{p,q} B_{p,\mathbf{s}}^3(u) B_{q,\mathbf{t}}^3(v). \tag{5.2}$$

We take the interpolation conditions to be the mn point conditions given by (5.1) plus the following boundary conditions

$$\frac{\partial f}{\partial u}(u_l, v_j) = z_{l,j}^{1,0}, \qquad l = 1, m, \qquad j = 1, \ldots, n,$$

$$\frac{\partial f}{\partial v}(u_i, v_l) = z_{i,l}^{0,1}, \qquad l = 1, n, \qquad i = 1, \ldots, m, \tag{5.3}$$

$$\frac{\partial^2 f}{\partial u \partial v}(u_i, v_j) = z_{i,j}^{1,1}, \qquad i = 1, \dot{m}, \qquad j = 1, n,$$

It is convenient to organize the z data in an $(m + 2) \times (n \times 2)$ matrix $\mathbf{F}$ as shown in Fig. 5.1.

For the choice of the z's we can use difference approximations as in the previous section. In particular we set

$z_{1,1}^{1,1}$	$z_{1,1}^{1,0}$	$\cdots$	$z_{1,n}^{1,0}$	$z_{1,n}^{1,1}$
$z_{1,1}^{0,1}$	$z_{1,1}$	$\cdots$	$z_{1,n}$	$z_{1,n}^{0,1}$
$\vdots$	$\vdots$	$z_{i,j}$	$\vdots$	$\vdots$
$z_{m,1}^{0,1}$	$z_{m,1}$	$\cdots$	$z_{m,n}$	$z_{m,n}^{0,1}$
$z_{m,1}^{1,1}$	$z_{m,1}^{1,0}$	$\cdots$	$z_{m,n}^{1,0}$	$z_{m,n}^{1,1}$

Figure 5.1. The data matrix for complete bicubic spline interpolation

$$z_{i,j}^{1,0} = \frac{\partial f}{\partial u}(u_i, v_j) = \frac{z_{i+1,j} - z_{i,j}}{u_{i+1} - u_i}, \qquad z_{i,j}^{0,1} = \frac{\partial f}{\partial v}(u_i, v_j) = \frac{z_{i,j+1} - z_{i,j}}{v_{i+1} - v_i}$$

$$z_{i,j}^{1,1} = \frac{\partial^2 f}{\partial u \partial v}(u_i, v_j) = \frac{z_{i+1,j+1} + z_{i,j} - z_{i+1,j} - z_{i,j+1}}{(u_{i+1} - u_i)(v_{j+1} - v_j)}.$$

There is a simple way to fill the data matrix $\mathbf{F}$ with the correct values around the boundary. Let $z_1, \ldots, z_m$ be the rows of $\mathbf{Z} = (z_{i,j})$. We then set

$$\mathbf{H} = \begin{bmatrix} (z_2 - z_1)/(u_2 - u_1) \\ z_1 \\ \vdots \\ z_m \\ (z_m - z_{m-1})/(u_m - u_{m-1}) \end{bmatrix} = (h_1 \ldots h_n),$$

and obtain the correct data matrix in Fig. 5.1 simply as

$$\mathbf{F} = \left(\frac{h_2 - h_1}{v_2 - v_1}, h_1, \ldots, h_n, \frac{h_n - h_{n-1}}{v_n - v_{n-1}} \right).$$

Here $h_1, \ldots, h_n$ are the n columns of the $(m+2) \times n$ matrix $\mathbf{H}$. The unknown coefficients $c_{p,q}$ in (5.2) can now be found using standard tensor product techniques.

We now present several examples showing the effect of using the tension parameters to control shape. In the first example we sample data from the test function

$$f(x, y) = h(x) + h(y) \quad \text{where} \quad h(x) = \begin{cases} e^{-10(x-1/4)}, & \text{if } x > 1/4, \\ 1, & \text{otherwise.} \end{cases} \tag{5.4}$$

Figure 5.2 shows a graph of this cutoff exponential function.

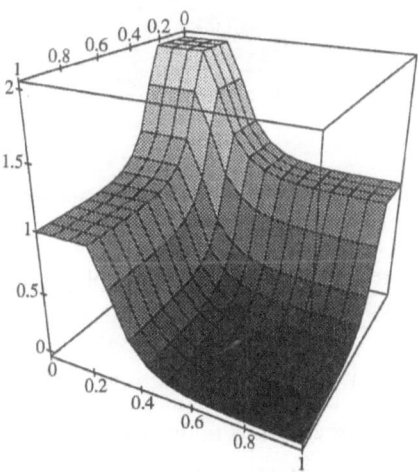

Figure 5.2. The function from (5.4) on the unit square

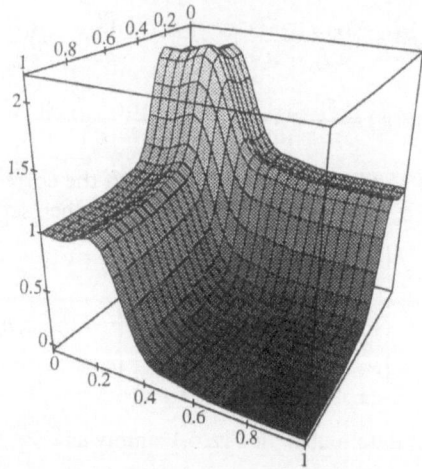

Figure 5.3. Bicubic spline approximation with no tension, to the function in Fig. 5.2

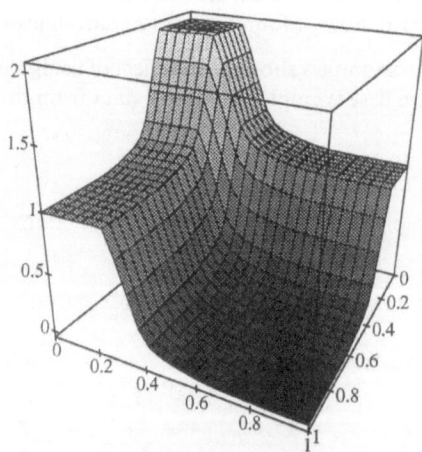

Figure 5.4. Same as Figure 5.3, but with tension parameters given by (5.5)

In Fig. 5.3 we show the result of complete bicubic spline interpolation to this data
using zero tension and the 9×9 grid defined by

$$u = v = \left(0, \frac{1}{8}, \frac{2}{8}, \frac{3}{8}, \ldots, \frac{8}{8}\right), \qquad z_{i,j} = f(u_i, v_j).$$

The interpolant has problems adjusting to the flat top part.

We can improve the appearance of this fit by introducing tension parameters. Figure
5.4 shows bicubic interpolation on the same grid using the two tension vectors

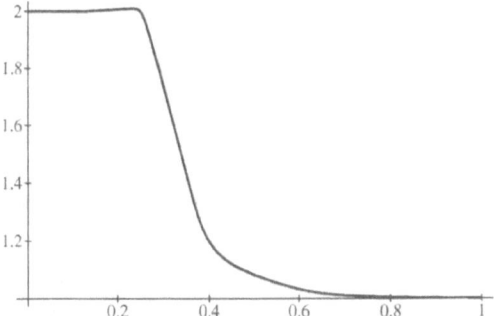

Figure 5.5. The v-curve at $u = 0$ of the surface in Fig. 5.4

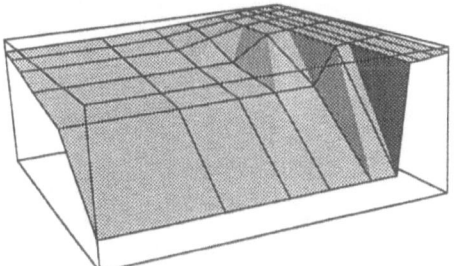

Figure 5.6. The CF data

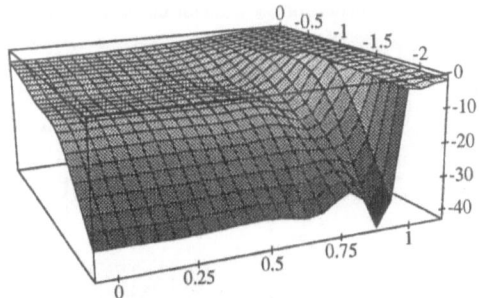

Figure 5.7. Complete bicubic C^2 interpolation to the CF data

$$\rho = \sigma = (200, 200, 100, 30, 20, 10, 0, 0). \tag{5.5}$$

A curve on the surface is shown in Fig. 5.5. Note that surface and the curve are C^2 smooth while the original surface is only C^0.

The data for the next example is on a 10×6 grid and is taken from Carlson and Fritsch's paper on monotone piecewise bicubic interpolation [4]. The exact data is shown in Fig. 5.6. The result of bicubic interpolation using no tension is shown in Fig. 5.7. The flat part followed by a rapid increase makes it difficult to obtain a

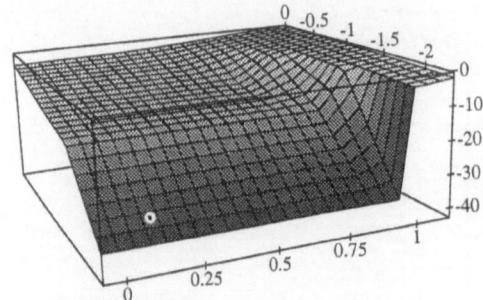

Figure 5.8. Interpolation with high tension

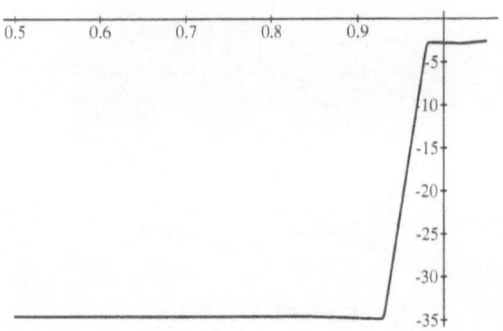

Figure 5.9. A u-curve at $v = -2.3$ for the surface in Fig. 5.8

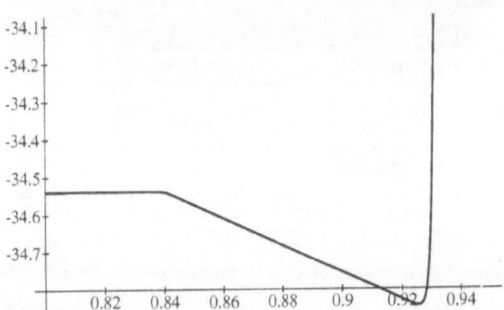

Figure 5.10. Part of the curve in Fig. 5.9 magnified

smooth fit which is monotone in both the u and v direction. In [4] the authors succeeded in obtaining a monotone C^1 fit to this data. Let us now try a C^2 exponential tension spline fit. Figure 5.8 shows the result of using uniform tension $\rho = \sigma = 1000$ on each subrectangle. It is difficult to tell from the plot in Fig. 5.8 whether the surface is monotone in each coordinate direction or not. To check this graphically we have plotted the u-curve at $v = -2.3$ in Fig. 5.9 and 5.10. The part of this curve around $u = 0.9$ shows that this u-curve has a small dip. Thus the fit in Fig. 5.8 is not quite monotone in the u direction. It would be interesting to see whether one of the schemes for automatically choosing the tension parameters [9, 16, 21, 22, 23] could be adapted to work for this example.

References

[1] Barsky, B. A.: Computer graphics and geometric modelling using beta-splines. Berlin: Springer 1988.

[2] Boehm, W.: Inserting new knots into B-spline curves. Comput. Aided Design *12*, 199–201 (1980).

[3] de Boor, C., Pinkus, A.: A backward error analysis for totally positive linear systems. Numer. Math. *27*, 485–490 (1977).

[4] Carlson, R. E., Fritsch, F. N.: Monotone piecewise bicubic interpolation. SIAM J. Numer. Anal. *22*, 386–400 (1985).

[5] Cinquin, Ph.: Splines unidimensionelles sous tension et bidimensionelles paramétrées: deux applications médicales, Thesis, Université de Saint-Etienne, 1981.

[6] Cline, A. K.: Scalar- and planar-valued curve fitting using splines under tension. Comm. ACM *17*, 218–223 (1974).

[7] Foley, T.: Local control of interval tension using weighted splines. Computer-Aided Geom. Design *3*, 281–294 (1986).

[8] Goldman, R. N.: Recursive triangles. In: Dahmen, W., Gasca, M., Micchelli, C. A. (eds.) Computation of curves and surfaces, pp. 27–72. Dordrecht: Kluwer Academic Publishers 1990.

[9] Heß, W., Schmidt, J. W.: Convexity preserving interpolation with exponential splines. Computing *36*, 335–342 (1986).

[10] Kincaid, D., Cheney, W.: Numerical analysis. Pacific Grove: Brooks/Cole Publishing Company 1991.

[11] Koch, P. E., Lyche, T.: Exponential B-splines in tension. In: Chui, C. K., Schumaker, L. L., Ward, J. D. (eds.) Approximation theory VI, pp. 361–364. New York: Academic Press 1989.

[12] Koch, P. E., Lyche, T.: Construction of exponential tension B-splines of arbitrary order. In: Laurent, P.-J., Le Méhauté, A., Schumaker, L. L. (eds.) Curves and surfaces, pp. 255–258. New York: Academic Press 1991.

[13] Koch, P. E., Lyche, T.: Calculating with exponential B-splines in tension (preprint).

[14] Kulkarni, R., Laurent, P.-J.: Q-splines. Numerical Algorithms *1*, 45–74 (1991).

[15] Lyche, T.: A recurrence relation for Chebyshevian B-splines. Constr. Approx. *1*, 155–173 (1985).

[16] Lynch, R. W.: A method for choosing a tension factor for splines under tension interpolation, Thesis, Univ. of Texas, Austin, 1982.

[17] Nielson, G. M.: Some piecewise polynomial alternatives to splines under tension. In: Barnhill, R. E., Riesenfeld, R. F. (eds.) Computer aided geometric design, pp. 209–235. New York: Academic Press 1974.

[18] Nielson, G. M., Franke, R.: A method for construction of surfaces under tension. Rocky Mt. J. Math. *14*, 203–221 (1984).

[19] Pruess, S.: Properties of splines in tension. J. Approx. Th. *17*, 86–96 (1976).

[20] Pruess, S.: An algorithm for computing smoothing splines in tension. Computing *19*, 365–373 (1978).

[21] Renka, R. J.: Interpolatory tension splines with automatic selection of tension factors. SIAM J. Sci. Stat. Comp. *8*, 393–415 (1987).

[22] Rentrop, P.: An algorithm for the computation of the exponential spline. Numer. Math. *35*, 81–93 (1980).

[23] Sapidis, N. S., Kaklis, P. D.: An algorithm for constructing convexity and monotonicity-preserving splines in tension. Computer-Aided Geom. Design *5*, 127–137 (1988).

[24] Salkauskas, K.: C^1 splines for interpolation of rapidly varying data. Rocky Mt. J. Math. *14*, 239–250 (1984).

[25] Schaback, R.: Rational curve interpolation. In: Lyche, T., Schumaker, L. L. (eds.) Mathematical methods in computer aided geometric design II, pp. 517–535. New York: Academic Press 1992.

[26] Schumaker, L. L.: Spline functions: basic theory. New York: Wiley 1981.

[27] Schweikert, D.: An interpolation curve using a spline in tension. J. Math. Phys. *45*, 312–317 (1966).

[28] Späth, H.: Exponential spline interpolation. Computing *4*, 225–233 (1969).

[29] Späth, H.: Spline algorithms for curves and surfaces. Winnipeg, Canada: Utilitas Mathematica Publishing 1974.

Per Erik Koch
Norges Tekniske Høgskole
Institutt for matematiske fag
N-7034 Trondheim, Norway
pek@ imf.unit.no

Tom Lyche
Universitetet i Oslo
Institutt for informatikk
P.O. Box 1080, Blindern,
N-0316 Oslo 3, Norway
tom@ ifi.uio.no

Computing Suppl. 8, 191–210 (1993)

Computing
© Springer-Verlag 1993

A Characterization of an Affine Invariant Triangulation

G. M. Nielson, Tempe

Abstract. The well known and widely used Delauny triangulation which is the dual of Dirichlet tessellation is not affine invariant. This means, among other things, that the triangulation is dependent upon the choice of the coordinate axes used to represent the vertices. We present a new type of triangulation of planar point sets which is unaffected by translations, rotations or changes of scale. A certain type of generalized Dirichlet tessellation is shown to always have a triangulation as its dual. When a particular affine invariant norm is used to create the tessellation, an affine invariant triangulation is characterized as the dual.

Key words: Triangulation, tessellation, affine invariant scattered data, interpolation, scientific computing.

1. Introduction

Triangulations are of interest in many aspects of multivariate approximations and scientific computing; including the finite element method and scattered data interpolation. In this report, we describe a certain type of optimal triangulation of the convex hull of a set of data points which is invariant under affine transformations. In particular, the type of triangulation we present is invariant under scale transformation which means that the choice in the units used to represent the data does not affect the triangulation. A property heretofore not enjoyed by other triangulation schemes. Basically, the problem of triangulating planar data sets requires the partitioning of the convex hull into a collection of triangles with vertices from the data set. We assume that the vertices V_i, $i = 1, \ldots, n$ are distinct and that they do not all lie on a line. The convex hull of these vertices will be denoted by $CH\langle V_1, V_2, \ldots, V_n \rangle\rangle = CH\langle \vec{V} \rangle$. For the vertices V_i, V_j, V_k, this convex hull is a triangle which we denote by T_{ijk}. This is a closed set, and the open set consisting the interior of this triangle will be denoted by $\mathring{T}_{ijk}$.

Definition 1.1. Given the vertices V_i, $i = 1, \ldots, n$, we say a collection of triple indices $I_t = \{ijk: 1 \le i, j, k \le n, i \ne j \ne k \ne i\}$ represents a triangulation provided:

    i) $\mathring{T}_{ijk} \ne \phi$
    ii) $\mathring{T}_{abc} \cap \mathring{T}_{\alpha\beta\gamma} = \phi,$ $abc, \alpha\beta\gamma \in I_t,$ $abc \ne \alpha\beta\gamma$
    iii) $V_m \notin T_{ijk},$ $m \ne i, j, k$
    iv) $\bigcup_{ijk \in I_t} T_{ijk} = CH\langle \vec{V} \rangle$

In other words, we require $\{T_{ijk}: ijk \in I_t\}$ to be a collection of nondegenerate trian-

gles which are mutually exclusive and collectively exhaustive. For computer implementations, the order of the indices of a triple in I_t can be used to represent certain
information, but here we do not differentiate and so we say that two entries abc and
$\alpha\beta\gamma$ of I_t are equal if they are equal as sets, i.e. $abc = \alpha\beta\gamma$ provided $\{b, c, a\} = \{\gamma, \alpha, \beta\}$.
Also, for our discussion, the order of the list of triples that make up a triangulation
does not matter and so we say two triangulations are equal provided they are equal
as sets (or triples). It is well known and easy to verify that n_t, the number of entries
of I_t, satisfies:

$$n_t = 2n_i + n_b - 2$$

where n_b is the number of points on the boundary of the convex hull and n_i is the
number of interior points so that $n = n_i + n_b$.

For a particular data set, there are a number of possible triangulations. For certain
applications, a preference for one triangulation over the other may exist. Usually,
an optimal criteria attempts to avoid long thin triangles [11]. Two optimal criteria
have been discussed quite extensively: the max-min criteria of Lawson [16] and the
min-max criteria of Little and Barnhill [2]. Both of these optimal triangulations
have a similar method of characterization. Associated with each triangulation there
is a vector with n_t entries representing either the largest or smallest angle of each
triangle. The entries of each vector are ordered and then a lexicographic ordering
of the vectors is used to impose an ordering on the set of all triangulations. In the

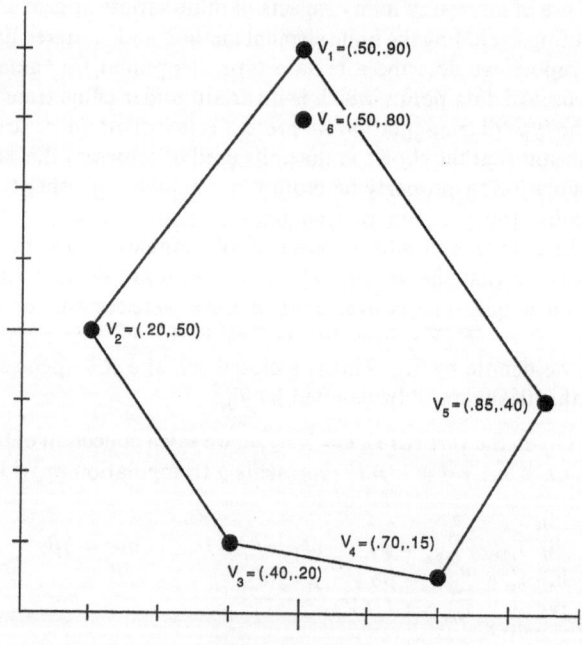

Figure 1. Six points

case of the min-max criteria, A_i is the largest angle of a triangle and the entries of each vector, A_t, are ordered so that

$$A_t = (A_1, A_2, \ldots, A_{n_i}), \qquad A_i \geq A_j, \qquad i < j.$$

The smallest of these vectors based on their lexicographic ordering associates with the optimal triangulation. In the case of the max-min criteria, a_i is the smallest angle and the entries of each vector are ordered the other way so that

$$a_t = (a_1, a_2, \ldots, a_{n_i}), \qquad a_i \leq a_j, \qquad i \leq j.$$

The largest of these vectors represents the optimal triangulation in max-min sense. The following example helps to illustrate these ideas. In Fig. 1, there are shown six (6) data points which have a total of ten (10) possible triangulations which are shown in Fig. 2. Based upon the max-min criteria we have the ordering:

$$\tau_1 < \tau_2 < \tau_5 < \tau_3 < \tau_0 < \tau_6 < \tau_7 < \tau_4 < \tau_8 < \tau_9,$$

and so τ_9 is the optimal triangulation in this sense. It is interesting to note that each triangulation is obtainable from one with a smaller associated vector by swapping the diagonal of a convex quadrilateral. In fact, this basic operation forms the basis of the algorithm of Lawson for computing the optimal triangulation. Lawson has shown [14] that any triangulation can be obtained from any other triangulation by

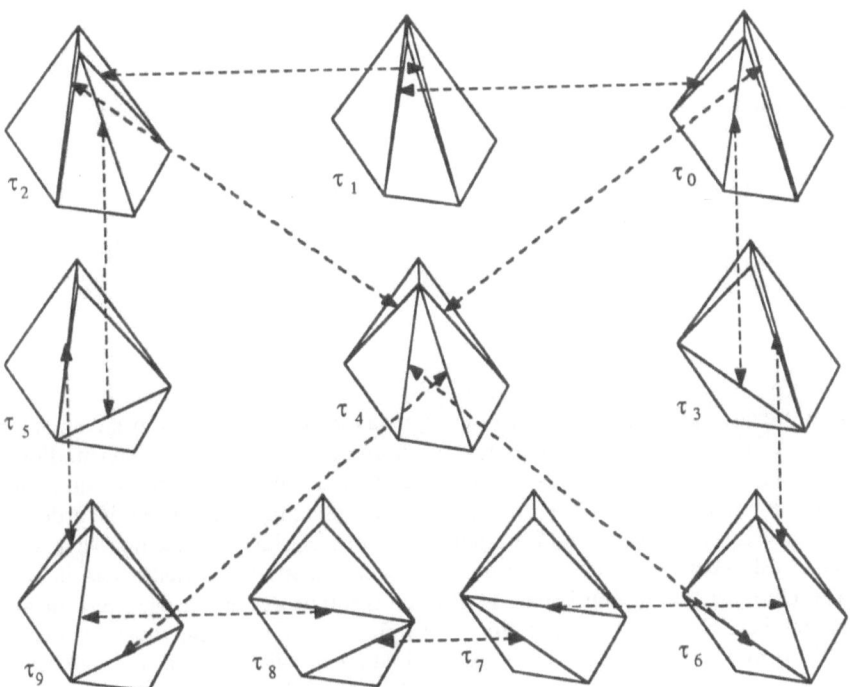

Figure 2. Ten triangulations of six points

a sequence of these operations. Furthermore, Lawson has proved that if the choice of the diagonal is made on the basis of the max-min criteria for the quadrilateral only, eventually the optimal triangulation will be obtained. In other words, for this criterion, a local optimum is a global optimum. It is interesting to note that this is not the case for the min-max criterion. This same example points this out. It turns out that based upon the min-max criteria, τ_4 is optimal and τ_8 is a local minimum. More details on this example and related results can be found in [21].

In many applications, the choice of the location of the coordinate axis or the choice of the units for measured data is rather arbitrary. Consequently, these choices should not affect the final results. Since the above criteria for an optimal triangulation is based upon angles only, it is clear that a translation or rotation of the data would not affect the triangulation. Unfortunately, a change of scale can affect matters. The following example shows this. In the left portion of Fig. 3, we show the optimal max-min triangulation of some data whose units of measurement are inches and seconds. In the right portion, we show the optimal triangulation of this same data except the units of measurement have been changed to feet and minutes. Even though both triangulations are optimal in the min-max sense, they are different. We consider this anomalous behavior to be a drawback and in this paper we describe a criterion for triangulations that is unaffected by this arbitrary choice.

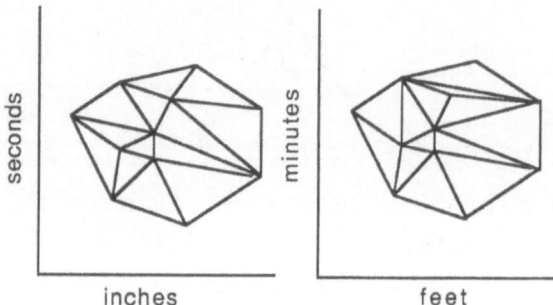

Figure 3. Different optimal triangulations for the same data

In a previous paper [20], we introduced a way of modifying certain methods for scattered data interpolation so as to be scale and rotation invariant. The modification consists of replacing the standard Euclidean norm with one that is invariant under affine transformations. In this paper we use this same metric in order to characterize an affine invariant triangulation. The basic idea is based upon a duality relationship between a certain optimal triangulation and the Dirichlet tessellation. The Dirichlet tessellation is based solely on distance and not angles. So our approach here is to form a tessellation of the plane using the affine invariant metric of [20] and use the triangulation which is dual to this tessellation as an affine invariant, optimal triangulation. The fact that all of his can be done is the subject of the remaining sections of this paper. In the next section, we introduce a generalization of the Dirichlet tessellation, based upon a fairly general norm and show that it will always yield a dual triangulation. In the following section, we incorporate an affine invariant norm and show some examples of the tessellations and triangulations that result.

2. A Generalized Dirichlet Tessellation and its Dual Triangulation

The Dirichlet tessellation based on the vertices V_i, $i = 1, \ldots, n$ is a partition of the plane into n regions R_i, $i = 1, \ldots, n$ called the Thiesen regions. The Thiessen region R_k consists of all points in the plane whose closest point among V_i, $i = 1, \ldots, n$ is V_k. A Dirichlet tessellation is usually illustrated by drawing the boundaries of the regions R_i, $i = 1, \ldots, n$. A example is shown in Fig. 4. There are several interesting properties that can be noted. The edges of the boundaries of the these regions consist of points which are equidistant to two vertices and are therefore perpendicular bisectors of two vertices. Points of intersection of edges are points that are equidistant to three (or more) points and are therefore the centers of great circles specified by these points. Two vertices which have Thiessen regions with a common edge are called neighbors and if neighbors are joined together a triangulation of the convex hull is obtained. This is often referred to as the Delaunay triangulation [23], [6] or the Thiessen triangulation [14], [25] [29]. A example based on the same points of Fig. 4 is given in Fig. 5. Lawson [14], as well as Sibson [27], has shown that this particular triangulation is the same as the optimal triangulation characterized by the max-min criterion. Each of the triangles of this optimal triangulation is characterized by the property that the interior of the circumscribing circle contains no vertices. The centers of these circumscribing circles are, of course, at the triads of the Dirichlet tessellation.

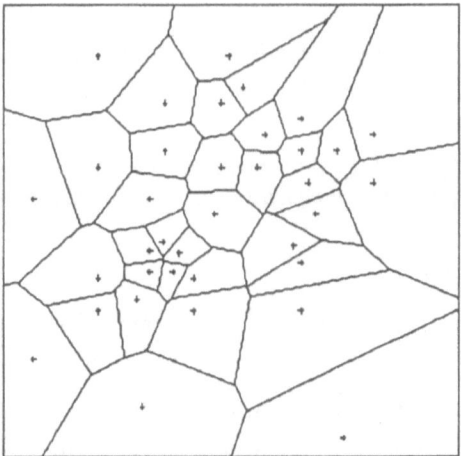

Figure 4. The Dirichlet tessellation of a set of points

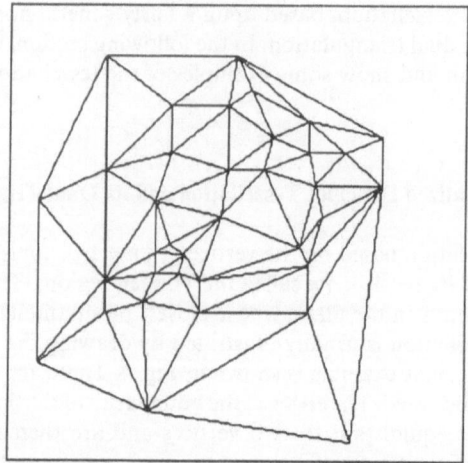

Figure 5. A triangulation of a set of points

We now proceed to generalize the duality relationship of the Dirichlet tessellation and Delaunay triangulation by considering a more general metric.

Any positive definite 2x2 matrix

$$A = \begin{pmatrix} A_{11} & A_{12} \\ A_{12} & A_{22} \end{pmatrix}$$

gives rise to an innerproduct

$$(V_1, V_2)_A = (x_1, y_1) A \begin{pmatrix} x_2 \\ y_2 \end{pmatrix}$$

and a resulting norm

$$\|V\|_A = \sqrt{A_{11}x^2 + 2A_{12}xy + A_{22}y^2}$$

which we refer to as the A-norm. The next lemma will lead to generalization of the notion of circumscribing circle.

Lemma 2.1. *If V_i, V_j and V_k are three vertices which are not collinear, then there is a unique point $C(V_i, V_j, V_k)$ which is equally distant, relative to the A-norm, from all three points.*

Proof. The conditions for a point

$$C(V_i, V_j, V_k) = (C_x(V_i, V_j, V_k), C_y(V_i, V_j, V_k))$$

to be equidistant from V_i, V_j and V_k are:

$$A_{11}(C_x - x_i)^2 + 2A_{12}(C_x - x_i)(C_y - y_i) + A_{22}(C_y - y_i)^2 = \rho^2$$
$$A_{11}(C_x - x_j)^2 + 2A_{12}(C_x - x_j)(C_y - y_j) + A_{22}(C_y - y_j)^2 = \rho^2$$
$$A_{11}(C_x - x_k)^2 + 2A_{12}(C_x - x_k)(C_y - y_k) + A_{22}(C_y - y_k)^2 = \rho^2.$$

The first equation minus the second equation and the first equation minus the third equation leads to

$$2[A_{11}(x_i - x_j) + A_{12}(y_i - y_j)]C_x + 2[A_{22}(y_i - y_j) + A_{12}(x_i - x_j)]C_y = C_1$$

$$2[A_{11}(x_i - x_k) + A_{12}(y_i - y_k)]C_x + 2[A_{22}(y_i - y_k) + A_{12}(x_i - x_k)]C_y = C_2$$

where

$$C_1 = A_{11}(x_i^2 - x_j^2) + 2A_{12}(x_iy_i - x_jy_j) + A_{22}(y_i^2 - y_j^2)$$

$$C_2 = A_{11}(x_i^2 - x_k^2) + 2A_{12}(x_iy_i - x_ky_k) + A_{22}(y_i^2 - y_k^2)$$

The coefficient matrix of this 2×2 system of equations is

$$2\begin{pmatrix} x_i - x_j & y_i - y_j \\ x_i - x_k & y_i - y_k \end{pmatrix} A.$$

The absolute value of the determinant of the first matrix is twice the area of T_{ijk} and so is not zero as long as the points V_i, V_j are V_k are not collinear. ∎

Definition 2.2. Let V_i, V_j and V_k be three non-collinear points. The set of points

$$E_A(V_i, V_j, V_k) = (V: \|V - C(V_i, V_j, V_k)\|_A \le \|V_j - C(V_i, V_j, V_k)\|_A)$$

will be called the circumscribing A-ellipse (c.f. Fig. 6).

As a result of this definition and Lemma 2.1, we have the following obvious result.

Lemma 2.3. *If V_a, V_b and V_c are any three non-collinear points on the boundary of the A-ellipse*

$$E_A(V_i, V_j, V_k),$$

then

$$E_A(V_a, V_b, V_c) = E_A(V_i, V_j, V_k).$$

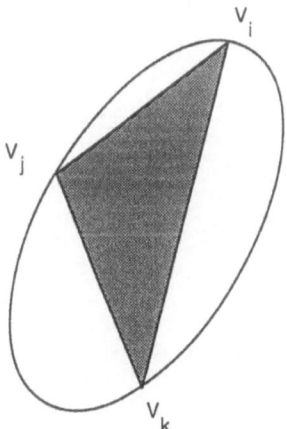

Figure 6. A circumscribing A-ellipse

We normally think of the information representing a tessellation as consisting of a list of points along with the edge connection information. Since the points of a tessellation are equidistant to (at least) three vertices, it would be equivalent to represent this information with a list of triples. Each one leading to a circumcenter which is computable by the equations of Lemma 2.1. The edge information is also represented by this list of triples for if two triples share a pair of indices, then there is an edge joining the two associated circumcenters. If a pair of indices ij of a particular triple are not shared by some other triple, then there is an infinite edge (ray) emanating from the center and in the direction of the perpendicular bisector of V_i and V_j. In the more general case the "perpendicular bisector" will consist of all points which are equal A-norm, distance to V_i and V_j and it has a direction vector given by

$$D = \begin{pmatrix} A_{12}(x_i - x_j) + A_{22}(y_i - y_j) \\ A_{11}(x_j - x_i) + A_{12}(y_j - y_i) \end{pmatrix}$$

which has the property that

$$(V_j - V_i, D)_A = 0,$$

or in other words, D is "perpendicular" to $V_j - V_i$ with respect to the A-innerproduct.

Definition 2.4. An A-norm tessellation consists of a list of triples $D_A = D_A(\vec{V})$ created in the following manner:

i) The triple $ijk \in D_A$ provided the set of vertices equidistant (using the A-norm) to $C(V_i, V_j, V_k)$ is exactly the three vertices V_i, V_j and V_k and the interior of $E_A(V_i, V_j, V_k)$ contains no other vertices.

ii) In the event that there are more that three vertices on the boundary of $E_A(V_i, V_j, V_k)$ (say, $V_{i_1}, V_{i_2}, \ldots, V_{i_m}$) and no other vertices in the interior of $E_A(V_i, V_j, V_k)$ we require that one of i, j or k be the minimum index of $i_1, i_2, \ldots, i_m$ and the remaining two vertices be contiguous neighbors on the boundary of E_{ijk}. This will account for $m - 1$ entries into D_A.

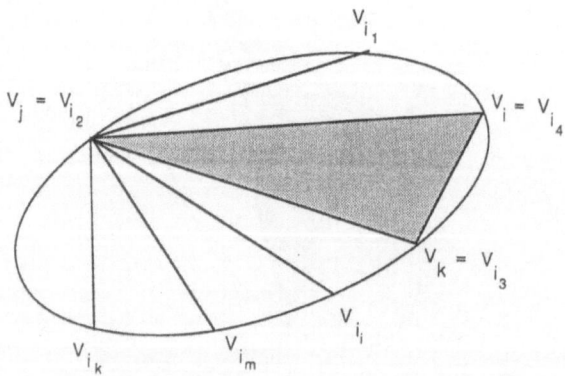

Figure 7. The arbitrary choice in the neutral case

This second condition of the above definition of a A-norm tessellation deals with the analog of what is often called the neutral case. We take an arbitrary, but well defined choice in this case. The triangles formed from the $m - 1$ entries of D_A created by $V_{i_1}, V_{i_2}, \ldots, V_{i_m}$ all lie in the A-ellipse with these points on its boundary and these tringles do not intersect. Figure 7 illustrates our choice.

It is now our task to show that a tessellation, as we have defined it, always gives rise to a triangulation. Once this is accomplished, we will have a way to define a triangulation which is entirely dependent upon distance and not angles. The following two lemmas will be needed.

Lemma 2.5. *Let $E_A(V_a, V_b, V_c)$ be an A-ellipse and assume that L is a line that intersects the interior of this ellipse (c.f. Fig. 8). Let H_+ and H_- be the two half-planes created by L. If $E_A(V_\alpha, V_\beta, V_\gamma)$ is another A-ellipse such that*

$$E_A(V_\alpha, V_\beta, V_\gamma) \cap L \subseteq E_A(V_a, V_b, V_c) \cap L$$

then either

$$E_A(V_\alpha, V_\beta, V_\gamma) \cap H_+ \subseteq E_A(V_a, V_b, V_c) \cap H_+$$

or

$$E_A(V_\alpha, V_\beta, V_\gamma) \cap H_- \subseteq E_A(V_a, V_b, V_c) \cap H_- .$$

More specifically, if

$$E_A(V_\alpha, V_\beta, V_\gamma) \cap L = E_A(V_a, V_b, V_c) \cap L$$

and $E_A(V_\alpha, V_\beta, V_\gamma) \neq E_A(V_a, V_b, V_c)$ then exactly one of the above conditions hold and the inclusion is the other way on the opposite side of L. That is, if

$$E_A(V_\alpha, V_\beta, V_\gamma) \cap H_+ \subseteq E_A(V_a, V_b, V_c) \cap H_+$$

then

$$E_A(V_\alpha, V_\beta, V_\gamma) \cap H_- \supseteq E_A(V_a, V_b, V_c) \cap H_-$$

or if

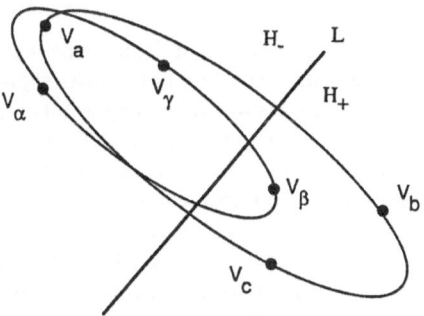

Figure 8. Two A-norm ellipses

$$E_A(V_\alpha, V_\beta, V_\gamma) \cap H_+ \supseteq E_A(V_a, V_b, V_c) \cap H_+$$

then

$$E_A(V_\alpha, V_\beta, V_\gamma) \cap H_- \subseteq E_A(V_a, V_b, V_c) \cap H_-.$$

Figure 9 illustrates this case.

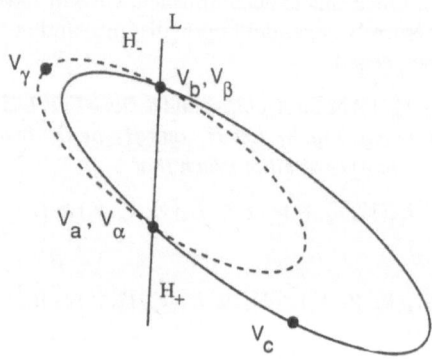

Figure 9. Two A-norm ellipses with two common points

Proof: We first prove the more specific result. Since

$$E_A(V_\alpha, V_\beta, V_\gamma) \cap L = E_A(V_a, V_b, V_c) \cap L$$

there are two points on the boundary of these two ellipses which are common. Without loss of generality, we may assume that these points are $V_a = V_\alpha$, and $V_b = V_\beta$ and so we have a situation like that of Fig. 9. It will suffice to show that the boundary of $E_A(V_\alpha, V_\beta, V_\gamma)$ is contained in $E_A(V_a, V_b, V_c)$ on one side of L, say H_+. Suppose that on H_+ the boundary of $E_A(V_\alpha, V_\beta, V_\gamma)$ contains points both in and out of $E_A(V_a, V_b, V_c)$, then it would have to be the case that there is a point other than $V_a = V_\alpha$ and $V_b = V_\beta$ where the boundary of $E_A(V_\alpha, V_\beta, V_\gamma)$ intersects the boundary of $E_A(V_a, V_b, V_c)$. But if these two A-ellipses share an additional point, then by Lemma 2.3 they would be exactly the same ellipse and so we have established the first part of this lemma.

Now to the proof of the more general result. If the two ellipses are nested, then the result is obvious. So let us assume that their boundaries share two points and again without loss of generality, let us assume that these points are $V_a = V_\alpha$ and $V_b = V_\beta$. Because of the assumption that $E_A(V_\alpha, V_\beta, V_\gamma) \cap L \subseteq E_A(V_a, V_b, V_c) \cap L$ we know that the line segment V_a to V_b does not intersect L or if it does, it is coincident with it and we have a case like above. Therefore, the intersection of L and both $E_A(V_\alpha, V_\beta, V_\gamma)$ and $E_A(V_a, V_b, V_c)$ must lie completely on one side of the line containing $V_a = V_\alpha$ and $V_b = V_\beta$. For the sake of this argument, let us say this is the side where $E_A(V_\alpha, V_\beta, V_\gamma) \subseteq E_A(V_a, V_b, V_c)$ which we will refer to as J_+. Since one of $H_+ \cap E_A(V_a, V_b, V_c)$ or $H_- \cap E_A(V_a, V_b, V_c)$ must be a subset of J_+, we may conclude that $E_A(V_\alpha, V_\beta, V_\gamma) \subseteq E_A(V_a, V_b, V_c)$ in H_+ or H_-. ∎

Lemma 2.6. *Let $ijk \in D_A$ and consider the edge e_{ij} joining V_i and V_j. If there are any vertices at all on the side of e_{ij} opposite from V_k then there is an entry $ijn \in D_A$ with V_n and V_k on opposite sides of e_{ij}.*

Proof: Let H_- and H_+ be the two half spaces created by the line containing the edge e_{ij}. Assume that $V_k \in H_-$ and let $V_{i_1}, V_{i_2}, \ldots, V_{i_m} \in H_+$ be those vertices opposite V_k. By Lemma 2.5, the sets $E_A(V_i, V_j, V_{i_j}) \cap H_+, j = 1, \ldots, m$ can be ordered by set inclusion. Let V_{i_k} associate with the smallest set so that

$$E_A(V_i, V_j, V_{i_k}) \cap H_+ \subseteq E_A(V_i, V_j, V_{i_j}) \cap H_+, \qquad j = 1, \ldots, m.$$

Now we claim that $iji_k \in D_A$. Because V_{i_k} associates with the smallest set, we know that the interior of $E_A(V_i, V_j, V_{i_k}) \cap H_+$ contains to vertices. Since $ijk \in D_A$ we know that $V_{i_k} \notin E_A(V_i, V_j, V_k) \cap H_+$ and so Lemma 2.5 implies that $E_A(V_i, V_j, V_{i_k}) \cap H_- \subset E_A(V_i, V_j, V_k) \cap H_-$ and so the interior of $E_A(V_i, V_j, V_{i_k})$ is void of vertices, which implies that $iji_k \in D_A$ and this completes the argument. ∎

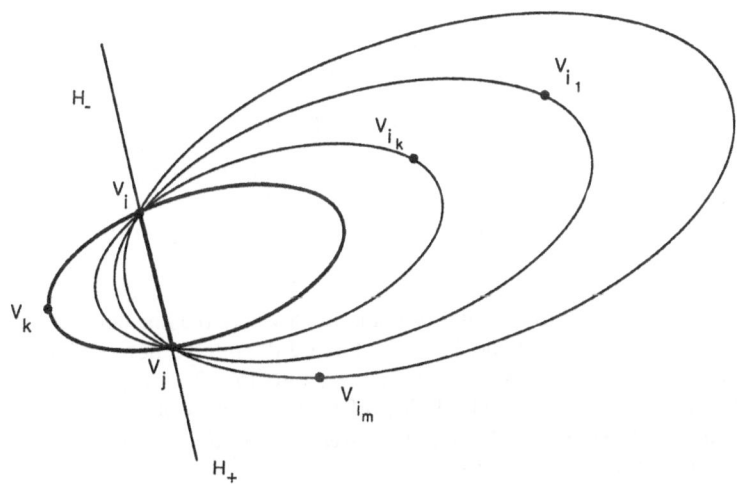

Figure 10. Nested A-ellipses

Theorem 2.7. *Let $D_A = D_{A(\vec{V})}$ be the A-norm Dirichlet tessellation for the collection of vertices $V_1, V_2, \ldots, V_n$ and let*

$$\tau = \{T_{ijk} : ijk \in D_A\}$$

be a collection of triangles represented by D_A, then τ is a triangulation of $CH\langle \vec{V} \rangle$.

Proof: We have to show that the conditions of Definition 1.1 are satisfied. We first show that the triangles of τ are mutually exclusive. Let abc and $\alpha\beta\gamma$ be two different triples of D_A. If $V_a, V_b, V_c, V_\alpha, V_\beta$ and V_γ all lie on the boundary of the same circumscribing A-ellipse, then the second condition of Definition 2.4 ensures that $\mathring{T}_{abc} \cap \mathring{T}_{\alpha\beta\gamma} = \phi$. If $\mathring{T}_{abc} \cap \mathring{E}_A(V_\alpha, V_\beta, V_\gamma) = \phi$ or $\mathring{E}_A(V_a, V_b, V_c) \cap \mathring{T}_{\alpha\beta\gamma} = \phi$ then again

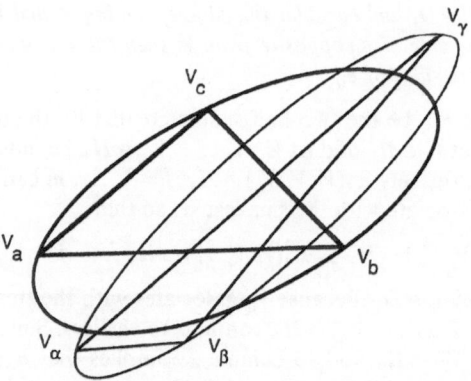

Figure 11. An impossible situation

$\mathring{T}_{abc} \cap \mathring{T}_{\alpha\beta\gamma} = \phi$ and we are finished. So let us assume for the moment that one of these two sets in not empty, say

$$\mathring{T}_{abc} \cap \mathring{E}_A(V_\alpha, V_\beta, V_\gamma) \neq \phi$$

Since we have eliminated the case of a common circumscribing ellipse, the conditions of D_A imply that V_a, V_b and $V_c \notin \mathring{E}_A(V_\alpha, V_\beta, V_\gamma)$. Therefore, it must be the case that some edge of T_{abc} (say e_{ab}) divides $E_A(V_\alpha, V_\beta, V_\gamma)$ (c.f. Fig. 11) and so

$$E_A(V_\alpha, V_\beta, V_\gamma) \cap e_{ab} \subseteq e_{ab}.$$

We now have a situation like that of Lemma 2.5 where the line containing V_a and V_b splits both $E_A(V_\alpha, V_\beta, V_\gamma)$ and $E_A(V_a, V_b, V_c)$ and the intersection of these two sets and the line are nested. Now, if we were to assume that $\mathring{T}_{\alpha\beta\gamma} \cap \mathring{T}_{abc} \neq \phi$ then there must be point of $\{V_\alpha, V_\beta, V_\gamma\}$ on both sides of e_{ab}. But because of Lemma 1 this would imply that at least one of V_α, V_β or V_γ were to be in $\mathring{E}_A(V_a, V_b, V_c)$ which is a contradiction and so we have that

$$\mathring{T}_{abc} \cap \mathring{T}_{\alpha\beta\gamma} = \phi.$$

Condition iii) of Definition 1.1 is quite obvious for if it were the case that some vertex V_m different from V_i, V_j and V_k were to lie in T_{ijk}, then it would certainly be the case that $V_m \in \mathring{E}_A(V_i, V_j, V_k)$ which contradicts the conditions for ijk to be in D_A.

Now we wish to show that every point of the convex hull lies is some triangle T_{ijk} for a triple $T_{ijk} \in D_A$. First we show that each edge of the boundary of the convex hull is included in the list of edges of τ:

$$E = \{ij: ijk \in D_A \text{ for some } k\}.$$

If e_{ij} is an edge of the convex hull, then there is one side of the line, L, containing e_{ij} which is void of vertices. Lets say this half space is H_-. Based upon Lemma 2.5 the sets $H_+ \cap E_A(V_i, V_j, V_l)$ with $V_l \in H_+$ can be ordered by set inclusion. Say $H_+ \cap E_A(V_i, V_j, V_k) \subset E_A(V_i, V_j, V_l)$ for $V_l \in H_+$ is the smallest (and it is unique). Then since H_- is void of vertices we know that $\mathring{E}_A(V_i, V_j, V_k)$ contains no other vertices and so ijk must be in D_A. If the smallest is not unique, the second property of the definition of D_A guarantees that there is at least one entry in D_A containing ij.

So far, we have shown that τ contains triangles that go all the way around the boundary of the convex hull. We need to show that the interior is completely covered with triangles. Let P be an arbitrary point of the convex hull and let L be a line containing P and not any vertices. Let E_L denote the set of all of the edges e_{ij}, $ij \in E$ which intersect L. From the argument of the previous paragraph, we know that there are at least two edges in E_L, namely the edges of the boundary of the convex hull. We let P_L and P_R be the points of the boundary edges which intersect L. We can order the edges of E_L based upon the distance of the intersection point to the point of intersection of a boundary edge, say P_L. Let e_{ij} be the edge with the largest distance which is less than (or equal to) the distance from P to P_L. Figure 12 illustrates this situation.

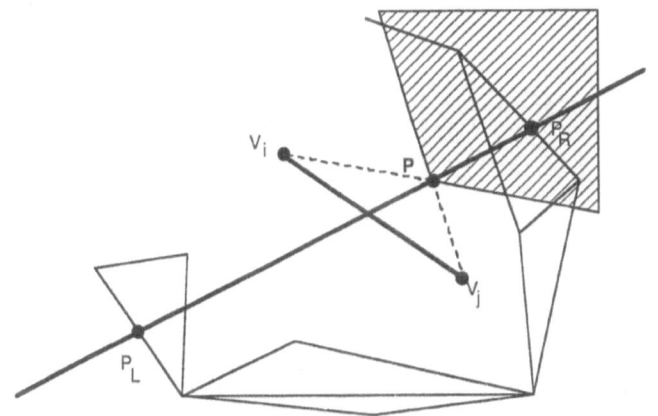

Figure 12. Edges intersecting a line containing P

Since $ij \in E$, there is some k such that $ijk \in D_A$. There are two cases to consider. Either V_k is on the same side of e_{ij} as P or not. If V_k is on the same side as P then due to the property that the point of intersection of e_{ij} with L is maximal, V_k must be on the opposite side of the line containing P and V_i from V_j and on the opposite of the edge containing P and V_j from V_i. This is the cross hatched region of Fig. 12. This implies that P lies in the triangle T_{ijk}. Now in the other case where V_k is on the side opposite P, Lemma 2.6 implies that there is a triple $ijn \in D_A$ and V_n is on the same side of e_{ij} as P. Similar arguments about the order of the intersection of the edges will imply that V_n must lie in the same region as above (the cross hatched region of Fig. 12) and so $P \in T_{ijn}$ and this completes the argument. ∎

Examples of A-norm, Dirichlet tessellations are shown in Fig. 13. Unit disks for the norms associated with each tessellation are shown in lower right. The triangulations dual to these tessellations are shown in Fig. 14.

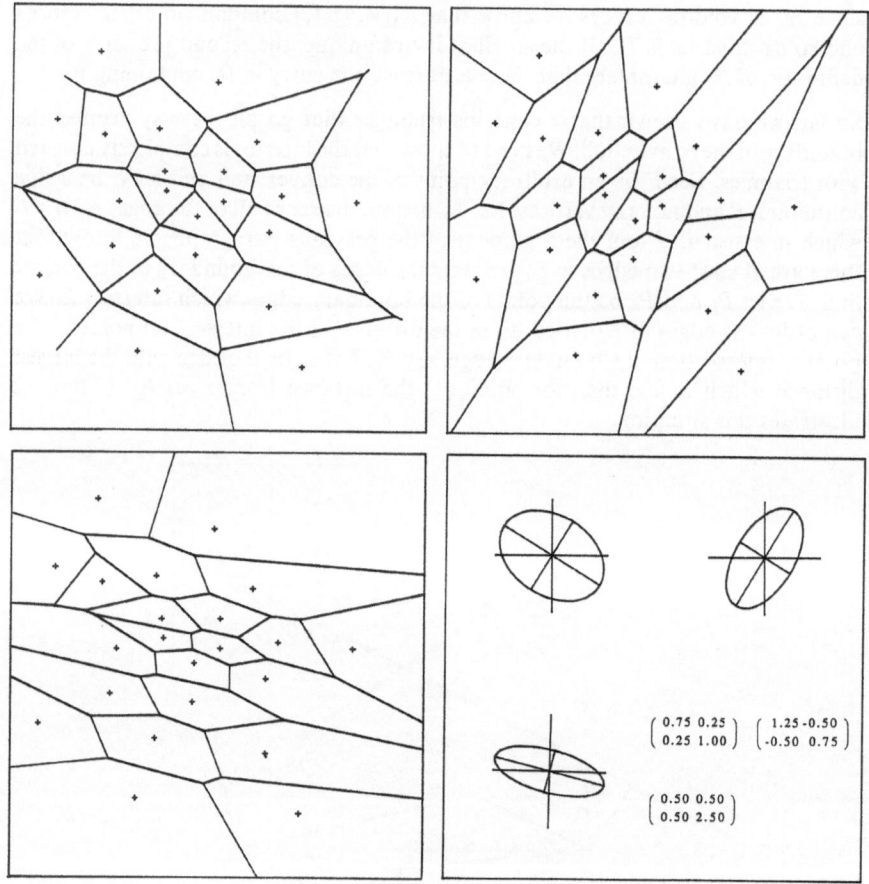

Figure 13. Some A-norm tessellations

3. Affine Invariant Triangulations

In a previous paper [20], we introduced the following innerproduct and associated norm:

$$(V, V)_{A(\vec{V})} = \|V\|^2_{A(\vec{V})} = (x, y) \begin{bmatrix} \dfrac{\Sigma_Y^2}{\Sigma_X^2 \Sigma_Y^2 - (\Sigma_{XY})^2} & \dfrac{-\Sigma_{XY}}{\Sigma_X^2 \Sigma_Y^2 - (\Sigma_{XY})^2} \\ \dfrac{-\Sigma_{XY}}{\Sigma_X^2 \Sigma_Y^2 - (\Sigma_{XY})^2} & \dfrac{\Sigma_X^2}{\Sigma_X^2 \Sigma_Y^2 - (\Sigma_{XY})^2} \end{bmatrix} \begin{pmatrix} x \\ y \end{pmatrix} \quad (1)$$

where

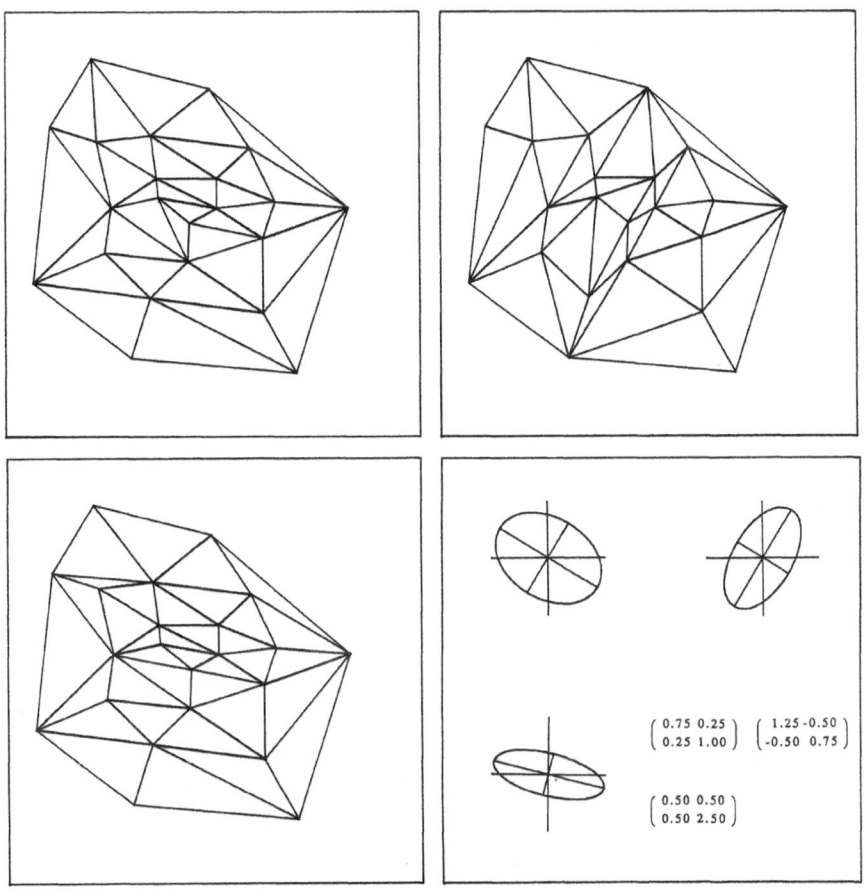

Figure 14. Dual triangulations

$$\Sigma_X^2 = \frac{\sum_{i=1}^n (x_i - \bar{x})^2}{n}, \qquad \bar{x} = \frac{\sum_{i=1}^n x_i}{n}$$

$$\Sigma_Y^2 = \frac{\sum_{i=1}^n (y_i - \bar{y})^2}{n}, \qquad \bar{y} = \frac{\sum_{i=1}^n y_i}{n}$$

and

$$\Sigma_{XY} = \frac{\sum_{i=1}^n (x_i - \bar{x})(y_i - \bar{y})}{n}$$

The fact that the vertices $V_i, i = 1, \ldots, n$ do not lie on a line is sufficient to guarantee that

$$\Sigma_X^2 \Sigma_Y^2 - (\Sigma_{XY})^2 \neq 0.$$

This norm has the property that it is invariant under affine transformations which means that

$$\|P - Q\|_{A(\vec{v})} = \|T(P) - T(Q)\|_{A(T(\vec{v}))}$$

for any two points P and Q and any affine transformation

$$T(V) = \begin{pmatrix} t_{11} & t_{12} \\ t_{21} & t_{22} \end{pmatrix} \begin{pmatrix} x \\ y \end{pmatrix} + \begin{pmatrix} c_1 \\ c_2 \end{pmatrix}.$$

Figure 15 further illustrates this property. Each of the data sets shown this figure are affine images of each other. Starting in the upper left and moving in a clockwise direction, the transformations are: counter clockwise rotation of 44°; a scaling in x by a factor of 2: a scaling in y by a factor of 0.4. The four ellipses in each figure represent points which are $\frac{1}{4}$, $\frac{1}{2}$, $\frac{3}{4}$ and 1 unit(s) from their center point.

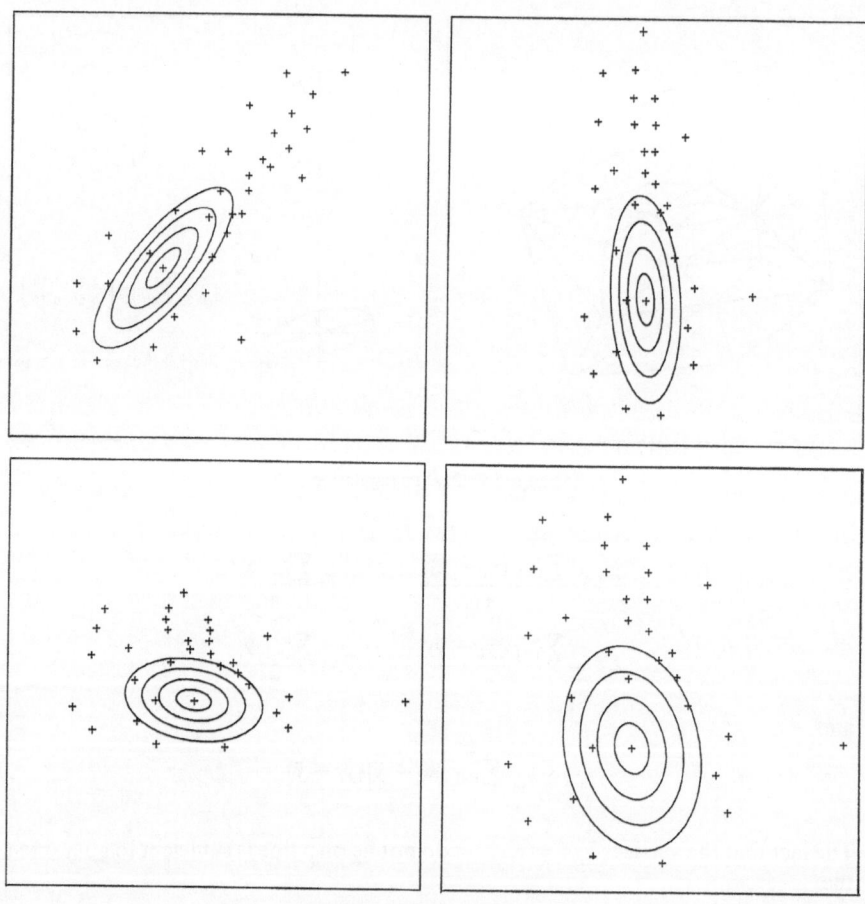

Figure 15. Affine invariant norm

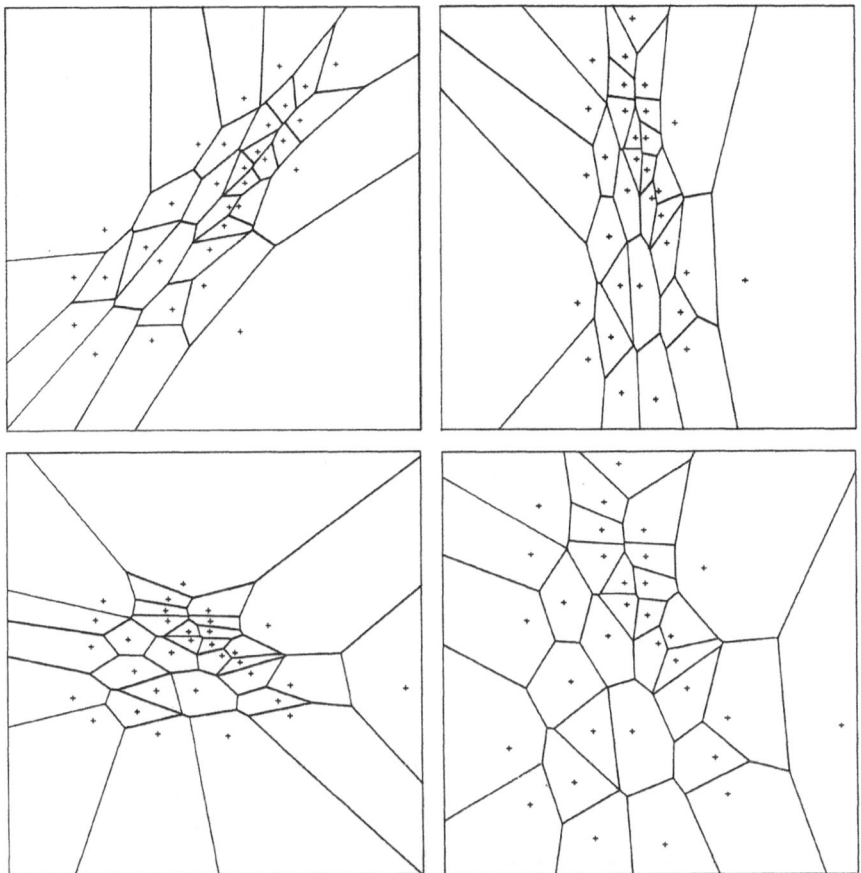

Figure 16. Tessellations based upon affine invariant metric

If we let $A(\vec{V})$ be the matrix of Eq. (1), then, based upon the notation of the previous section, $D_{A(\vec{V})}(\vec{V})$ represents the tessellation of the vertices when the norm is based upon $A(\vec{V})$. An example of the A-norm, Dirichlet tessellation in this case is shown in Fig. 16. The vertices are the same as in Fig. 15.

In the previous section, we proved that for any 2×2, positive definite matrix A, the list of triples $D_A(\vec{V})$ given by Definition 2.4 always represents a triangulation. Based upon the affine invariant properties of $A(\vec{V})$ of Eq. (1) we have the following result.

Theorem 3.1. *The triangulation represented by $D_{A(\vec{V})}(\vec{V})$ is invariant under affine transformations. That is*

$$D_{A(T(\vec{V}))}(T(\vec{V})) = D_{A(\vec{V})}(\vec{V})$$

for any affine transformation T.

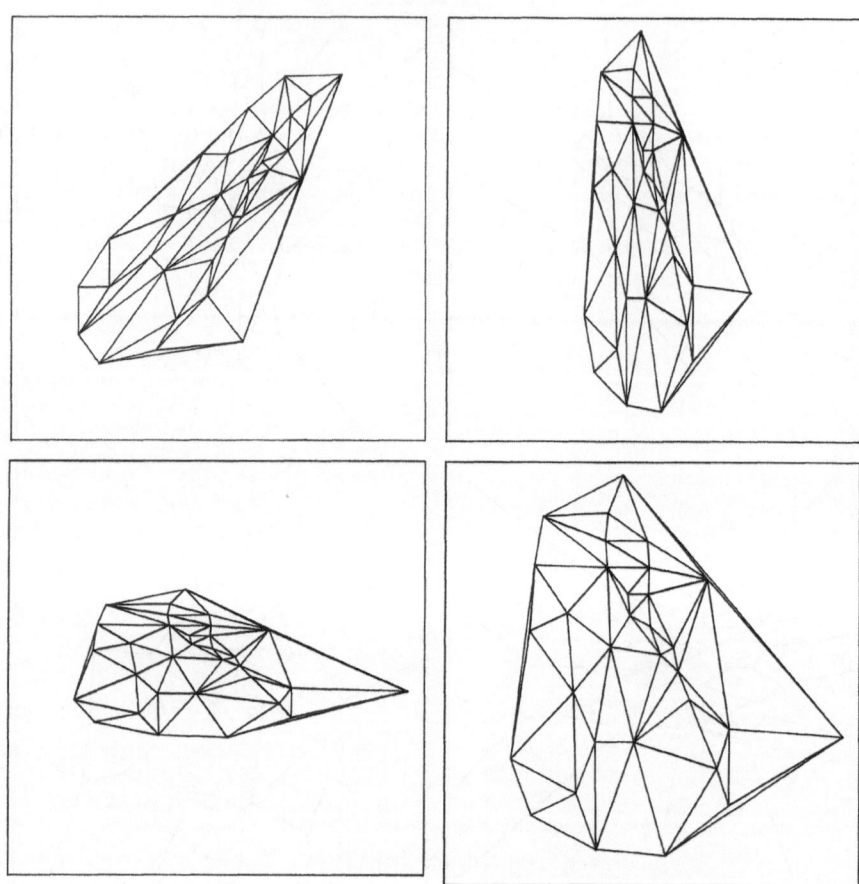

Figure 17. Triangulations from the tessellations of Fig. 16

In Fig. 17, we show the triangulations which are dual to the tessellations of Fig. 16. For comparison, we have included in Fig. 18 the triangulation of these vertices which results from the use of the standard Euclidean metric.

4. Remarks

It is general knowledge that the dual of the Dirichlet tessellation leads to a triangulation of the convex hull. A proof for points in E^2 and the Euclidean norm is given by Preperata and Shamos [23] and attributed to Delaunay [6]. A proof for E^n and the Euclidean norm (excluding the case of neutral points) was given many years ago by Rogers [24]. Both of these arguments use similar techniques to what we have used here for the more general norm. There seems to be no obstacles to generalizing our proof to the case of E^n. Also, since the affine invariant norm has an extension

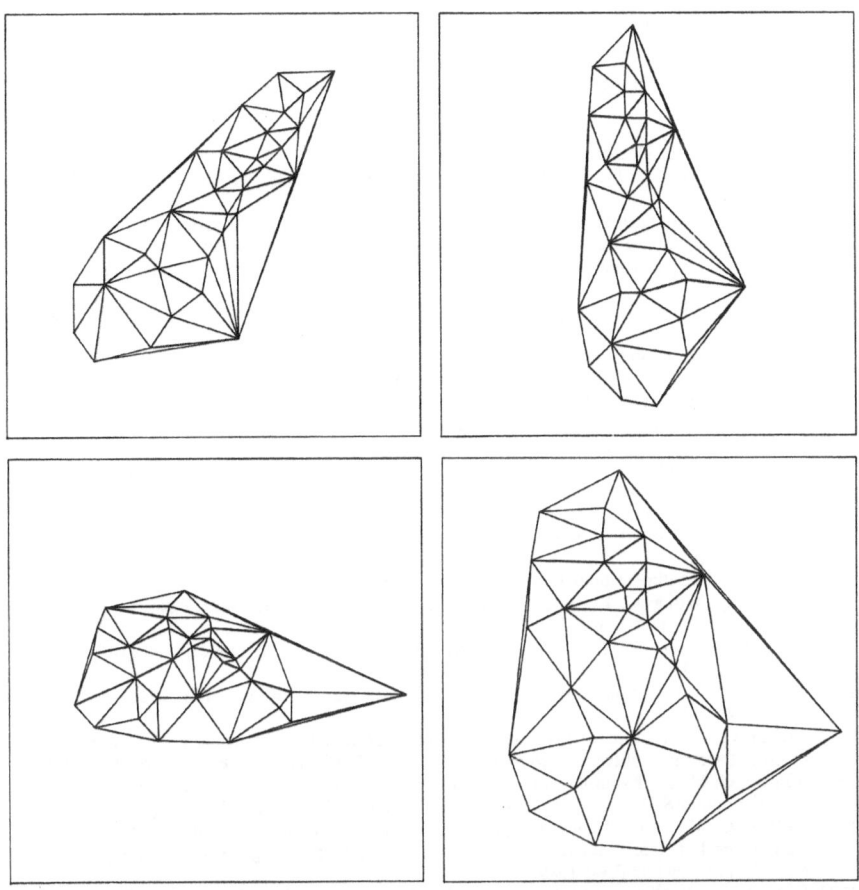

Figure 18. Triangulations based upon standard Euclidean distance

to E^n (c.f. Nielson [20]), an optimal affine invariant triangulation in E^n could be characterized. This topic will be taken up in more detail in a later paper.

Acknowledgments

Much of this work was accomplished while the author was a participatory guest scientist at Lawrence Livermore National Laboratory. The support of Fred Fritsch in this connection is greatly appreciated.

This research was supported by a grant from the North Atlantic Treaty Organization, NATO RG. 0097/88, and by a US Department of Energy research contract, DE-FG02-87ER25041, to Arizona State University.

References

[1] Akima, H.: A method of bivariate interpolation and smooth surface fitting for irregularly distributed data points. ACM Trans. Math. Software 4, 149–159 (1978).

[2] Barnhill, R. E., Little, F. F.: Three- and four-dimensional surfaces. Rocky Mt. J. Math. *14*, 77–102 (1984).

[3] Bowyer, A.: Computing Dirichlet tessellations. Computer J. *24*, 162–166 (1981).

[4] Brassel, K. E., Reif, D.: Procedure to generate Thiessen polygons. Geograph. Anal. *11*, 289–303 (1979).

[5] Cavendish, J. C.: Automatic triangulation of arbitrary planar domains for the finite element method. Int. J. for Numer. Methods in Engr. *8*, 679–696 (1974).

[6] Delaunay, B.: Sur la sphre vide. *Bull. Acad. Sci.* U.S.S.R.(VII), *Classe Sci. Mat. Nat.* Volume 7, 793–800 (1934).

[7] Forrest, R.: Computational Geometry. Proc. Royal Society London *321* Series 4, 187–195 (1971).

[8] Franke, R.: Scattered data interpolation: Tests of some methods. Math. Comp. *38*, 181–200 (1982).

[9] Gleue, J.: Triangulierung und Interpolation von im R^2 unregelmässig verteilten Daten. HMI B 357, 1981.

[10] Green, P. J., Sibson, R.: Computing Dirichlet tesselations in the plane. Computer J. *21*, 168–173 (1978).

[11] Gregory, J. A.: Error bounds for linear interpolation on triangles. In: Whiteman J. (ed.) The mathematics of finite elements and applications II, pp. 163–170. London: Academic Press 1975.

[12] Guibas, L., Stolfi, J.: Primitives for the manipulation of general subdivisions and the computation of Voronoi diagrams. ACM Trans. Graphics *4*, 74–123 (1985).

[13] Hermeline, F.: Triangulation automatique d'un polyedre in dimension n. RAIRO Anal. Numer. *76*, 211–242 (1982).

[14] Lawson, C. L.: Software for C^1 surface interpolation. In: Rice J. R. (ed.) Mathematical software III, pp. 161–194. New York: Academic Press 1977.

[15] Lawson, C. L.: Properties of n-dimensional triangulations. Computer Aided Geometric Design *3*, 231–246 (1986).

[16] Lee, D. T.: Two dimensional Voronoi diagram in the L_p-metric. J. ACM *27*, 604–618 (1980).

[17] Lee, D. T., Wong, C. K.: Voronoi diagrams in L_1 (L_∞) metrics with 2-dimensional storage applications. SIAM J. Comput. *9*, 200–211 (1980).

[18] Lewis, B. A., Robinson, J. S.: Triangulation of planar regions with applications. Computer J. *21*, 324–332 (1978).

[19] Nelson, J. M.: A triangulation algorithm for arbitrary planar domains. Appl. Math. Modelling *2*, 151–159 (1978).

[20] Nielson, G. M.: Coordinate free scattered data interpolation. In: Schumaker, L. L., Chui, C., Utreras F. (eds.) Topics in multivariate approximation, pp. 175–184. New York: Academic Press 1987.

[21] Nielson, G. M.: An Example with a local minimum for the MinMax ordering of triangulations, ASU Computer Science Department Technical Report TR87-014 (1987), 10 pages.

[22] Nielson, G. M., Foley, T.: A survey of applications of an affine invariant metric. In: Lyche, T., Schumaker, L. L. (eds.) Mathematical methods in computer aided geometric design, pp. 445–467. New York: Academic Press 1989.

[23] Preparata, F. P., Shamos, M. I.: Computational geometry: an introduction. New York: Springer 1985.

[24] Rogers, C. A.: Packing and Covering. Cambridge: Cambridge University Press, 1964.

[25] Rhynsburger, D.: Analytic delineation of Thiessen polygons. Geograph. Anal. *5*, 133–144 (1973).

[26] Schumaker, L. L.: Triangulation methods. In: Schumaker, L. L., Chui, C., Utreras F. (eds.) Topics in multivariate approximation, pp. 219–232. New York: Academic Press 1987.

[27] Sibson, R.: Locally equiangular triangulations. Computer J. *21*, 243–245 (1978).

[28] Shamos, M. I.: Computational geometry, Ph.D. Dissertation, Yale University, 1978.

[29] Thiessen, A. H.; Precipitation averages for large areas. Mon. Wea. Rev. *39*, 1032–1034 (1911).

[30] Voronoi, G.: Nouvelles applications des paramàtres continus à la théorie des formes quadratiques, Deuxième Mémoire, Recherches sur les parallélloèdres primitifs. J. Reine Angew. Math. *134*, 198–287 (1908).

[31] Watson, D. F.: Computing the n-dimensional Delaunay tessellation with application to Voronoi polytopes. Comp. J. *24*, 167–172 (1981).

[32] Watson, D. F., Philp, G. M.: Systematic triangulations. Computer Vision, Graphics, and Image Processing *26*, 217–223 (1984).

Gregory M. Nielson
Computer Science
Arizona State University
Tempe, Arizona 85287-5406, U.S.A.
nielson@asuvax.eas.asu.edu

Computing Suppl. 8, 211–226 (1993)

A Data Structure for Representing and Efficient Querying Large Scenes of Geometric Objects: MB^* Trees

H. Noltemeier[1], **K. Verbarg**[2], and **C. Zirkelbach**[3], Würzburg

Abstract. We are concerned with the problem of partitioning complex scenes of geometric objects in order to support the solutions of proximity problems in general metric spaces with an efficiently computable distance function. We present a data structure called *Monotone Bisector* Tree* (MB^* Tree), which can be regarded as a divisive hierarchical approach of centralized clustering methods (compare [3] and [10]). We analyze some structural properties showing that MB^* Trees are a proper tool for a general representation of proximity information in complex scenes of geometric objects.

Given a scene of n objects in d-dimensional space and some Minkowski-metric. We additionally demand a general position of the objects and that the distance between a point and an object of the scene can be computed in constant time. We show that a MB^* Tree with logarithmic height can be constructed in optimal $O(n \log n)$ time using $O(n)$ space. This statement still holds if we demand that the cluster radii, which appear on a path from the root down to a leaf, should generate a geometrically decreasing sequence.

We report on extensive experimental results which show that MB^* Trees support a large variety of proximity queries by a single data structure efficiently.

Key words: Proximity queries, geometric data structures, clustering, computational geometry.

1. Introduction

The appropriate and efficient representation of proximity information in large sets of objects is a crucial problem in a wide range of applications [11]. We will not try to give an account of this field; we will mention only some examples directly related to our results. For instance consider a large scene of objects S (as points, line segments, polygons, ...) in the plane and a distance function d. Now, we want to move an arbitrary object o within this scene avoiding collisions with the objects of the scene. If the scene is huge, it is useful to find at first all objects which might become dangerous to o and hand them over to the motion-planning algorithm. So, we need a data structure and an algorithm to retrieve the set

---

[1] e-mail: noltemei@informatik.uni-wuerzburg.dbp.de
[2] e-mail: verbarg@informatik.uni-wuerzburg.dbp.de
[3] e-mail: zirkelba@informatik.uni-wuerzburg.dbp.de
Lehrstuhl für Informatik I, Universität Würzburg, Am Hubland, 8700 Würzburg, Germany, Fax: +49 931 888 4600.
This work was supported by the Deutsche Forschungsgemeinschaft (DFG) under contract (No 88/6-4).

$$\{s \in S \,|\, d(o, s) \leq MAXDIST\}$$

efficiently, where $MAXDIST$ denotes a suitable limitation of proximity.

One approach is the use of cluster centers [3]: given an arbitrary space E, a distance function $d: E \times E \to \mathbb{R}_+^0$ satisfying the triangle inequality and a finite set $S \subset E$ of objects. After selecting a subset $E' \subset E$ with $k \geq 2$ elements, each $s \in S$ is related to its *cluster center* i.e. its nearest neighbor in E'. Thus, the partition is carried out by the (generalized) Voronoi-diagram [12] of the cluster centers. This provides a partitioning method which is sensitive to the underlying distance function d, because any separation sheet is given by the bisector of two cluster centers. This separation scheme, of course, is more flexible than the separation by hyperplanes which are parallel to the axis, which is the concept of the k-d-tree and the Quadtree [1] [13] [9] and thus, for example, it is possible to separate natural groupings. In general, a higher degree of flexibility in choosing the separation sheet allows to carry out the separation more sensitive to the spatial position of the objects. This is also the idea of the polygon-tree [14] and the cell-tree [5], however those and related data structures don't take care of the underlying metric. With this intention, for every cluster center $e \in E'$ let

$$RADIUS(e) := \max\{d(e, s) \,|\, s \in S \text{ and } e \text{ is the cluster center of } s\}$$

denote the corresponding *cluster radius*. The recursive use of this partitioning method generates a k-nary tree which is suitable for representing proximity properties by means of the cluster centers and the corresponding radii [6]. For example, if we search for the objects in the neighborhood of $p \in E$ in S, the search process can prune the subtree of the actual cluster center e if $d(p, e) - RADIUS(e) \geq MAXDIST$ holds. On the other hand, the whole subtree can be accepted, if $d(p, e) + RADIUS(e) \leq MAXDIST$ holds. If we search for the nearest neighbor of $p \in E$ in S the search process can prune the subtree of the actual cluster center e if $d(p, e) - RADIUS(e) \geq DACTUAL$ holds with $DACTUAL$ denoting the distance of p to its actual nearest neighbor in S. The first example can be regarded as an absolute neighborhood query and the latter one as a relative neighborhood query.

Here we notice, if a cluster center represents a cluster element, then the cluster radius estimates the loss of spatial information. Therefore, the generation of representative cluster centers (cluster centers with small radii) is a general goal of this clustering technique.

Early attempts date back to Kalantari and McDonald [7]. They use a data structure, called *bisector tree*, which can be regarded as a straightforward generalization of binary search trees (which support nearest neighbor search in $\mathbb{R}^1$) to normed spaces. Unfortunately, those early approaches failed in the following **two criteria** which are necessary to support proximity queries efficiently:

1. The radii of the elements stored in nodes which appear on a path from the root down to a leaf should generate a monotonously decreasing sequence. So, we do not permit *eccentric sons* i.e. successors in the tree that have larger radii than the current center.
2. The height of the tree should be $O(\log|S|)$ to achieve small subtrees.

Now, this paper is organized as follows. To overcome the disadvantages caused by eccentric sons, we present a new type of the bisector tree—the so-called *Monotone Bisector Tree* (*MB* Tree)—for partitioning large sets of points in section 2. Afterwards, in section 3, we generalize this approach to complex scenes of "higher" geometric objects introducing *MB* Trees*. In section 4, we summarize experimental results and finally, in section 5 point out goals of further research.

2. *MB* Trees

Let E be an arbitrary space with distance function d and $S \subset E$ a finite set of n points with $s_1 \in S$. Now, a *Monotone Bisector Tree* $MBT(S, s_1)$ is a binary tree having the following features:

$1 \leq |S| \leq 2$: $MBT(S, s_1)$ consists of a single node containing the elements of S.
$\quad |S| > 2$: The root w of $MBT(S, s_1)$ contains s_1 and a point $s_2 \in S \setminus \{s_1\}$.
$\qquad$ The subtrees of w are $MBT(S_1, s_1)$ and $MBT(S_2, s_2)$ with

$$\{s \in S | d(s, s_i) < d(s, s_j)\} \subseteq S_i \subseteq \{s \in S | d(s, s_i) \leq d(s, s_j)\}$$

where $i \neq j$, $S_1 \cup S_2 = S$ and $S_1 \cap S_2 = \varnothing$.

Now the construction of the tree in its basic steps is as follows: Starting with a given element $s_1 \in S$ we recursively choose a second splitting element $s_2 \in S$. Then we relate each $s \in S$ to its cluster center s_1 or s_2, respectively, which provides a partition in the sense given above. These two cluster centers are reached down to their corresponding subtrees. So, each node (except of the root) contains exactly one redundant element—which preserves linear storage (see Fig. 1 and Fig. 2, which show two partitions in their first few steps under different metrics). To achieve a good start, the first cluster center should be chosen carefully, for instance in the

Figure 1. Monotone Bisector Tree of a set of 400 points under L_1-metric; Height: 3

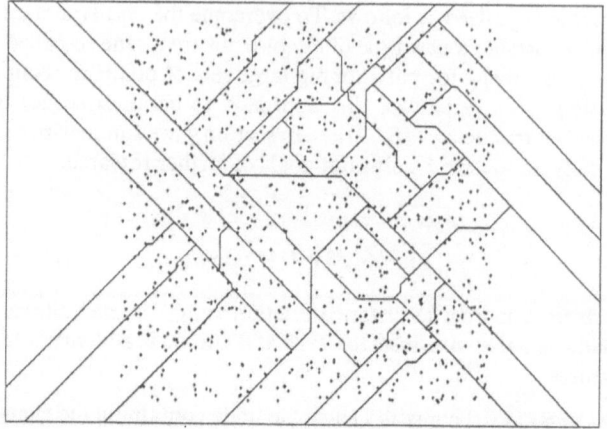

Figure 2. Monotone Bisector Tree of a set of 1000 points under L_∞-metric; Height: 4

euclidean case the nearest neighbor of the center of the smallest enclosing circle which can be computed in linear time [8].

The proposed method removes the disadvantages of bisector trees by an adaptive sequential insertion of new cluster points in those areas, where the greatest error (with respect to the two criteria demanded in the introduction) occurs. If we choose those splitting elements carefully, we gain a partition of the scene which is sensitive to the spatial position of the elements in S and to the distance function d, because the partition is carried out by the bisector of the two cluster centers of a node. This is the basic idea which enables us to prove the following theorem.

Theorem 1

1. *MB Trees do not have eccentric sons. The radii of the elements stored in nodes which appear on a path from the root to a leaf generate a monotonously decreasing sequence (justifying the term:* monotone *tree).*
2. *Let $S \subset \mathbb{R}^d$ be a finite set of n points. For any L_p-metric $(1 \le p \le \infty)$ a MB Tree with logarithmic height can be constructed in optimal $O(n \log n)$ time and $O(n)$ storage.*
3. *Let $S \subset \mathbb{R}^2$ be a finite set of n points. For any* convex distance function[1] *a MB Tree with logarithmic height can be constructed in optimal $O(n \log n)$ time using $O(n)$ storage.*

The last two statements still hold if we demand that the radii of clusters, which appear on a path from the root down to a leaf, should generate a geometrically decreasing sequence.

---

[1] A distance function based on an expanding convex shape is called *convex distance function* and was first defined by Minkowski. Indeed, in some references it is called the *Minkowski distance function* [2].

We sketch the proof in appendix A. The complete proof can be found in [15].

However, there are simple examples demonstrating, that it is impossible to generate balanced Monotone Bisector Trees for a set of "higher" objects (as line segments, polygons, ...). That leads to a generalization of this data structure in the following section.

3. *MB** Trees

Let $S \neq \emptyset$ be a finite set of objects, $E \neq \emptyset$ a (not necessarily finite) set of *splitting elements* with $e_1 \in E$ and $d: E \times S \rightarrow \mathbb{R}_+^0$ a distance function. A *Monotone Bisector* Tree* $MBT^*(S, e_1)$ is a binary tree having the following properties:

$1 \leq |S| \leq 2$: $MBT^*(S, e_1)$ consists of a single node containing the elements of S.
$\quad |S| > 2$: The root w of $MBT^*(S, e_1)$ contains e_1 and a point $e_2 \in E \backslash \{e_1\}$.
$\qquad$ The subtrees of w are $MBT^*(S_1, e_1)$ and $MBT^*(S_2, e_2)$ with

$$\{s \in S | d(e_i, s) < d(e_j, s)\} \subseteq S_i \subseteq \{s \in S | d(e_i, s) \leq d(e_j, s)\}$$

$$\text{where } i \neq j, S_1 \neq \emptyset, S_2 \neq \emptyset, S_1 \cup S_2 = S \text{ and } S_1 \cap S_2 = \emptyset.$$

Here, we apply our idea by reaching down the cluster centers to their corresponding subtrees. Notice, that a *MB* Tree is a *MB** Tree where $S = E$. Now we select the cluster centers from the set E of splitting elements and store the objects of S in the buckets of the tree (see Fig. 3 for an example).

In addition to theorem 1, we can prove

Figure 3. Monotone Bisector* Tree of a set of 300 points, line segments and polygons under L_2-metric (cluster centers are marked by a ◆); Height: 3

Theorem 2

1. *MB* Trees do not have eccentric sons. The radii of the elements stored in nodes which appear on a path from the root to a leaf generate a monotonously decreasing sequence.*
2. *Let S be a finite set of n convex objects in d-dimensional space and $E = \mathbb{R}^d$ the set of splitting elements. We demand that the distance between a point and an object of S can be computed in constant time.*
   (a) *For any* symmetrical convex distance function *(for instance any L_p-Metric), a MB* Tree with logarithmic height and geometrically decreasing radii can be constructed in optimal $O(n \log n)$ time and $O(n)$ storage.*
   (b) *If we additionally demand that at most k objects of S intersect any common hyperplane, the objects of the scene S need not to be convex to hold* (a).

For the proof we refer to [16]. In addition we should remark, that under the conditions of statement 2 the quality of balance doesn't depend on the dimension.

4. Applications and Experimental Results

We have implemented both variants of the Monotone Tree with success. We ran the performance tests on SUN workstations under UNIX using C implementations. To show the practicability of our algorithms we have examined six different types of scenes illustrated in Fig. 4. While A-E are artificial data the data of F are of natural origin, received from the "Institut für Astronomie, Universität Würzburg". The objects of the artificial scenes are points, line segments, circles and convex polygons with bounded size. The overlap (the summarized area of the objects divided by the area of the whole scene) is 0.06. In the case of F we have only point data.

A shows a scene of disjoint objects under uniform distribution.
B shows a scene of objects under uniform distribution.
C shows a scene of objects distributed along a sinus curve.
D shows a scene of objects distributed along a diagonal line.
E shows a scene of objects distributed to seven clusters.
F shows a picture of two galaxies (10000 data points).

For every type of scenes (except F) we have run experiments where the number of objects ranged between some hundred and 20000 (compare appendix B). Using $O(n \log n)$ preprocessing time we achieved logarithmic height. In any case the height of the tree was by factor 2 within the lower bound. We should remark that the lower bound can be achieved, if we don't care about the development of the cluster radii.

The general concept of information retrieval, as described in the introduction, enables us to support a large variety of proximity queries by a single data structure. Our experiments show this for:

- nearest-neighbor queries
- fixed-radius-near neighbor-queries

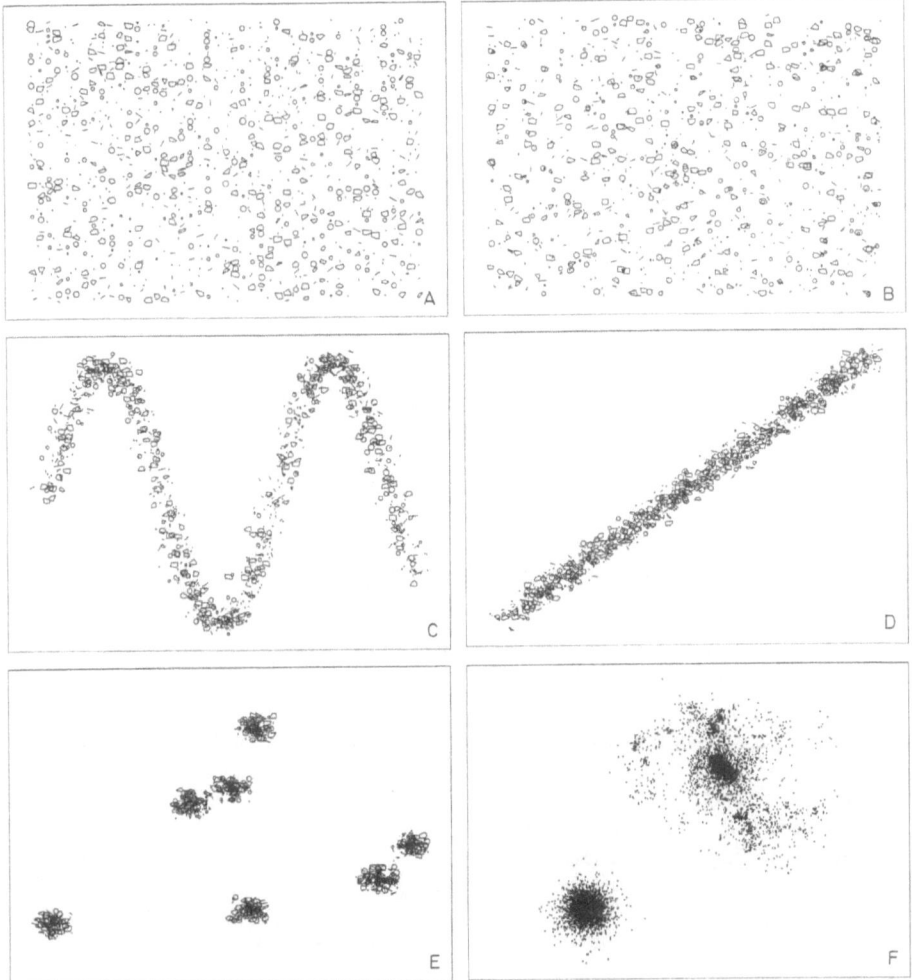

Figure 4.

- ray-shooting queries
- range queries
- points/objects in polygon retrieval
- objects hitting polygon retrieval
- objects hitting curve retrieval
- hidden-line/surface queries
- special problems of motion planning

Performing the list of proximity queries we achieve an average query time of $\log n + k$ where k is the size of the output. So we found out in general that the query

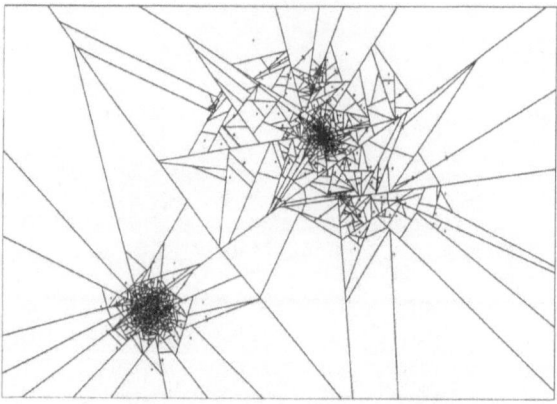

Figure 5. Monotone Bisector* Tree of scene F

cost is the complexity of the output if the size of the output is large with respect to the height of the tree. In appendix C we demonstrate this result by performing a constant set of 500 queries for each of the following types:

- nearest-neighbor query
  The query-points are randomly distributed in the scene.
- fixed-radius-near-neighbor query
  The query-points are randomly distributed and the radii are randomly chosen in the range from 0 to $\frac{1}{5}$ of the diameter of the scene. The whole query-scene is shown in Fig. 6.
- visibility query
  We used a set of unlimited cones.
- objects hitting curve retrieval

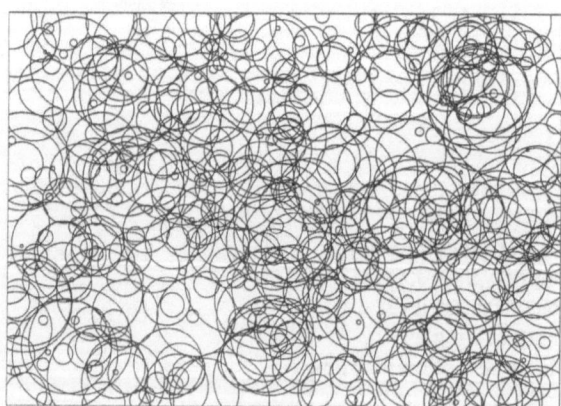

Figure 6.

The curves consist of five edges, where the length of each edge is limited by $\frac{1}{10}$ of the diameter of the scene. The whole query-scene is shown in Fig. 7.

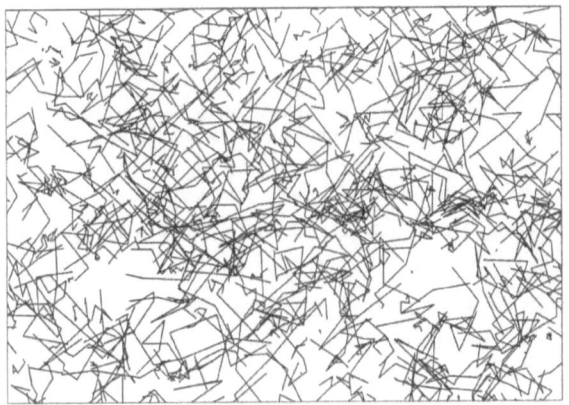

Figure 7.

In summary the expansive experiments indicate that by the single data structure MB^* Tree we cannot only represent complex scenes of objects in a comprehensive way but additionally can support a large variety of types of queries in a very efficient way.

5. Further Directions of Research

At next a more rigorous **analysis** of certain kinds of queries will follow. The efficiency of our data structure in a scene with a high degree of overlapping objects is still unsettled today. Performance tests in this field are one goal of experimental research.

In near future, we will turn our attention to further applications: imagine that an objects is very complex and the computation of its position is expensive. If we are able to find a cheap estimation our concept still works, if we define the cluster radius as follows:

$$RADIUS_\varepsilon(e) := RADIUS(e) + \varepsilon,$$

where ε denotes the maximum error that occurs in the cluster. If ε is small with respect to the whole scene, this idea enables us to work efficiently under uncertainties. We also will set the decomposition of large problems to our work with applications in image analysis and image understanding.

Appendix A

We sketch the proof of theorem 1 in chapter 2 in a sequence of lemmas and refer to the exhaustive proof in [15].

Theorem 1

1. *MB Trees do not have eccentric sons. The radii of the elements stored in nodes which appear on a path from the root to a leaf generate a monotonously decreasing sequence.*
2. *Let $S \subset \mathbb{R}^d$ be a finite set of n points. For any L_p-metric $(1 \le p \le \infty)$ a MB Tree with logarithmic height can be constructed in optimal $O(n \log n)$ time and $O(n)$ storage.*
3. *Let $S \subset \mathbb{R}^2$ be a finite set of n points. For any convex distance function a MB Tree with logarithmic height can be constructed in optimal $O(n \log n)$ time using $O(n)$ storage.*

Proof: The first statement is immediately implied by the structure of the tree. The goal to achieve logarithmic height using $O(n \log n)$ preprocessing time can be reduced to the following

Problem: Given a finite set S of points in $\mathbb{R}^d$. Find in $O(|S|)$ time two points $a, b \in S$ such that each of the two half-spaces defined by the bisector of a and b contains at least a constant fraction of S.

Lemma 1.1: *For any given convex distance function d, all possible Voronoi regions are star-shaped.*

The proof can be found in [2].

Let $\|b - a\|$ denote the euclidean distance of $a \in \mathbb{R}^d$ and $b \in \mathbb{R}^d$.

For $p \in \mathbb{R}^d$ and $\delta \ge 0$ we denote the ball $K_\delta(p) := \{x \in \mathbb{R}^d | \|p - x\| \le \delta\}$.

For $a, b \in \mathbb{R}^d$ and a convex distance function d let $H(a, b) := \{x \in M | d(a, x) \le d(b, x)\}$ denote the half-space associated to the point a, given by the bisector of a and b.

Lemma 1.2: *For every convex distance function d there exists an $\varepsilon > 0$ such that for every $a, b \in \mathbb{R}^d$ $K_{\varepsilon \|a-b\|}(a) \subset H(a, b)$ holds.*

This depends on the fact that in any convex distance function the speed has an upper and a lower non-negative bound. $\left(\text{If } \lambda_1 \text{ denotes its infimum and } \lambda_2 \text{ its supremum we can choose } \varepsilon := \varepsilon(d) := \dfrac{\lambda_1}{\lambda_1 + \lambda_2}\right)$. Let in the following denote $\alpha := \alpha(d) := \arcsin(\varepsilon)$.

The combination of lemma 1.1 and lemma 1.2 implies immediately

Lemma 1.3: *For every $a, b, x \in \mathbb{R}^d$ the following statement holds:*

$$\{\lambda x + (1 - \lambda)b | 0 \le \lambda \le 1\} \cap K_{\varepsilon \|a-b\|}(a) \ne \varnothing \quad \Rightarrow \quad x \in H(a, b)$$

Let $\angle(v_1, v_2)$ denote the *(positive) angle given by the two vectors $v_1 \in \mathbb{R}^d$ and $v_2 \in \mathbb{R}^d$. For $a \in \mathbb{R}^d$, $v \in \mathbb{R}^d$ $(\|v\| = 1)$ and $0 \le \beta \le \pi$ we define the cone*

$$W(a, v, \beta) := \left\{x \in \mathbb{R}^d | \angle(v, x - a) \le \frac{\beta}{2}\right\}$$

and the double-cone

$$DW(a, v, \beta) := W(a, v, \beta) \cup W(a, -v, \beta).$$

Applying lemma 1.3 (see Fig. A-1) we gain

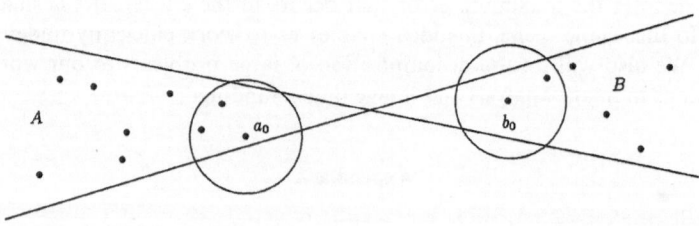

Figure A-1

Lemma 1.4: *Given a double-cone $DW(p, v, \alpha) \subset \mathbb{R}^d$ and two finite point sets $A, B \subset \mathbb{R}^d$ where*

$$A \subset W(p, v, \alpha) \qquad and \qquad B \subset W(p, -v, \alpha).$$

If we choose $a_0 \in A$ and $b_0 \in B$ such that

$$a_0 v = \min\{av | a \in A\}$$
$$b_0 v = \max\{bv | b \in B\}$$

then $A \subset H(a_0, b_0)$ and $B \subset H(b_0, a_0)$ holds.

In the remainder of the proof we still have to show that it is possible to construct in linear time a double-cone with angle $\leq \alpha$ where each cone contains at least a constant fraction of S. In the plane this is possible for every $0 < \alpha \leq \pi$. Up to now we solve this problem in $\mathbb{R}^d$ only for $\alpha \geq \dfrac{\pi}{2}$ which reduces the generality of convex distance functions to the case of L_p-metrics. We investigate at first the plane case (statement 3):

Lemma 1.5: *Given a double-wedge $DW(p, v, \beta) \subset \mathbb{R}^2$ and two finite sets $A, B \subset \mathbb{R}^2$ where*

$$A \subset W(p, v, \beta) \qquad and \qquad B \subset W(p, -v, \beta)$$

then we find in $O(|A| + |B|)$ time a double-wedge $DW(p', v', \beta/2)$ where

$$|W(p', v', \beta/2) \cap A| \geq \lceil |A|/2 \rceil \qquad and \qquad |W(p', -v', \beta/2) \cap B| \geq \lceil |B|/2 \rceil.$$

Let $DW(p, v, \beta)$ be given by the two lines g_1 and g_2 (see Fig. A-2). We apply a median-cut to the set A using an oriented line g with direction v. Now the double-wedge defined by g_1 and g or the double wedge defined by g_2 and g is a solution.

In a first step we separate the set S by the vertical line which is given by the median of the first component. Thus we achieve a double-wedge with angle π. The recursive application of lemma 1.5 to the resulting double-wedges will provide a double-wedge with angle $\leq \alpha$ where each wedge contains at least a constant fraction of S. Due to the fact that the median can be found in linear time we solve this problem in $O(n) + O\left(\dfrac{n}{2}\right) + O\left(\dfrac{n}{4}\right) + \cdots = O(n)$ time. This completes the proof of the plane case (statement 3).

Now we turn our attention to the higher-dimensional case (statement 2). Let $\|b - a\|_p$ $(1 \leq p \leq \infty)$ denote the L_p-distance of $a \in \mathbb{R}^d$ and $b \in \mathbb{R}^d$. With the help of the *Binomial Formula* we can show:

Lemma 1.6: *Let $a, b, c \in \mathbb{R}^d$ with $\|c - a\|_p \geq \|c - b\|_p$. Then each*

$$x \in Q := \{(x_1, \ldots, x_d)^T \in \mathbb{R}^d | a_i < b_i \Rightarrow x_i \geq c_i \text{ and } a_i > b_i \Rightarrow x_i \leq c_i\}$$

has the following property: $\|x - a\|_p \geq \|x - b\|_p$.

Lemma 1.7: *Let $S \subset \mathbb{R}^d$ $(d \in \mathbb{N}_+)$ be a set of n points. Then we can find in $O(n)$ time three points $p^1, p^2, p^3 \in \mathbb{R}^d$ and two sets of indices $I_1, I_2 \subset \{1, \ldots, d\}$, such that*

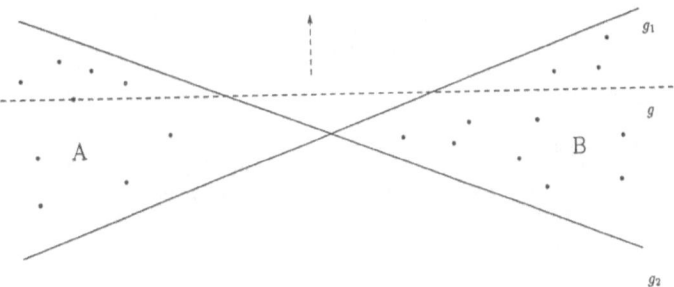

Figure A-2

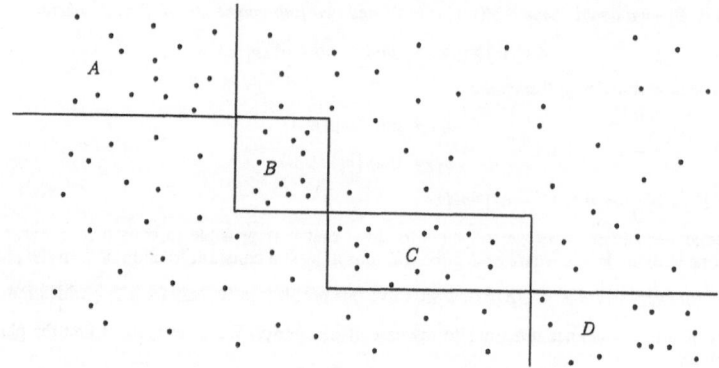

Figure A-3

$$I_1 \cup I_2 = \{1,\dots,d\}, \qquad I_1 \cap I_2 = \varnothing$$
$$p_i^1 \le p_i^2 \le p_i^3 \qquad \textit{if} \qquad i \in I_1$$
$$p_j^1 \ge p_j^2 \ge p_j^3 \qquad \textit{if} \qquad j \in I_2$$

and each set

$$A := \{x \in S \mid x_i \le p_i^1 \text{ if } i \in I_1 \text{ and } x_j \ge p_j^1 \text{ if } j \in I_2\}$$
$$B := \{x \in S \mid p_i^1 \le x_i \le p_i^2 \text{ if } i \in I_1 \text{ and } p_j^2 \le x_j \le p_j^1 \text{ if } j \in I_2\}$$
$$C := \{x \in S \mid p_i^2 \le x_i \le p_i^3 \text{ if } i \in I_1 \text{ and } p_j^3 \le x_j \le p_j^2 \text{ if } j \in I_2\}$$
$$D := \{x \in S \mid x_i \ge p_i^3 \text{ if } i \in I_1 \text{ and } x_j \le p_j^3 \text{ if } j \in I_2\}$$

contains at least a constant fraction of S (see Fig. A-3).

For the exhaustive prove of this lemma which is an improvement of Heusinger's lemma 3.3 in [6] we refer to [15]. If we choose $a \in B$ and $b \in C$ lemma 1.6 guarantees that $A \subset H(a,b)$ and $D \subset H(b,a)$ holds, which completes the prove of the second statement.

Due to the fact that in any convex distance function the unit sphere can be covered by a constant number of equal sized smaller spheres it is possible to reduce a cluster radius by a factor $0 < q < 1$ after a constant number of k (depending on q and the dimension) splitting steps. The alternating application of balancing and reducing the cluster radii doesn't touch the complexities and provides a tree with logarithmic height and a development of the cluster radii in the sense of a geometrically decreasing sequence.

Appendix B

The results of our tree-constructing algorithm are as follows:

Figure B-1 shows the height of the trees and the preprocessing time versus the number of objects. Using $O(n \log n)$ preprocessing we achieve logarithmic height (compare the dashed lines). In the case of scene D, the algorithm is able to make use of the one-dimensional structure of this scene, which reduces the preprocessing costs by a constant factor.

Figure B-2 gives an example of the development of the clusterradii in scene F. The two solid lines show the maximum and the average cluster radii versus the level in the tree. The crosses mark the average balance versus the level in the tree. The balance b for a single node is defined as $b := \dfrac{|l - r|}{l + r}$, where l is

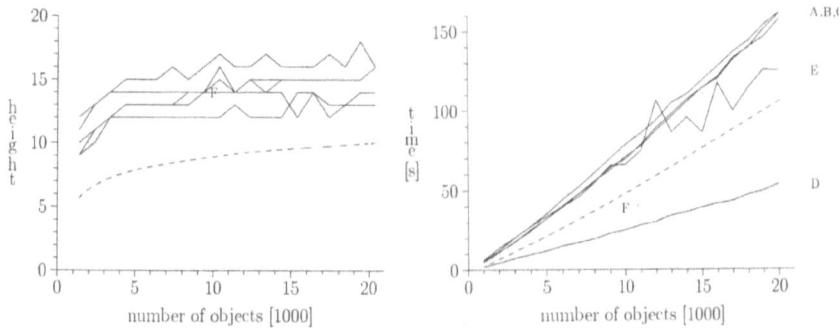

Figure B-1

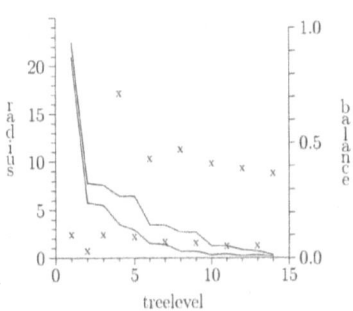

Figure B-2

the number of objects in the left subtree and r the number of objects in the right subtree. We notice a trade-off between the decrease of the radii and the quality of the balance. That depends on the algorithm which satisfies in each splitting-step at least one of the two criteria. To satisfy both criteria the algorithm is forced to be adaptive to the spatial character of the scene (compare Figs. 1, 2, 3 with Fig. 5).

Appendix C

We have performed a constant set of 500 queries for each of the following types:

- nearest-neighbor query (Fig. C-1)
  The query-points are randomly distributed in the scene.
- fixed-radius-near-neighbor query (Fig. C-2)
  The query-points are randomly distributed and the radii are randomly chosen in the range from 0 to $\frac{1}{3}$ of the diameter of the scene.
- visibility query (Fig. C-3)
  We used a set of unlimited cones.
- objects hitting curve retrieval (Fig. C-4)
  The curves consist of five edges, where the length of each edge is limited by $\frac{1}{10}$ of the diameter of the scene.

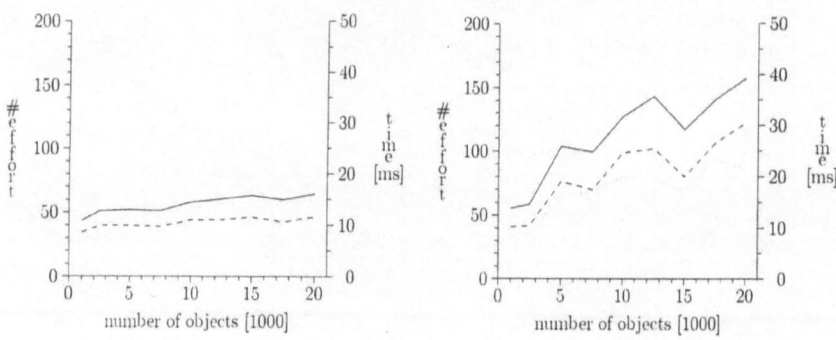

Figure C-1

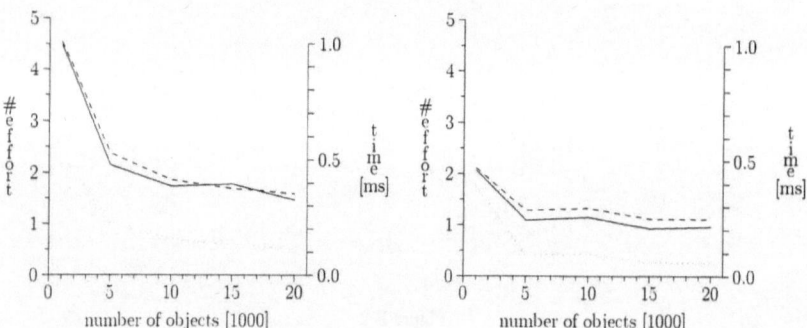

Figure C-2

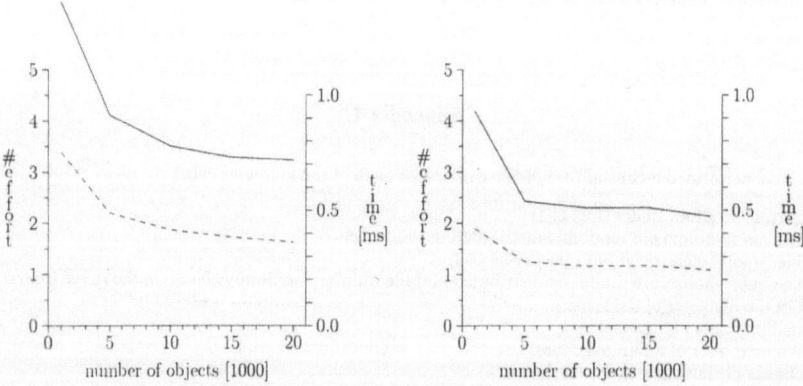

Figure C-3

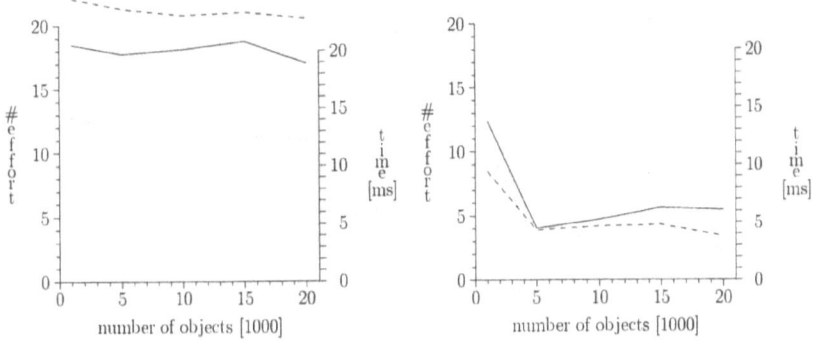

Figure C-4

The average results for each query are shown on the left series for scene A and on the right series for scene E. The solid curve reports the quotient of the running time and the number of the reported objects. To achieve a performance measure which is independent of the hardware we also used the number of the touched nodes (dotted curve) and the number of the touched objects (dashed curve) instead of the running time. So we can say, roughly speaking, that the curves tell us the amortised costs of each reported object.

References

[1] Bentley, J. L.: Multidimensional binary search trees used for associative searching. Communications of the ACM *18* (9), 509–517 (1975).

[2] Chew, L. P., Drysdale III, R. L.: Voronoi diagrams based on convex distance functions. 1st ACM Symposium on Computational Geometry, Baltimore, Maryland, 1985.

[3] Dehne, F., Noltemeier, H.: A computational geometry approach to clustering problems. Proceedings of the 1st ACM Symposium on Computational Geometry, Baltimore, Maryland, 1985.

[4] Edelsbrunner, H.: Algorithms in combinatorial geometry. Berlin, Heidelberg: Springer 1987 (EATCS Monographs in Computer Science, 10).

[5] Günther, O.: Efficient structures for geometric data management. (ed. G. Goos, J. Hartmanis). Berlin, Heidelberg: Springer 1988 (LNCS 337).

[6] Heusinger, H.: Clusterverfahren für Mengen geometrischer Objekte. Report, Universität Würzburg, 1989.

[7] Kalantari, I., McDonald G.: A data structure and an algorithm for the nearest point problem. IEEE Transactions on Software Engineering *SE-9* (5), 446–454 (1983).

[8] Megiddo, N.: Linear-time algorithms for linear programming in $\mathbb{R}^3$ and related problems. SIAM Journal on Comput. *12*, 759–776 (1983).

[9] Murtagh, F.: A survey of recent advances in hierarchical clustering algorithms. The Computer Journal *26* (4), 354–359 (1983).

[10] Noltemeier, H.: Voronoi trees and applications. In: H. Imai (ed.) Discrete algorithms and complexity. (Proceedings), Fukuoka/Japan, 1989.

[11] Noltemeier, H.: Layout of flexible manufacturing systems-selected problems. DIMACS—Workshop on Applications of Combinatorial Optimization in Science and Technology (COST), New Brunswick, New Jersey, 1991.

[12] Preparata, F. P., Shamos, M. I.: Computational geometry—an introduction. New York: Springer 1985.

[13] Samet, H.: The quadtree and related hierarchical data structures. ACM Computing Surveys *16*, 187–260 (1984).

[14] Willard, D. E.: Polygon retrieval. SIAM J. Comput. *11* (1), 149–165 (1982).

[15] Zirkelbach, C.: Partitionierung mit Bisektoren. Techn. Report, Universität Würzburg, 1990.
[16] Zirkelbach, C.: Monotone Bisektor* Bäume unter Minkowski-Metrik. Techn. Report, Universität Würzburg, 1991.

Prof. Dr. H. Noltemeier
Lehrstuhl für Informatik I
Universität Würzburg
Am Hubland,
D-W-8700 Würzburg,
Federal Republic of Germany

Computing Suppl. 8, 227–239 (1993)

Computing
© Springer-Verlag 1993

Properties of Local Coordinates Based on Dirichlet Tessellations

B. Piper, New York

Abstract. Local coordinates based on the Dirichlet tessellation (Voronoi diagram) provide a means to express a point as a linear combination of certain fixed points by using ratios of areas (or volumes) of certain regions. We add insight into the structure of the local coordinates by proving some of their basic properties by deriving formulas for their gradients from some simple geometry and proving the properties of smoothness and linear precision.

Key words: Dirichlet tessellation, Voronoi diagram.

1. Introduction

Local coordinates based on the Dirichlet tessellation (Voronoi diagram) were introduced in [Sibson 1981] where they were used to form C^0 and C^1 interpolants to scattered data in $\mathscr{R}^s$. More recently, in [Farin 1990], further properties of the local coordinates were explored and they were used them to construct a new C^1 interpolant in $\mathscr{R}^2$. In the present paper, we add insight into the structure of the local coordinates by proving some of their basic properties and deriving their derivatives from some simple geometry. We begin by reviewing the relevant definitions.

Let $\mathscr{P} = \{\mathbf{p}_j\}_{j=1}^n$ be a set of distinct points in $\mathscr{R}^s$ and let

$$\mathscr{T}_k = \{\mathbf{x}: \|\mathbf{x} - \mathbf{p}_k\| < \|\mathbf{x} - \mathbf{p}_j\| \qquad j = 1,\ldots, k-1, k+1,\ldots,n\}.$$

The region $\mathscr{T}_k$ is called the Dirichlet tile associated with $\mathbf{p}_k$ and consists of the region closer to $\mathbf{p}_k$ than any other points of $\mathscr{P}$. By definition, the tile $\mathscr{T}_k$ is the intersection of the half spaces $\{\mathbf{x}: \|\mathbf{x} - \mathbf{p}_k\| < \|\mathbf{x} - \mathbf{p}_j\|\}$ and hence is a convex polytope. Two points of $\mathscr{P}$ are called (strong) Dirichlet neighbors if the intersection of the closure of their tiles is a region with non zero $s - 1$ dimensional volume. The corners of the tile $\mathscr{T}_k$ are points where at least s perpendicular bisectors of the form $\{\mathbf{x}: \|\mathbf{x} - \mathbf{p}_k\| = \|\mathbf{x} - \mathbf{p}_j\|\}$ intersect and hence are centers of spheres through $\mathbf{p}_k$ and certain collections of s of its Dirichlet neighbors.

Let $\mathscr{C}$ denote the interior of the convex hull of the point set $\mathscr{P}$ and let $\mathbf{p}$ be a point in $\mathscr{C}\backslash\mathscr{P}$. Let $\mathscr{T}_{\mathbf{p}} = \{\mathbf{x}: \|\mathbf{x} - \mathbf{p}\| < \|\mathbf{x} - \mathbf{p}_j\| \, j = 1,\ldots,n\}$ be the tile closer to $\mathbf{p}$ than any point of $\mathscr{P}$ and define $\lambda_k(\mathbf{p})$ to be the volume of the intersection of the "old" tile $\mathscr{T}_k$ (before the insertion of $\mathbf{p}$) and the "new" tile $\mathscr{T}_{\mathbf{p}}$ as shown in figure 1 for the case $s = 2$. Thus $\lambda_k(\mathbf{p})$ is the volume of the subtile

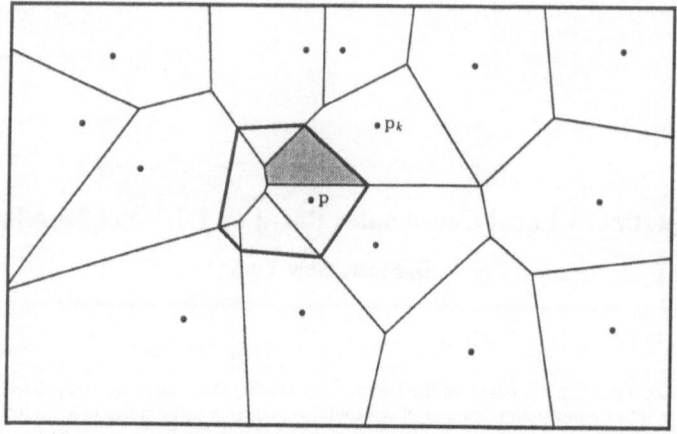

Figure 1. The area of the shaded region is $\lambda_k(\mathbf{p})$

$$\{\mathbf{x}: \|\mathbf{x} - \mathbf{p}_k\| < \|\mathbf{x} - \mathbf{p}_j\|, \qquad j = 1, \ldots, k - 1, k + 1, \ldots, n$$

$$\|\mathbf{x} - \mathbf{p}\| < \|\mathbf{x} - \mathbf{p}_k\|\}$$

and since this is bounded whenever $\mathbf{p} \in \mathscr{C}\backslash\mathscr{P}$, the function λ_k is well defined on $\mathscr{C}\backslash\mathscr{P}$. The local coordinates of $\mathbf{p} \in \mathscr{C}$ are then defined to be

$$u_k(\mathbf{p}) = \begin{cases} \lambda_k(\mathbf{p})/V(\mathbf{p}) & \text{if } \mathbf{p} \in \mathscr{C}\backslash\mathscr{P} \\ \delta_{k,j} & \text{if } \mathbf{p} = \mathbf{p}_j \in \mathscr{P} \end{cases}$$

where $V(\mathbf{p})$ is the volume of tile $\mathscr{T}_{\mathbf{p}}$. The function u_k are clearly nonnegative and sum to one because the subtiles for $k = 1, \ldots, n$ partition $\mathscr{T}_{\mathbf{p}}$ and hence $\sum \lambda_k(\mathbf{p}) = V(\mathbf{p})$. The functions u_k are called "local" coordinates because $u_k(\mathbf{p})$ depends only on the Dirichlet neighbors of $\mathbf{p}$. The functions u_k are coordinates of $\mathbf{p}$ because they satisfy the "local coordinates property" that

$$\sum_{j=1}^{n} u_j(\mathbf{p})\mathbf{p}_j = \mathbf{p}, \qquad \forall \mathbf{p} \in \mathscr{C},$$

as proved originally in [Sibson 1981].

The C^0 natural neighbor interpolant to the scattered data $\{\mathbf{p}_j, w_j\}_{j=1}^{n}$ was defined by Sibson as

$$N(\mathbf{p}) = \sum_{j=1}^{n} u_j(\mathbf{p})w_j.$$

In one dimension the interpolant reduces to piecewise linear interpolation (see [Sibson 1981]) but in higher dimensions it has a more complex algebraic structure. It has linear precision because the functions, u_j, have the local coordinates property and is continuously differentiable with respect to $\mathbf{p}$ for all $\mathbf{p} \in \mathscr{C}\backslash\mathscr{P}$ and continuous for all $\mathbf{p} \in \mathscr{C}$. Further, the function N has the desirable property that it depends continuously on the input "data sites" $\mathbf{p}_i$. In fact, if $\mathbf{p}$ is fixed in $\mathscr{C}\backslash\mathscr{P}$, then u_k is continuously differentiable with respect to $\mathbf{p}_j \in \mathscr{P}$ provided we restrict $\mathbf{p}_j$ to be not equal to $\mathbf{p}$ and in $\mathscr{C}\backslash\{\mathscr{P}\backslash\{\mathbf{p}_j\}\}$. The smoothness properties cited above were first noted in [Sibson 1981] and explained further in [Farin 1990].

In this paper, we will study the properties of local coordinates restricted to $\mathscr{C}$ and for $s \geq 2$. We give some simple proofs of the smoothness properties in Sections 2 and 3 and contribute a new geometric interpretation for the gradient of λ_k in Section 4 and a new proof of the local coordinates property in Section V.

2. Smoothness of u_k

In this section, we will prove that the function u_k is continuously differentiable on $\mathscr{C}\backslash\mathscr{P}$ and continuous on $\mathscr{C}$ and, in the process, reveal some of the algebraic structure of λ_k. The proof essentially follows the idea of [Farin 1990] but includes the necessary generalizations for the s dimensional case.

We will need some additional terminology. A hyperplane will refer to an $s - 1$ dimensional affine subspace of $\mathscr{R}^s$ and a facet of a tile will be a region of non-zero $s - 1$ dimensional volume that consists of the intersection of the boundary of the tile with some hyperplane. Each of the facets of a Dirichlet tile is in a hyperplane which is the perpendicular bisector between two points of $\mathscr{P}$.

Consider u_k restricted to $\mathscr{C}\backslash\mathscr{P}$. Since the denominator $V(\mathbf{p})$ does not vanish on $\mathscr{C}\backslash\mathscr{P}$, it follows that u_k is continuously differentiable on $\mathscr{C}\backslash\mathscr{P}$ if and only if $\lambda_k(\mathbf{p})$ is continuously differentiable on $\mathscr{C}\backslash\mathscr{P}$, so to prove differentiability it is sufficient to consider λ_k.

We first show that λ_k has a particularly simple form when restricted to lines through $\mathbf{p}_k$. Let $\vec{\mathbf{n}}$ be any unit vector and define the ray $\mathbf{l}_k$ by $\mathbf{l}_k(t) = \mathbf{p}_k + t\vec{\mathbf{n}}$ for $t \geq 0$. If $\mathbf{p}_k$ is on the boundary of $\mathscr{C}$ we restrict $\vec{\mathbf{n}}$ so that $\mathbf{l}_k(t) \in \mathscr{C}$ for some $t > 0$. For each t consider the hyperplane (a line for $s = 2$) that passes through $\mathbf{l}_k(t)$ and is perpendicular to the ray $\mathbf{l}_k$. Define a function $L_k(t)$ so that it is zero for values of t where this hyperplane does not intersect the closure of $\mathscr{T}_k$ and otherwise define $L_k(t)$ to be the $s - 1$ dimensional volume of the intersection of this hyperplane with the closure of $\mathscr{T}_k$. Let c be the smallest positive value such that $L_k(t)$ is zero for any $t > c$. Thus, $\mathbf{l}_k(c)$ marks the end point of the perpendicular projection of the tile $\mathscr{T}_k$ onto the line $\mathbf{l}_k$. Figure 2 illustrates the situation for $s = 2$.

Since the boundary of the tile $\mathscr{T}_k$ consists of portions of hyperplanes, an $s - 1$ dimensional volume computation gives that L_k is a piecewise polynomial function of degree $s - 1$ that is continuous for $t \in (0, c)$. The function L_k may or may not be continuous at $t = c$ but as can be seen in Fig. 2,

$$\lim_{t \to c} L_k(t) = 0$$

provided that the perpendicular bisector of $\mathbf{l}_k(2c)$ and $\mathbf{p}_k$ does not contain a facet of

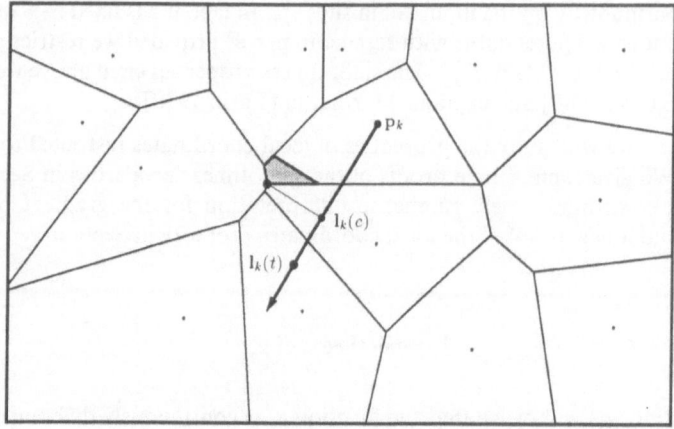

Figure 2. The area of the shaded region is $\lambda_k(\mathbf{l}_k(t)) = \int_{t/2}^{c} L_k(s)\,ds$

$\mathcal{T}_k$ which will not happen so long as $\mathbf{l}_k(2c)$ is not a point in the set $\mathcal{P}$. This implies that L_k is continuous for those $t \geq 0$ such that $\mathbf{l}_k(2t) \notin \mathcal{P}$.

It is clear from the definition of λ_k that for $\mathbf{l}_k(t) \in \mathcal{C}\backslash\mathcal{P}$

$$\lambda_k(\mathbf{l}_k(t)) = \begin{cases} \int\int_{t/2}^{c} L_k(s)\,ds & \text{if } 0 < t < 2c; \\ 0 & \text{if } t \geq 2c. \end{cases}$$

By the fundamental theorem of calculus, $\lambda_k(\mathbf{l}_k(t))$ is a continuously differentiable piecewise polynomial function of degree s whenever $\mathbf{l}_k(t) \in \mathcal{C}\backslash\mathcal{P}$.

We now need to look further at the structure of λ_k as a function of s variables. The function λ_k is the s dimensional volume of a subtile with vertices that are either vertices of the tile $\mathcal{T}_k$ or vertices of the intersection of $\mathcal{T}_k$ with the perpendicular bisector of the variable point $\mathbf{p}$ and the fixed point $\mathbf{p}_k$. If $\mathbf{p}$ is restricted to a region such that this perpendicular bisector intersects the same set of facets of the tile $\mathcal{T}_k$, then there are a fixed number of vertices of the subtile and a straightforward computation shows they are each rational (or constant) functions of $\mathbf{p}$ on this region. Thus, the function λ_k is piecewise rational and it changes from one rational function to another exactly when $\mathbf{p}$ is positioned so that the perpendicular bisector of $\mathbf{p}$ and $\mathbf{p}_k$ contains a vertex of $\mathcal{T}_k$ which occurs when $\mathbf{p}$ is on a sphere through $\mathbf{p}_k$ centered at such a vertex. Therefore the boundaries that divide λ_k into different rational functions are spheres that pass through the point $\mathbf{p}_k$ and are centered at vertices of $\mathcal{T}_k$ as shown in Fig. 3 for the case $s = 2$. Note also that the support of the function λ_k is in the union of the disks with boundary through $\mathbf{p}$, centered at the vertices of $\mathcal{T}_k$. (This was shown in [Farin 1990] with a figure similar to that of Fig. 3.)

Now, since λ_k is continuously differentiable in regions where it is one rational function, it suffices to show that some cross boundary derivative of λ_k is continuous across the spheres through $\mathbf{p}_k$ centered at vertices of $\mathcal{T}_k$. Let $\mathbf{q} \in \mathcal{C}\backslash\mathcal{P}$ be a point on

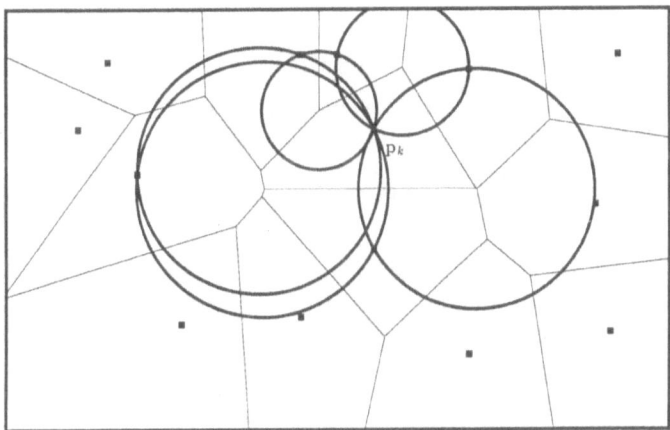

Figure 3. The boundaries of the rational pieces of the function $\lambda_k(\mathbf{p})$ are circular arcs

such a sphere and let $\mathbf{l}_k(t)$ be the ray starting at $\mathbf{p}_k$ that passes through $\mathbf{q}$ for some $t > 0$. Since we know that $\lambda_k(\mathbf{l}_k(t))$ is continuously differentiable for $\mathbf{l}_k(t) \in \mathscr{C}\backslash\mathscr{P}$, the directional derivative at $\mathbf{q}$ in the direction of $\mathbf{l}_k$ is continuous. We may take this direction as the direction of the cross boundary derivative provided that $\mathbf{l}_k$ is not tangent to the sphere at $\mathbf{q}$. But since $\mathbf{l}_k$ passes through both $\mathbf{p}_k$ and $\mathbf{q} \notin \mathscr{P}$, it cannot be tangent to the sphere at $\mathbf{q}$ and we conclude that λ_k is continuously differentiable everywhere in $\mathscr{C}\backslash\mathscr{P}$.

In general, the function λ_k cannot be defined continuously at $\mathbf{p}_k$ [Farin, 1990]. However, the function u_k is continuous at the points of $\mathscr{P}$ since from the definition of λ_j it follows that

$$\lim_{\mathbf{p}\to\mathbf{p}_k} u_j(\mathbf{p}) = \lim_{\mathbf{p}\to\mathbf{p}_k} \lambda_j(\mathbf{p}) = 0$$

whenever $j \neq k$. Then, since $\sum u_j(\mathbf{p}) = 1$, we must have that

$$\lim_{\mathbf{p}\to\mathbf{p}_k} u_k(\mathbf{p}) = 1$$

so u_k is continuous everywhere in $\mathscr{C}$. In general, the function u_k is not differentiable at the points of $\mathscr{P}$.

3. Smoothness with Respect to $\mathbf{p}_j$

Fix $\mathbf{p}$ and all points of $\mathscr{P}$ except for some $\mathbf{p}_j$ and let $\mathscr{Q} = \mathscr{P} \cup \{\mathbf{p}\}\backslash\{\mathbf{p}_j\}$. We claim that $\lambda_k(\mathbf{p})$ is a continuously differentiable function of $\mathbf{p}_j$ provided $\mathbf{p}_j \in \mathscr{C}\backslash\mathscr{Q}$. To prove this, we express $\lambda_k(\mathbf{p})$ in terms of subtile volumes that we know are continuously differentiable with respect to $\mathbf{p}_j$.

In the formula below, the superscript on λ_k indicates the point set used to form the Dirichlet tessellation into which the point ($\mathbf{p}$ or $\mathbf{p}_j$) is inserted to form a subtile. Let

$\mathcal{Q}$ be as above and set $\mathcal{O} = \mathcal{P} \setminus \{\mathbf{p}_j\}$. For $j \neq k$ we claim that

$$\lambda_k^{\mathcal{P}}(\mathbf{p}) = \lambda_k^{\mathcal{O}}(\mathbf{p}) - [\lambda_k^{\mathcal{O}}(\mathbf{p}_j) - \lambda_k^{\mathcal{Q}}(\mathbf{p}_j)]. \tag{1}$$

This formula shows how λ_k depends on $\mathbf{p}_j$. In particular, $\lambda_k^{\mathcal{O}}(\mathbf{p})$ is constant and the other two terms are continuously differentiable functions of $\mathbf{p}_j$ since they are just subtile volumes and these are continuously differentiable as shown in Section 2. It follows that $\lambda_k(\mathbf{p})$ is continuously differentiable with respect to the $\mathbf{p}_j$ provided that $\mathbf{p}_j \in \mathscr{C} \setminus \mathcal{Q}$.

We discovered Eq. (1) by examining 2 dimensional pictures, but it can be proved directly by using the definitions of the various subtiles. To do this, recall the definition of subtiles to see that the subtile with volume $\lambda_k^{\mathcal{O}}(\mathbf{p}_j)$ is

$$\mathcal{T}_{\mathbf{p}_j}^{\mathcal{O}} = \{\mathbf{x} \colon \|\mathbf{x} - \mathbf{p}_k\| < \|\mathbf{x} - \mathbf{p}_i\|, \quad i \neq j, k, \quad \|\mathbf{x} - \mathbf{p}_j\| < \|\mathbf{x} - \mathbf{p}_k\|\}.$$

and the subtile with volume $\lambda_k^{\mathcal{Q}}(\mathbf{p}_j)$ is

$$\mathcal{T}_{\mathbf{p}_j}^{\mathcal{Q}} = \{\mathbf{x} \colon \|\mathbf{x} - \mathbf{p}_k\| < \|\mathbf{x} - \mathbf{p}_i\|, \quad i \neq j, k, \quad \|\mathbf{x} - \mathbf{p}_k\| < \|\mathbf{x} - \mathbf{p}\|,$$
$$\|\mathbf{x} - \mathbf{p}_j\| < \|\mathbf{x} - \mathbf{p}_k\|\}.$$

Since $\mathcal{T}_{\mathbf{p}_j}^{\mathcal{Q}} \subset \mathcal{T}_{\mathbf{p}_j}^{\mathcal{O}}$ the quantity $\lambda_k^{\mathcal{O}}(\mathbf{p}_j) - \lambda_k^{\mathcal{Q}}(\mathbf{p}_j)$, is the volume of the region

$$\mathcal{T}_1 = \{\mathbf{x} \colon \|\mathbf{x} - \mathbf{p}_k\| < \|\mathbf{x} - \mathbf{p}_i\|, \quad i \neq j, k, \quad \|\mathbf{x} - \mathbf{p}_j\| < \|\mathbf{x} - \mathbf{p}_k\|,$$
$$\|\mathbf{x} - \mathbf{p}_k\| \geq \|\mathbf{x} - \mathbf{p}\|\}.$$

Now the subtile with volume $\lambda_k^{\mathcal{O}}(\mathbf{p})$ is

$$\mathcal{T}_{\mathbf{p}}^{\mathcal{O}} = \{\mathbf{x} \colon \|\mathbf{x} - \mathbf{p}_k\| < \|\mathbf{x} - \mathbf{p}_i\|, \quad i \neq j, k, \quad \|\mathbf{x} - \mathbf{p}\| < \|\mathbf{x} - \mathbf{p}_k\|\}.$$

Since the interior of $\mathcal{T}_1$ is a subset of $\mathcal{T}_{\mathbf{p}}^{\mathcal{O}}$ the quantity on the right hand side of equation (1) is the volume of the region

$$\{\mathbf{x} \colon \|\mathbf{x} - \mathbf{p}_k\| < \|\mathbf{x} - \mathbf{p}_i\|, \quad i \neq j, k, \quad \|\mathbf{x} - \mathbf{p}\| \leq \|\mathbf{x} - \mathbf{p}_k\|, \quad \|\mathbf{x} - \mathbf{p}_j\| \geq \|\mathbf{x} - \mathbf{p}_k\|\}$$

and the interior of this region is precisely the subtile with volume $\lambda_k^{\mathcal{P}}(\mathbf{p})$ which proves equation (1).

A similar proof may be used to establish that for $j = k$,

$$\lambda_j^{\mathcal{P}}(\mathbf{p}) = \sum_{\substack{i=1 \\ i \neq k}}^{n} \lambda_i^{\mathcal{O}}(\mathbf{p}_j) - \left[\sum_{\substack{i=1 \\ i \neq k}}^{n} \lambda_i^{\mathcal{Q}}(\mathbf{p}_j) + \lambda_{\mathbf{p}}^{\mathcal{Q}}(\mathbf{p}_j) \right]$$

and thus, λ_j is also continuously differentiable with respect to $\mathbf{p}_j$ for $\mathbf{p}_j \in \mathscr{C} \setminus \mathcal{Q}$.

4. The Gradient of λ_k

We first develop a formula for the gradient in the case when the dimension, s, is 2 and then extend it to higher dimensions.

Denote the m vertices of the tile $\mathcal{T}_{\mathbf{p}}$ as $\{\mathbf{a}_i\}_{i=1}^{m}$ arranged in counterclockwise order around $\mathbf{p}$ (see Fig. 4). Index arithmetic will be treated cyclically so that $0 = m$,

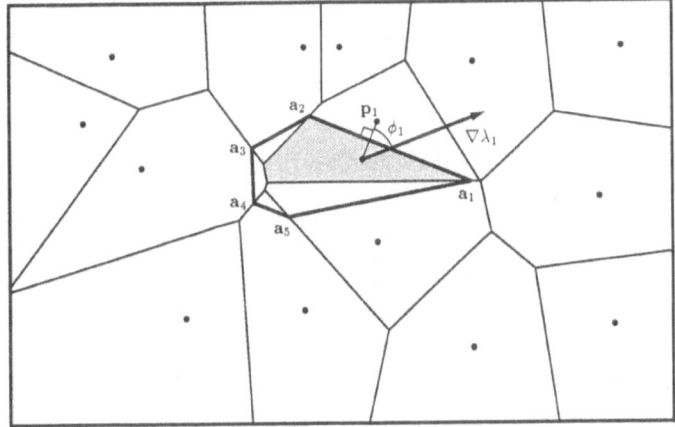

Figure 4. The gradient of λ_1 at $\mathbf{p}$ is the vector $(\vec{v}_1 + \vec{v}_2)/2$ times the scale factor f_1/d_1

$m + 1 = 1$ and for convenience, reorder the point set $\mathscr{P}$ so that $\mathbf{a}_i$ is the center of the circle through the points $\mathbf{p}_{i-1}$, $\mathbf{p}$ and $\mathbf{p}_i$. Let $f_i = \|\mathbf{a}_{i+1} - \mathbf{a}_i\|$, $d_i = \|\mathbf{p}_i - \mathbf{p}\|$ and $\vec{v}_i = \mathbf{a}_i - \mathbf{p}$.

We claim that whenever $\mathbf{p} \in \mathscr{C}\backslash\mathscr{P}$ is in the support of λ_k the gradient is given by

$$\nabla\lambda_k(\mathbf{p}) = \frac{f_k}{d_k}\left(\frac{\mathbf{a}_k + \mathbf{a}_{k+1}}{2} - \mathbf{p}\right) = \frac{f_k}{d_k}\left(\frac{\vec{v}_k + \vec{v}_{k+1}}{2}\right). \tag{2}$$

Thus, the direction of the gradient is from $\mathbf{p}$ to the midpoint of the edge $\mathbf{a}_k\mathbf{a}_{k+1}$ and a simple trigonometric argument shows that the length of the gradient is $f_k/(2\cos(\phi_k))$ where ϕ_k is the angle between the vector $\mathbf{p}_k - \mathbf{p}$ and the gradient vector $\nabla\lambda_k$ as shown in Fig. 4 for $k = 1$.

To prove Eq. (2) we first find the directional derivatives of λ_k in two distinct directions. One simple direction to take would be directly away from the point $\mathbf{p}_k$ because, as shown in Section 2, the function λ_k is piecewise quadratic on lines through $\mathbf{p}_k$. However, we will not use this direction because one direction is not sufficient to calculate the gradient and it will be simpler to choose two other directions as described below.

The idea is to restrict λ_k to a path so that as a point moves along this path, only one of the vertices of the subtile changes (as opposed to two in general). The two paths we choose will be circles centered at $\mathbf{a}_k$ and $\mathbf{a}_{k+1}$ and the directional derivatives will be taken in directions tangential to these circles.

Fix $\mathbf{p} \in \mathscr{C}\backslash\mathscr{P}$ and a point $\mathbf{p}_k \in \mathscr{P}$ so that $\mathbf{p}$ is in the support of λ_k. Consider the circle centered at $\mathbf{a}_k$ that passes through $\mathbf{p}$, $\mathbf{p}_{k-1}$ and $\mathbf{p}_k$. Parameterize this circle counterclockwise as $\mathbf{r}(t)$ so that $\mathbf{r}(0) = \mathbf{p}$ and t is the angle between $\mathbf{p} - \mathbf{a}_k$ and $\mathbf{r}(t) - \mathbf{a}_k$. It is clear that $\mathbf{r}'(0) = \vec{v}_k^{\perp}$ where $\vec{v}_k^{\perp}$ is the vector orthogonal to $\vec{v}_k = (x_k, y_k)$ given by $\vec{v}_k^{\perp} = (-y_k, x_k)$.

We will use some geometry to compute the derivative of $g(t) = \lambda_k(\mathbf{r}(t))$ at $t = 0$. By definition, this derivative is given by

$$g'(0) = \lim_{t \to 0} \frac{g(t) - g(0)}{t}.$$

For sufficiently small values of $t > 0$, $\mathbf{r}(t) \neq \mathbf{p}_k$ and the subtile associated with $\mathbf{r}(t)$, $\mathcal{T}_{\mathbf{r}(t)} = \{\mathbf{x} \colon \|\mathbf{x} - \mathbf{r}(t)\| < \|\mathbf{x} - \mathbf{p}_j\| \; \forall j = 1, \ldots, n\}$, is well defined and has the vertex $\mathbf{a}_k$. The quantity $g(t) - g(0)$ is the difference of two areas and is the area of the shaded region shown in Fig. 5. For sufficiently small t this region is always a triangle so its area for such t is given by $f_k \hat{f}_k(t) \sin(\theta(t))/2$ where $\hat{f}_k(t)$ is the length of the edge of $\mathcal{T}_{\mathbf{r}(t)}$ in $\mathcal{T}_k$ and $\theta(t)$ is the angle shown in Fig. 6 which is an enlargement of the circle

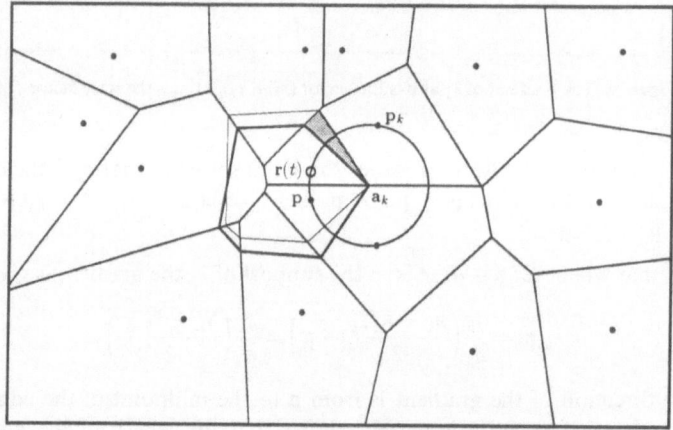

Figure 5. The area of the shaded region is $g(t) - g(0)$

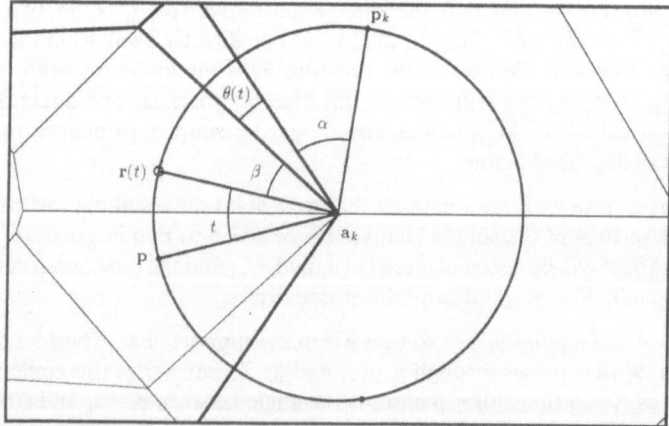

Figure 6. By perpendicular bisectors $\theta(t) + \alpha = \beta + t$ and $\theta(t) + \beta = \alpha$. Thus $\theta(t) = t/2$

shown in Fig. 5. By the simple geometric argument illustrated in Fig. 6, $\theta(t) = t/2$. Also it is clear that as h tends to zero, $\hat{f}_k(t)$ tends to f_k. Hence

$$g'(0) = \lim_{t \to 0} \frac{g(t) - g(0)}{t} = \lim_{t \to 0} \frac{f_k \hat{f}_k(t) \sin(t/2)}{2t} = \frac{f_k^2}{4}.$$

Now since $g(t) = \lambda_k(\mathbf{r}(t))$ the chain rule gives us that

$$g'(0) = \nabla \lambda_k(\mathbf{r}(0)) \cdot \mathbf{r}'(0) = \nabla \lambda_k(\mathbf{r}(0)) \cdot \vec{\mathbf{v}}_k^{\perp}.$$

so the gradient must satisfy

$$\nabla \lambda_k(\mathbf{p}) \cdot \vec{\mathbf{v}}_k^{\perp} = \nabla \lambda_k(\mathbf{r}(0)) \cdot \vec{\mathbf{v}}_k^{\perp} = \frac{f_k^2}{4}. \tag{3}$$

A similar argument for a circle centered at $\mathbf{a}_{k+1}$ shows the gradient must also satisfy

$$\nabla \lambda_k(\mathbf{p}) \cdot \vec{\mathbf{v}}_{k+1}^{\perp} = \nabla \lambda_k(\mathbf{r}(0)) \cdot \vec{\mathbf{v}}_{k+1}^{\perp} = \frac{-f_k^2}{4} \tag{4}$$

where the sign is reversed due to the fact that on this circle, $g(h) - g(0)$ is negative for $h > 0$.

Hence we have two directional derivatives of λ_k from which a formula for the gradient may be found. To establish that Eq. (2) provides such a formula we take the dot product of the right hand side of Eq. (2) with $\vec{\mathbf{v}}_k^{\perp}$ to get

$$\frac{f_k}{2d_k}(\vec{\mathbf{v}}_k + \vec{\mathbf{v}}_{k+1}) \cdot \vec{\mathbf{v}}_k^{\perp} = \frac{f_k}{2d_k}(\vec{\mathbf{v}}_{k+1} \cdot \vec{\mathbf{v}}_k^{\perp}) \tag{5}$$

Note now that $\vec{\mathbf{v}}_{k+1} \cdot \vec{\mathbf{v}}_k^{\perp}$ is twice the area of the triangle with vertices $\mathbf{p}$, $\mathbf{a}_k$, $\mathbf{a}_{k+1}$. Alternatively, since $\mathbf{p} - \mathbf{p}_k$ is perpendicular to $\mathbf{a}_{k+1} - \mathbf{a}_k$, twice the area of this triangle is also given by $f_k d_k/2$. Substitute $\vec{\mathbf{v}}_{k+1} \cdot \vec{\mathbf{v}}_k^{\perp} = d_k f_k/2$ into the right hand side of Eq. (5) to get

$$\nabla \lambda_k(\mathbf{r}(0)) \cdot \vec{\mathbf{v}}_k^{\perp} = \frac{f_k^2}{4}$$

as required by Eq. (3). A similar argument shows that Eq. (4) is also satisfied by the formula for the gradient given in Eq. (2). This proves that equation (2) gives the gradient of λ_k at $\mathbf{p}$.

For general dimension s we claim that the gradient of λ_k is

$$\nabla \lambda_k(\mathbf{p}) = \frac{f_k}{d_k}(\mathbf{c}_k - \mathbf{p}) \tag{6}$$

where $d_k = \|\mathbf{p}_k - \mathbf{p}\|$ as before, f_k is the $s - 1$ dimensional volume of the facet $\mathscr{F}_k$ of $\mathscr{T}_\mathbf{p}$ that is contained in $\mathscr{T}_k$, and $\mathbf{c}_k$ is the centroid of $\mathscr{F}_k$. This is the centroid of the entire facet and not just the vertices on the facet and may be expressed as

$$\mathbf{c}_k = (1/f_k) \int_{\mathbf{q} \in \mathscr{F}_k} (\mathbf{q} - \mathbf{p}) \, dq.$$

Substituting, we obtain the alternative formula,

$$\nabla \lambda_k(\mathbf{p}) = (1/d_k) \int_{\mathbf{q} \in \mathscr{F}_k} (\mathbf{q} - \mathbf{p}) \, dq, \tag{7}$$

which will be used in Section V to prove the local coordinates property.

To prove Eq. (6), we first need a lemma that generalizes the property that we used above that related $\bar{\mathbf{v}}_k^\perp \cdot \bar{\mathbf{v}}_{k+1}$ to $f_k d_k/2$. In this lemma and what follows a k face of an s dimensional simplex will refer to a region of non-zero k dimensional volume that consists of the intersection of the boundary of the simplex with a k dimensional affine subspace of $\mathscr{R}^s$. A facet is thus an $s - 1$ face. The distance from a point to a k face will always be taken to be the perpendicular distance from the point to the k dimensional affine subspace that contains the k face.

Lemma: *Let σ be a s dimensional simplex with vertices $\mathbf{b}_1, \ldots, \mathbf{b}_{s+1}$. Let $\mathscr{E}$ be the $s - 2$ face of σ not containing $\mathbf{b}_1$ or $\mathbf{b}_2$. Let H_j be the distance from $\mathbf{b}_j$ to the opposite facet and let h_j be the distance from $\mathbf{b}_j$ to $\mathscr{E}$. Then $H_1 h_2 = H_2 h_1$.*

Proof: If $V_\mathscr{E}$ is the $s - 2$ dimensional volume of $\mathscr{E}$ then $h_1 V_\mathscr{E}/(s - 1)$ is the $s - 1$ dimensional volume of the facet opposite $\mathbf{b}_2$ and hence $h_1 H_2 V_\mathscr{E}/[(s)(s - 1)]$ is the volume of σ. Similarly, by first finding the volume of the facet opposite $\mathbf{b}_1$, the volume of σ is also given by $H_1 h_2 V_\mathscr{E}/[(s)(s - 1)]$ and setting these two formulas equal to one another gives the desired result. □

Now, for the proof of Eq. (6), triangulate the vertices of $\mathscr{F}_k$ into $s - 1$ dimensional simplices $\tau_k^i, i = 1, \ldots, m_k$ and partition $\mathscr{T}_k$ into regions $\mathscr{T}_k^i, i = 1, \ldots, m_k$ where each $\mathscr{T}_k^i$ is an s dimensional polytope containing τ_k^i in its interior. Having fixed this partition, define

$$\lambda_k^i(\mathbf{p}) = \text{Area of } (\mathscr{T}_\mathbf{p} \cap \mathscr{T}_k^i).$$

We claim that for $\mathbf{p} \in \mathscr{C} \backslash \mathscr{P}$ that is in the support of λ_k the formula for the gradient is given by

$$\nabla \lambda_k^i(\mathbf{p}) = \frac{f_k^i}{d_k} (\mathbf{c}_k^i - \mathbf{p}), \tag{8}$$

where f_k^i is the $s - 1$ dimensional volume of τ_k^i and $\mathbf{c}_k^i$ is the centroid of τ_k^i. Then since $\lambda_k(\mathbf{p}) = \sum_{i=1}^m \lambda_k^i(\mathbf{p})$, Eq. (6) is obtained by summing the formula in Eq. (8) over i to get

$$\nabla \lambda_k(\mathbf{p}) = \sum_{i=1}^{m_k} \nabla \lambda_k^i(\mathbf{p}) = \frac{f_k}{d_k} \sum_{i=1}^{m_k} \frac{f_k^i}{f_k} (\mathbf{c}_k^i - \mathbf{p}) = \frac{f_k}{d_k} (\mathbf{c}_k - \mathbf{p})$$

and hence it is sufficient to prove Eq. (8). (Note that since the centroid of each τ_k^i is just the average of its vertices, the above summation provides a simple way of actually computing the gradient.)

To prove Eq. (8), fix k and i and let σ be the simplex with vertex $\mathbf{p} = \mathbf{b}_1$ and the vertices of τ_k^i and let $\mathbf{b}_2$ be any vertex of τ_k^i. Let $\mathscr{E}$ be (as in the lemma) the $s - 2$ face opposite $\mathbf{b}_1$ and $\mathbf{b}_2$. We will find a directional derivative along a circular path $\mathbf{r}(t)$ that is in a 2 dimensional affine subspace perpendicular to $\mathscr{E}$ and containing $\mathbf{p}$. The circular path will be chosen so as to contain $\mathbf{p}$ and will be centered at the point where $\mathscr{E}$ projects onto this two dimensional affine subspace. As in the two dimensional case, parameterize the circular path by the angle t and choose the parameterization so that $\mathbf{r}(0) = \mathbf{p}$ and further require that $\mathbf{r}'(0) \cdot (\mathbf{b}_2 - \mathbf{p})$ is positive (so that the directional derivative will be positive). If $g(t) = \lambda_k^i(\mathbf{r}(t))$, for small $t > 0$, the quantity $g(t) - g(0)$ is the volume of a simplex with facet τ_k^i and height $\sin(t/2)h_2(t)$ where $h_2(t)$ is the distance from $\mathscr{E}$ to the vertex of this simplex that is not on τ_k^i. Letting t approach zero and using an argument similar to the 2-dimensional case it follows that

$$\nabla \lambda_k^i(\mathbf{p}) \cdot \vec{\mathbf{v}} = g'(0) = \lim_{t \to 0} \frac{f_k^i \sin(t/2)h_2(t)}{2st} = \frac{f_k^i h_2}{2s} \tag{9}$$

where $\vec{\mathbf{v}} = \mathbf{r}'(0)$ and f_k^i is the $s - 1$ dimensional volume of τ_k^i. The quantity h_2 is the limit as t approaches zero of $h_2(t)$ and is hence the distance from $\mathbf{b}_2$ to $\mathscr{E}$.

To establish that the formula given in Eq. (8) is the gradient we must show that the dot product with $\vec{\mathbf{v}}$ does in fact give the right hand side of Eq. (9). Now,

$$\mathbf{c}_k^i - \mathbf{p} = \sum_{j=2}^{s} (\mathbf{b}_j - \mathbf{p})/s$$

and $\vec{\mathbf{v}}$ is the vector perpendicular to the hyperplane containing $\mathscr{E}$ and $\mathbf{p}$ which is in fact the hyperplane containing the facet opposite $\mathbf{b}_2$ in σ and hence $\vec{\mathbf{v}}$ is perpendicular to all of the terms in the above sum except for $j = 2$ so

$$\vec{\mathbf{v}} \cdot (\mathbf{c}_k^i - \mathbf{p}) = \vec{\mathbf{v}} \cdot (\mathbf{b}_2 - \mathbf{p}) = H_2 \|\vec{\mathbf{v}}\|$$

where H_2 is the distance from $\mathbf{b}_2$ to the opposite facet in σ. Also, by the parameterization of the circle, the vector $\vec{\mathbf{v}}$ has length h_1 which is the distance from $\mathbf{p} = \mathbf{b}_1$ to $\mathscr{E}$ and hence $\vec{\mathbf{v}} \cdot (\mathbf{b}_2 - \mathbf{p}) = H_2 h_1$. Thus, the dot product of $\vec{\mathbf{v}}$ with the formula in Eq. (8) becomes

$$\frac{f_k^i [(\mathbf{b}_2 - \mathbf{p}) \cdot \vec{\mathbf{v}}]}{d_k s} = \frac{f_k^i H_2 h_1}{2 H_1 s}$$

which by the lemma is

$$\frac{f_k^i H_1 h_2}{2 H_1 s} = \frac{f_k^i h_2}{2s}$$

and since this works for $\mathbf{b}_2$ being any of the s vertices of τ_k^i, Eq. (8) must give the gradient.

Remark: Given the gradients of the λ_k, the gradients of u_k may be found by using the quotient rule.

5. The Local Coordinates Property

For any fixed point $\mathbf{p} \in \mathscr{C}$ we wish to prove that

$$\sum_{k=0}^{n} u_k(\mathbf{p}) \mathbf{p}_k = \mathbf{p}.$$

This was proven originally in [Sibson 1980]. We will present a new proof using the formula for the gradient of λ_k and Stokes theorem. This proof adds further insight into the nature of λ_k.

Note first that by definition, the local coordinates property is satisfied at all points of $\mathscr{P}$, so we will restrict our discussion to points in $\mathscr{C}\backslash\mathscr{P}$. Recall that for $\mathbf{p} \in \mathscr{C}\backslash\mathscr{P}$

$$u_k(\mathbf{p}) = \frac{\lambda_k(\mathbf{p})}{\sum_{j=1}^n \lambda_j(\mathbf{p})}.$$

Since $\lambda_k(\mathbf{p}) = u_k(\mathbf{p}) = 0$ for all k such that $\mathbf{p}_k$ is not a Dirichlet neighbor of $\mathbf{p}$, we only need show that

$$\sum_{i=1}^m u_i(\mathbf{p})\mathbf{p}_i = \mathbf{p}$$

where (by reordering) $\{\mathbf{p}_i\}_{i=1}^m$ are the Dirichlet neighbors of $\mathbf{p}$ (i.e. $\lambda_i(\mathbf{p}) \neq 0$, $i = 1,\ldots,m$). By substituting in the definition of u_i, clearing the denominator and subtracting the right hand side we find that we must prove that

$$\sum_{i=1}^m \lambda_i(\mathbf{p})(\mathbf{p}_i - \mathbf{p}) = \mathbf{0}.$$

Let

$$\mathbf{H}(\mathbf{p}) = \sum_{i=1}^m \lambda_i(\mathbf{p})(\mathbf{p}_i - \mathbf{p})$$

and note that H maps $\mathscr{C}\backslash\mathscr{P}$ into $\mathscr{R}^s$.

To prove $\mathbf{H} = \mathbf{0}$ we first show that H is constant by showing that the derivative, $\mathbf{H}'$, is the zero $s \times s$ matrix. In the notation used below, $\mathbf{q}$ will be a dummy variable, $\mathbf{p}_i - \mathbf{p}, \mathbf{q} - \mathbf{p}$ and $\nabla\lambda_k$ will be row vectors, $(\cdot)^t$ will denote the transpose of a vector, and I will be the $s \times s$ identity matrix. Using the product rule on each component of H and rewriting in matrix form gives

$$\mathbf{H}'(\mathbf{p}) = \sum_{i=1}^m (\nabla\lambda_i(\mathbf{p}))^t(\mathbf{p}_i - \mathbf{p}) - \lambda_i(\mathbf{p})\mathbf{I}$$

Recalling that $\sum_{i=1}^m \lambda_i(\mathbf{p})$ is the volume $V(\mathbf{p})$ of tile $\mathscr{T}_\mathbf{p}$ and substituting in formula (7) for the gradient, the above expression becomes

$$\mathbf{H}'(\mathbf{p}) = \sum_{i=1}^m \int_{\mathbf{q} \in \mathscr{F}_i} (\mathbf{q} - \mathbf{p})^t \, dq \, \frac{\mathbf{p}_i - \mathbf{p}}{d_i} - V(\mathbf{p})\mathbf{I}$$

where $\mathscr{F}_i$ is the facet of $\mathscr{T}_\mathbf{p}$ that is in $\mathscr{T}_i$. Hence

$$\mathbf{H}'(\mathbf{p}) = \int_{\mathbf{q} \in \partial\mathscr{T}_\mathbf{p}} (\mathbf{q} - \mathbf{p})^t \, \vec{\mathbf{n}} \, dq - V(\mathbf{p})\mathbf{I}$$

where $\vec{\mathbf{n}}$ is the outward pointing unit normal to the polytope $\mathscr{T}_\mathbf{p}$.

Now, let $\vec{\mathbf{e}}_r$ be the r-th (row) vector in the standard basis for $\mathscr{R}^s$ and set

$$\vec{m}_{rc} = [(q - p) \cdot \vec{e}_r]\, \vec{e}_c \qquad r = 1, \ldots, s, \qquad c = 1, \ldots, s.$$

Then the entry in the r-th row and c-th column of $\mathbf{H}'(\mathbf{p})$ is

$$\int_{q \in \partial \mathcal{T}_p} \vec{m}_{rc} \cdot \vec{n}\, dq - V(\mathbf{p})\delta_{rc}$$

which by Stokes' Theorem equals

$$\int_{q \in \mathcal{T}_p} \nabla \cdot \vec{m}_{rc}\, dq - V(\mathbf{p})\delta_{rc}$$

where $\nabla = (\partial_1, \partial_2, \ldots, \partial_s)$ is the gradient operator (this gradient is taken with respect to $\mathbf{q}$). Since $\nabla \cdot \vec{m}_{rc} = \delta_{rc}$,

$$\mathbf{H}'(\mathbf{p}) = \int_{q \in \mathcal{T}_p} dq\,\mathbf{I} - V(\mathbf{p})\mathbf{I} = 0.$$

Since $\mathbf{p}$ is arbitrary, the derivative vanishes for all $\mathbf{p}$ and $\mathbf{H}(\mathbf{p})$ is constant on $\mathcal{C} \backslash \mathcal{P}$.

Now consider

$$\lim_{\mathbf{p} \to \mathbf{p}_j} \mathbf{H}(\mathbf{p}) = \sum_{i=1}^{m} \lim_{\mathbf{p} \to \mathbf{p}_j} \lambda_i(\mathbf{p})(\mathbf{p}_i - \mathbf{p})$$

for some $\mathbf{p}_j \in \mathcal{P}$. In the sum, $\mathbf{p}_j - \mathbf{p}$ approaches zero and for $i \neq j$, $\lambda_i(\mathbf{p})$ approaches zero so each term in the sum approaches zero. Since $\mathbf{H}(\mathbf{p})$ is constant, $\mathbf{H}(\mathbf{p}) = 0$ in $\mathcal{C} \backslash \mathcal{P}$. This proves the local coordinates property.

References

[1] Farin, G.: Surfaces over Dirichlet tessellations. Computer Aided Geometric Design 7(1–4), 281–292 (1990).
[2] Sibson, R.: A vector identity for the Dirichlet tessellations. Math. Proc. Cambridge Philos. Soc. 87, 151–155 (1980).
[3] Sibson, R.: A brief description of the natural neighbour interpolant. In: Barnett, D. V. (ed.) Interpolation Multivariate Data. New York: Wiley 1981.

B. Piper
Department of Mathematical Sciences
Rensselaer Polytechnic Institute
Troy, NY 12180
USA

Computing Suppl. 8, 241–250 (1993)

Automated Feature Recognition and its Role in Product Modelling

M. J. Pratt, Bedford

Abstract. The paper discusses several issues which will need to be addressed in the development of effective feature-based systems for integrated computer-aided design, analysis and manufacture. Design features are often significantly different from the features appropriate for applications downstream of design. This implies that, even in a system permitting design by features, some form of automated recognition of application features will be needed. It is pointed out that this is easier from a model which already contains design feature information than from a pure geometric model, the basis of most research to date on automated feature recognition. Closely related to feature recognition is the process of feature validation, necessary for checking that features created by the designer are not invalidated as the result of subsequent design operations. The paper concludes by making a case for a feature definition language which will permit the integrated system to be configured flexibly to meet the precise requirements of any organisation using it. This will also have strong implications on the development of standards for the transfer of CAD information between different systems.

Key words: Form features, integrated CAD/CAM, feature recognition, feature model transmutation, feature validation, feature definition language.

1. Introduction

Much has been written in the past few years on the topic of form features. These are local geometric configurations on the surface of a part which have significance at some stage during the life-cycle of the part. For example, the designer creates form features for reasons concerned with functionality, while manufacturing features are key elements in determining the operations needed in a manufacturing plan. A review of form features and some of their applications is given in Pratt (1991).

One of the major research areas concerning form features (or simply features, for short) is that of automated feature recognition. The approach usually taken (see the recent review paper by Shah, 1991) is based on the prior existence of a pure geometric model of the part. Feature recognition is then the process of interrogating the data structure of the model in order to identify the presence and nature of the features relating to some process downstream of design. The object of doing this is to provide a basis for the automation of that downstream process. The most popular application area is that of the manufacturing features of machined metal parts, when the relevant feature types include pockets, slots, grooves, bosses and the like. Each of these may only be manufactured in a limited number of ways, the actual choice depending upon available manufacturing resources and possibly also some non-geometric information associated with the model, concerning surface finish or engineering tolerances. The geometric modellers used are in most cases solid modellers of the boundary representation type. It has also proved possible, however, to identify features automatically from 2-dimensional drawing data; Meeran & Pratt (1993) discuss this in the context of machining features and Swift (1987) in the context of features for automated assembly.

Automated feature recognition from geometric models has a history of more than ten years, but recently the idea of design-by-features has come into vogue. This permits the designer actually to create his model in terms of (for example) pockets, slots and grooves, defined upon some initially generated base shape. Some workers have taken the view that this approach to design obviates the need for automated feature recognition, since it is only necessary to require the designer to work in terms of manufacturing features in order to obtain without further ado the basis of a manufacturing plan. Unfortunately this is a naive view; the designer's features are concerned with functionality of the part, and are by no means always coincident with manufacturing features, as will be shown shortly. Further, there are many other potential applications of form features during the life-cycle of a part, and forcing the designer to use manufacturing features still leaves the requirement for automated recognition of these other classes of features. Although design-by-features promises to have many virtues in speeding up the design process and facilitating engineering changes it nevertheless appears not to remove the need for some form of automated feature recognition.

2. Feature Model Transmutation

In what follows the assumption will be made that the initial product model is created by a design-by-features system rather than by a pure geometric modeller. If this is so, then information about the designer's features is present in the model at the outset. The recognition of features for other, downstream, applications may therefore draw not only on the geometric information but also on the design feature data. It is the present writer's belief that the presence of this latter data will significantly ease the task of identifying features for manufacturing and other post-design activities. The extent to which this is true will of course depend upon the appropriateness and completeness of the design feature data. The process of starting with a feature model for one application and generating from it a feature model for a different application has been called *feature mapping* or *feature transformation* (Magleby 1988, Shah 1988), but both 'mapping' and 'transformation' are words having precise mathematical meanings which are not quite appropriate in the present context. Features do not map one into another and are not transformed into one another by the kind of process envisaged. Rather, the features are decomposed into their component elements which are then reassembled in different groupings. The term which will be used here for this operation is *feature model transmutation*. The remainder of this section gives three simple examples to show how feature-related information may be used for this purpose.

2.1 Machined Wing Rib

The first is a familiar example (see Fig. 1), but is worth airing once again. It concerns a $2\frac{1}{2}$D part machined from the solid, whose peripheral shape is specified from aerodynamic considerations and is therefore not under the control of the designer. In order to transform his given 2D shape into a 3D part he will need to add thickness to the shape, to create the *web*, then a peripheral *flange*, to which the wing skin will be bonded or riveted, and some transverse *stiffeners* to give the part structural rigidity. A final stage will probably be the addition of *fillets* where the stiffeners join the flange. The designer's features, with their defining parameters, are thus (i) the peripheral shape (probably specified in terms of spline coefficients), (ii) the web (thickness h), (iii) the flange (height H, thickness T_2), (iv) the stiffeners (height H, thickness T_1) and (v) the fillets (radius R). The machining features are of course the top face, the peripheral shape, and three pockets. How is it then possible to derive the second feature model from the first? One straightforward answer is based on simple labelling of design feature elements. For example, the designer's flange, stiffeners and fillets all have wall faces; if these are automatically labelled as such when the features are first created by the system, then the first step in the identification of the peripheral shape and the pockets is a search for closed loops of connected 'wall' faces. The 'top' faces of all these features also become a single top face of the part as a whole. The process planning system can now proceed on the basis of these machining features, together with any associated surface finish or tolerance data in the model, to plan and sequence the machining operations. The radii of the corner fillets will have implications on the cutter size for both the roughing and finishing cuts of the pocket walls.

This example is idealised in several respects in the interests of clarifying the main principles. In practice matters are likely to be a good deal less simple; one example of added complexity arises when the thicknesses of the flange and/or stiffener features is small. Then the material may deflect under the machining forces applied so that the finished part is not within tolerance of the design specification. This

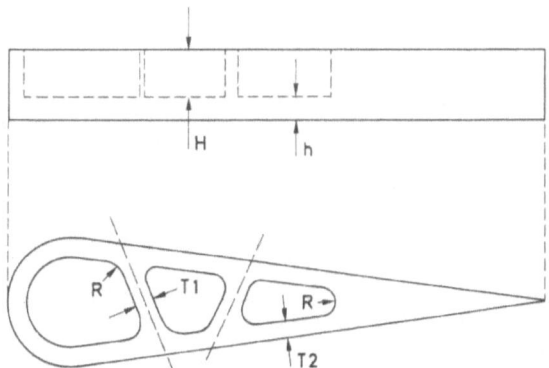

Figure 1. Machined wing rib, showing details of designer's features

problem can be avoided by making the finishing cuts very shallow. In this example, then, the process planning should not be based purely on the machining features (pockets) but should also take the designer's features (stiffeners, flange) into account as secondary information. Such situations occur frequently in practice.

2.2 Machined Block with Rectangular Boss

This part is illustrated in Fig. 2. The designer has based the model on a rectangular block; one smaller block has been removed to create a slot, and another has been added to create the protruding boss. Both of these design features exist to fulfil some functional purpose. The slot feature is also a machining feature, since it defines a volume of material which must be machined away to produce the functional slot. The boss is another matter, however. In this case we have a protrusion, and material must be machined away to reveal the feature in the finished part. The machining feature corresponding to the boss is therefore some volume of material which surrounds the boss. What are its bounding surfaces? Clearly these include the side and top faces of the boss itself, but the other bounding surfaces are rather more difficult to define. They may be related to the surfaces of the original billet of material from which the part is being machined, or possibly to surfaces previously machined at some intermediate stage in the manufacture; this depends upon the order in which the various machining operations are performed. We may therefore have a chicken-and-egg situation where precise details of the machining features cannot be determined until the sequence of operations is known, while the sequencing cannot be determined until the features are known. However this problem is solved (and few solutions have yet been proposed) it is clear that the information in the initial product model, containing only geometry and design feature data, will need to be

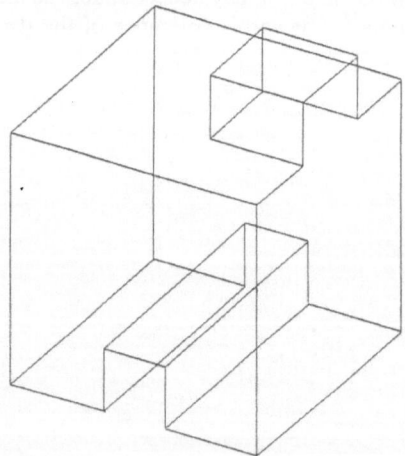

Figure 2. Machined part having slot and boss features

supplemented by other contextual information, at the very least the dimensions of the original billet. All this takes us rather beyond the notion of a form feature as it is generally understood, but the need for this wider concept was recognised by CAM-I (1982) several years ago, and the name *delta-volume* was given to it.

It is clear that the face-labelling stratagem suggested in the previous example is also useful in the construction of delta-volumes. Wall faces of the boss become wall faces of the delta-volume, but now also the surfaces of the 'base' face (from which the boss emerges) and the 'top' face become the surfaces of the bottom and top faces of the desired volume. The other faces of the delta-volume, it should be reiterated, cannot in general be determined in terms of the part geometry alone, and this is the primary distinction between the form feature and the delta-volume.

2.3 Features for Finite Element Analysis

The next example is taken from a benchmark test on automatic finite element (FE) mesh generation (CAM-I 1985). The object subjected to analysis is the one shown in Fig. 3, and the prescribed loading conditions are as follows:

(i) The component is mounted on a cylindrical pillar fitting tightly through the large hole at one end.
(ii) A longitudinal force is applied via a pin fitting the small hole at the other end.

Figure 3 also shows the mesh used for the analysis, with a plot of the resulting stresses superimposed upon it. Note that the four tapering upright pillars on the

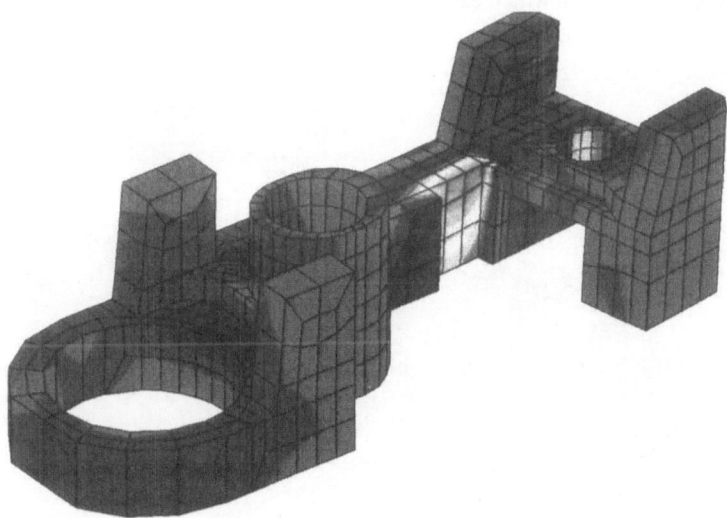

Figure 3. Result of finite element stress analysis of aircraft part, showing subdivision into elements (lighter shade indicates higher stress)

part are very lightly stressed. From the designer's point of view they are present because they are needed for some functional purpose—this is, in fact, a real part taken from an aircraft door locking mechanism. On the other hand, for this particular FE analysis the presence or absence of the four pillar features has no significant effect on the result. The neglect of these features would lead to a much simpler mesh model, and in consequence to a smaller computational problem and a faster and probably more accurate solution. The manual preparation of FE mesh models is a very meticulous and time-consuming process, and some method for the automatic detection of features having no significance in the analysis would be very valuable. Note, however, (and this is very important) that the presence of the pillar features may be crucial to the structural analysis of the same part under different loading conditions, such as a compressive load between their tops and the base of the object. Once again we have an example of the context dependency of features; the significance for FE analysis of the pillars cannot be decided on the information in the designer's model alone, but only when the loading conditions are known. For the present example, the results of neglecting the four pillars is shown in Fig. 4. Note the considerable simplification in the mesh, and also the fact that in the most highly stressed part of the object the stress distribution is indistinguishable from that of the initial full model.

How can the system automatically recognise that the pillar features may be neglected in the desired stress calculation? If the features considered are restricted to design features, which are already present in the product model, some form of initial 'quick and dirty' stress analysis may form the basis of a method, allowing the simplification of the model on which the full accurate analysis is later performed. However, it is possible that the features which may appropriately be neglected are not design features, in which case the problem assumes an added dimension. These

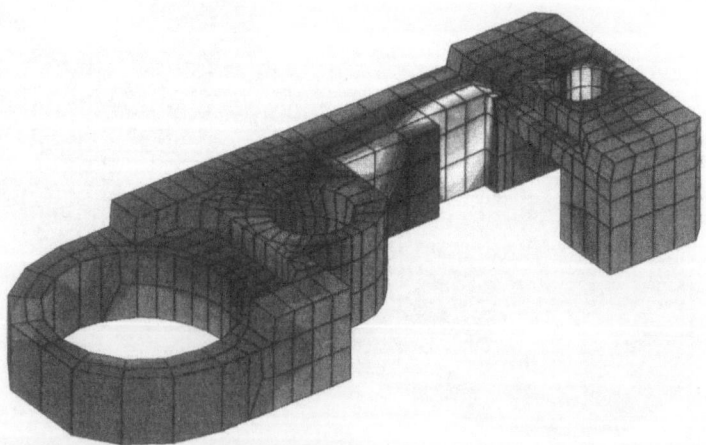

Figure 4. The part of Fig. 3 simplified by omission of pillar features having negligible effect on the stress calculation

features must first be recognised, then the system must check that they exist in a lightly stressed region, and finally they must be deleted from the model.

There are other kinds of features which will also be important for finite element analysis. It is customary to obtain further simplification of the mesh model by idealising certain regions of the full 3D model as plates or shells (meshable in terms of 2D elements) or beams (meshable in terms of 1D elements). The automatic recognition of such regions is therefore also a requirement. A 'shell' feature, for example, may be characterised by the existence in the model of two curved parallel faces whose separation is small compared with their dimensions. Further, in a situation where both the model and the loading conditions are symmetric it is only necessary to analyse part of the model, which allows additional simplification. The automatic detection of symmetries is thus also desirable.

3. Feature Validation during Design

Features belong to classes, and feature classes are distinguished from each other by sets of rules. In a design-by-features system all features invoked by the designer will be created by the system in conformance with those rules, which are concerned not only with the elements making up the feature and their interrelationships but also with the manner in which the feature is installed on the part model (Pratt, 1991).

Example: Blind Cylindrical Hole in Planar Face
Rules (subset only):

Hole bottom face may be planar, conical or hemispherical; if planar it must be perpendicular to the hole axis, and otherwise it must be axially symmetric about the hole axis.

Only wall faces of the hole may intersect the part surface when the hole feature is installed on the model. The intersection curve must be a single circular or elliptical closed loop.

If, for instance, the intersection curve has two circular loops the second rule is broken; the hole is a through hole. Alternatively, if the bottom face intersects the part surface the hole once again is not blind.

Every time a feature is created the appropriate rules should be checked to ensure that the feature is validly positioned and oriented on the part. The process of rule checking is essentially the same process as that of feature recognition, except that in this case the system knows exactly what kind of feature it is trying to recognize; if it fails, the designer has done something wrong.

It is also possible that when the designer creates a feature on the part he destroys the validity of some previously existing feature. Consider the example shown in Figure 5. Here the designer has started with a rectangular block, and has created a perfectly valid cylindrical blind hole in the side of the block. Next he creates a rectangular pocket in the top face of the block. However, this renders the previous hole feature invalid, since what was a blind hole has now become a through hole,

EXAMPLE - Valid blind hole created in block:

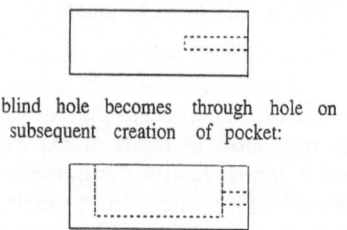

blind hole becomes through hole on
subsequent creation of pocket:

Figure 5. Blind hole feature rendered invalid by subsequent creation of a pocket feature

conforming to a different set of rules. The system must therefore detect, on the installation of any feature on the model, whether it interacts with any existing features and if so whether it renders them invalid. Once again this requires application of the feature class rules.

The examples given in this section indicate that feature validation and revalidation will be required throughout the process of design-by-features. Both processes, since they involve the checking of feature elements against sets of rules, use some of the basic processes of feature recognition. As previously, however, the situations described have been oversimplified in the interests of clarity. In practice the rules defining a class of features for design purposes will be of various types. The most important ones will be concerned with ensuring the functionality of the feature, but others may be concerned with manufacturability or conformance to company standards, for example. Feature validation and revalidation during design has the primary intention of ensuring that features created and modified during the design process possess and retain their intended functionality. Checking of the *functionality* rules is therefore the most crucial requirement; their violation will render a design unacceptable. Should the system detect that other types of rules are violated at any stage then the outcome might simply be an advisory message to the designer, indicating that a feature is now expensive to manufacture or does not conform to company standards. There may be good reasons for either situation, in which case the design may be acceptable. On the other hand the system advice may prompt the designer to seek alternative, more satisfactory solutions for the achievement of the desired functionality.

4. Feature Definition Language

The feature-based design systems of the future will need to be configurable to meet the needs of a wide variety of manufacturing organisations. It must be possible to define precisely those classes of design features which are most meaningful to the designers in a particular company, and also the classes of manufacturing features which are most appropriate for the manufacturing facilities available. The system must therefore provide some form of Feature Definition Language which, however it appears to the user of the system, has the effect internally of creating sets of rules for the definition of the desired feature classes. The rules for each feature class should be grouped into subsets according to their purpose, as indicated in the last paragraph. They will be used not only in the creation and installation of features, and for their validation and revalidation as described above, but also, in the case of manufacturing or other application features, for automated feature recognition through the transmutation process discussed earlier. System configuration, through the medium of this language, will not be one of the functions of the designer; it will be a higher-level responsibility based on decisions collectively taken within the company.

Some progress towards the flexible definition and recognition of features has been reported by Shah & Rogers (1988a, b) and by van Houten et al. (1989). The desirability of this approach calls into question current efforts within the ISO STEP standard development (ISO 1988) towards the standardisation of feature definitions; what is really needed is a standard means for defining features, not a standard set of ready-defined features.

5. Summary and Conclusions

Although automated feature recognition from pure geometric models has been demonstrated by numerous research teams, significant problems remain to be solved, particularly in the elucidation of interactions between features. However, the future appears to lie with systems providing the capability for design-by-features, which implies that feature model transmutation as defined earlier will become increasingly important. Not much research on this topic has yet been published. Some elements of 'pure' feature recognition from geometry and topology will remain a significant requirement in the context of feature validation and revalidation during design.

One of the most interesting challenges facing us is that of finding mechanisms for the flexible definition of feature classes in configurable systems. The definitions must be easily created, and must provide for the requirements not only of design-by-features but also any desired approaches to automated feature recognition or feature model transmutation. A preliminary intensive study is needed to discover what are the most appropriate data requirements and methods for the latter purpose. An eye must be kept firmly on CAD/CAM data transfer standards during this research, and if necessary the standards must be influenced to bring them into line with the emerging new developments in product modelling.

Acknowledgement

The author is indebted to FEQS Ltd, Oakington, Cambridge, England, who provided the illustrations used in Figs. 3 and 4.

References

[1] CAM-I: Design of an advanced numerical control processor. Report R-82-ANC-01, Computer Aided Manufacturing International Inc., Arlington, Texas, 1982.
[2] CAM-I: Solid modeling applications—the real payback. Proc. 3rd CAM-I Geometric Modeling Seminar, March 1985, Nashville, Tennessee; Computer Aided Manufacturing International Inc., Arlington, Texas, 1985.
[3] ISO: Industrial automation systems—exchange of product modelling data—Representation and Format Description (PDES/STEP Version 1.0), ISO Draft Proposal DP10303 (expected to be ratified as a Draft International Standard in 1992). National Institute of Science & Technology (NIST), Gaithersburg, Maryland, 1988.
[4] Magleby, S. P.: Design by functional features for aircraft structure. PhD thesis, University of Wisconsin-Madison, 1988.
[5] Meeran, S., Pratt, M. J.: Automated feature recognition from 2D drawings. Computer Aided Design 25, 1 (1993).
[6] Pratt, M. J.: Aspects of form feature modelling. In: Hagen, H., Roller, D. (eds.) Geometric Modeling: Methods and Applications. Berlin, Heidelberg: Springer 1991.
[7] Shah, J. J.: Feature transformations between application-specific feature spaces. Computer Aided Engineering Journal 5, 6, 247–255 (1988).
[8] Shah, J. J.: Assessment of features technology. Computer Aided Design 23, 5, 331–343 (1991).
[9] Shah, J. J., Rogers, M. T.: Functional requirements and conceptual design of the feature-based modelling system. Computer Aided Engineering Journal 5, 1, 9–15 (1988a).
[10] Shah, J. J., Rogers, M. T.: Feature-based modeling shell: design and implementation. Proc. ASME Computers in Engineering Conf., San Francisco, July/Aug 1988; American Society of Mechanical Engineers, 1988b.
[11] Swift, K. G.: Knowledge-based design for manufacture, Kogan Page, 1987.
[12] van Houten, F. J. A. M., van't Erve, A. H., Kals, H. J. J.: PART: A Feature-based CAPP System. Proc. 21st CIRP International Seminar on Manufacturing Systems, June 1989, Stockholm, Sweden, 1989.

Dr. M. J. Pratt
21 High Street
Carlton, Bedford MK43 HA
England

Computing Suppl. 8, 251–258 (1993)

Computing
© Springer-Verlag 1993

Approximate C^r-Blending with Tensor Product Polynomials

H. Prautzsch, Karlsruhe

Abstract. An algorithm is presented which generates a 4-sided C^r surface. The surface is piecewise polynomial of degree $2r + 1$ and interpolates arbitrary C^k boundary data within a given tolerance. The number of polynomial pieces depends on the subdivision strategy used. Therefore it can be controlled. The method developed in this paper has the advantage over known solutions that it generates no singularities and generates a non-rational surface versus a rational one.

Key words: Subdivision, blending, Bézier surfaces, approximation, C^r.

1. Introduction

A common problem in CAGD is the following interpolation problem. A C^r surface patch $\mathbf{p}(u, v)$, $(u, v) \in [0, 1]^2$, is to be constructed from its boundaries, i.e., $\dfrac{\partial^\rho}{\partial u^\rho} \mathbf{p}(i, v)$ and $\dfrac{\partial^\rho}{\partial v^\rho} \mathbf{p}(u, i)$ are given for $i = 0, 1$ and $\rho = 0, 1, \ldots, r$ while $\mathbf{p}$ is unknown otherwise.

Difficulties arise if higher mixed partial derivatives are incompatible at the corners, e.g., if

$$\lim_{u \to 0} \frac{\partial^{r+1}}{\partial u^r \partial v} \mathbf{p}(u, 0) \neq \lim_{v \to 0} \frac{\partial^{r+1}}{\partial u^r \partial v} \mathbf{p}(0, v).$$

Known solutions [2, 5] are therefore rational with singularities at the corners. In [3] a simpler method was proposed for cubic boundary curves. The method gives an approximation to the interpolation problem which can be refined iteratively such that the limiting surface consists of infinitely many bicubic pieces and solves the interpolation problem exactly. A similar method exists for the construction of triangular piecewise bicubic patches with prescribed boundaries and cross boundary derivatives [4].

In this paper, the method given in [3] is formulated for higher smoothness degrees. Also new parameters are introduced which can be used to get a solution consisting of fewer polynomial pieces at the same approximation error.

The central part of this paper is the algorithm given in Section 3. The algorithm itself is based on subdivision techniques. Therefore one should use the Bernstein-

Bézier representation (see, e.g., [1]) for numerical computations. Here, however, a slightly different representation of polynomials is used in order to pave the way for the subsequent analysis in Sections 6 and 7.

2. Notation

Throughout the paper, the small letters a, b, d, e with and without sub- and superscripts denote bivariate polynomials of degree $(2r + 1, 2r + 1)$, the letters p, q piecewise bivariate polynomials. A polynomial $a(u, v)$ can be given by its derivatives

$$D(a, \mathbf{x}) = \left[\frac{\partial^{i+j}}{\partial u^i \partial v^j} a(\mathbf{x}) \right], \qquad (i,j) \in \{0, 1, \ldots, r\}^2, \qquad (2.1)$$

at $\mathbf{x} = (0,0), (1,0), (0,1), (1,1)$. In particular there are polynomials e_{ij} for which the matrix of their derivatives

$$E_{ij} = \begin{bmatrix} D(e_{ij}, (0,0)) & D(e_{ij}, (0,1)) \\ D(e_{ij}, (1,0)) & D(e_{ij}, (1,1)) \end{bmatrix}$$

has only zero entries besides $E_{ij}[i,j] = 1$. Note that

$$\{e_{ij} | i,j = 0, 1, \ldots, 2r + 1\} \qquad (2.2)$$

is a basis for the space of all bivariate polynomials up to degree $(2r + 1, 2r + 1)$. It is the Hermite two point Taylor tensor product basis.

In CAGD applications, one usually deals with vector valued functions. Hence, $\mathbf{p}$, $\mathbf{q}$, $\mathbf{a}$, $\mathbf{b}$, ... may map into some $\mathbb{R}^d$. Then the entries of $D(a, \mathbf{x})$ are vectors in $\mathbb{R}^d$. Vectors, points, and vector valued functions are denoted by bold letters.

3. The Method

Let $\mathbf{f} \in C^r[0,1]^2$ be the function whose boundaries and cross boundary derivatives up to order r are to be interpolated. The boundaries and cross boundary derivatives of $\mathbf{f}$ are assumed to be polynomial of degree $\leq 2r + 1$.

Although there is no polynomial solution in general, one can always interpolate opposite boundaries of $\mathbf{f}$ by a polynomial. Thus let $\mathbf{a}$ and $\mathbf{b}$ be two bivariate polynomials of degree $\leq (2r + 1, 2r + 1)$ such that $\mathbf{a}(u, v)$ agrees with $\mathbf{f}(u, v)$ and all its v-partials up to order r for all $(u, v) \in [0, 1] \times \{0, 1\}$ and $\mathbf{b}(u, v)$ agrees with $\mathbf{f}(u, v)$ and all its u-partials up to order r for all $(u, v) \in \{0, 1\} \times [0, 1]$.

The procedure BLEND below needs $\mathbf{a}$ and $\mathbf{b}$ as input and renders a piecewise polynomial surface which solves the interpolation problem within a given tolerance ε. BLEND is a recursive procedure with two integer parameter m and n where m is incremented by one during each call of BLEND and where n defines the maximum recursion level. In Section 7 it is shown how the number of recursion levels $n - m$ can be computed from the tolerance ε. Each recursive call of BLEND generates a piece of the solution. The parameter interval of this piece is given by $\mathbf{x} = (x_1, x_2)$ and $\mathbf{y} = (y_1, y_2)$, i.e., $[\mathbf{x}, \mathbf{y}] = [\min\{x_1, y_1\}, \max\{x_1, y_1\}] \times [\min\{x_2, y_2\}, \max\{x_2, y_2\}]$ is the parameter interval.

BLEND $(\mathbf{a}, \mathbf{b}, m, n, \mathbf{x}, \mathbf{y})$

begin
    if $m \geq n$ or $\mathbf{a} = \mathbf{b}$
    then render $\frac{1}{2}[\mathbf{a} + \mathbf{b}](u, v)$ for $(u, v) \in [\mathbf{x}, \mathbf{y}]$
    else begin
        Increment m by 1.
        The point $\mathbf{z} = \gamma\mathbf{y} + (1 - \gamma)\mathbf{x}$ divides the interval $[\mathbf{x}, \mathbf{y}]$ into
        four subintervals F_i, $i = 1, 2, 3, 4$, cf. Fig. 1.

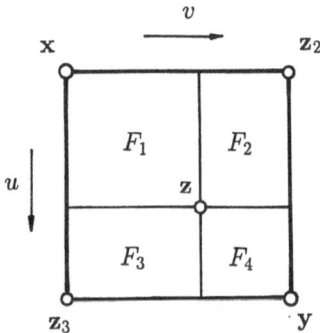

Figure 1. The subintervals F_i

For each F_i define polynomials $\mathbf{a}_i^{(m)}$ such that in the corners of F_i, $D(\mathbf{a}_i^{(m)}, \cdot)$ agrees with $D(\mathbf{a}, \cdot)$, $D(\mathbf{b}, \cdot)$ or $D(\frac{1}{2}[\mathbf{a} + \mathbf{b}], \cdot)$ if the corner is marked by $+$, $-$, or o, respectively in Fig. 2.

Similarly define polynomials $\mathbf{b}_i^{(m)}$ via Fig. 3.

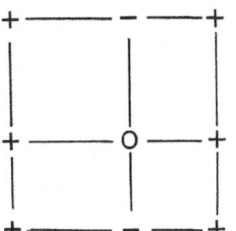
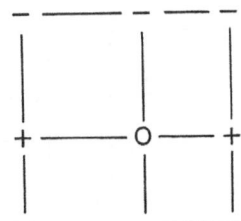

Figure 2. Defining the $\mathbf{a}_i^{(m)}$ **Figure 3.** Defining the $\mathbf{b}_i^{(m)}$

Let $\mathbf{z}_1 = (x_1, x_2) = \mathbf{x}$
$\qquad \mathbf{z}_4 = (y_1, y_2) = \mathbf{y}$
$\qquad \mathbf{z}_2 = (x_1, y_2)$
$\qquad \mathbf{z}_3 = (y_1, x_2)$
For $i = 1, 2, 3, 4$
$\qquad$ execute BLEND $(\mathbf{a}_i^{(m)}, \mathbf{b}_i^{(m)}, m, n, \mathbf{z}_i, \mathbf{z})$
$\quad$ **end** of else
end of the procedure.

4. Discussion

As mentioned earlier for practical computations with BLEND, it is advantageous to represent the polynomials in their Bernstein-Bézier form. Then the $\mathbf{a}_i$ and $\mathbf{b}_i$ can be determined by a subdivision and subsequent averaging and interchanging of Bézier points, cf [3]. In this form, BLEND computes only convex linear combinations.

In order to prepare for the analysis in the following Sections we observe a few facts about $\mathbf{p}_n(u, v)$, the piecewise polynomial surface over $[0, 1]^2$ rendered by BLEND $(\mathbf{a}, \mathbf{b}, 0, n, \mathbf{o}, \mathbf{e})$, where $\mathbf{o} = (0, 0)$ and $\mathbf{e} = (1, 1)$. The piecewise polynomial surface

$$\mathbf{q}_a(u, v) := \mathbf{a}_i^{(1)}(u, v), \qquad (u, v) \in F_i,$$

is C^r continuous and interpolates $\mathbf{a}$ and its v-partials derivatives up to order r at all $(u, v) \in [0, 1] \times \{0, 1\}$. Similarly,

$$\mathbf{q}_b(u, v) := \mathbf{b}_i^{(1)}(u, v), \qquad (u, v) \in F_i,$$

is a C^r surface interpolating $\mathbf{b}$ and its cross boundary derivatives up to order r at all $(u, v) \in \{0, 1\} \times [0, 1]$.

During each call of BLEND with $m \geq 1$, $D(\mathbf{a}^{(m)}, \cdot)$ and $D(\mathbf{b}^{(m)}, \cdot)$ agree at three corners of $[\mathbf{x}, \mathbf{y}]$. Therefore three of the four pairs $\mathbf{a}_i^{(m+1)}, \mathbf{b}_i^{(m+1)}$ consist of identical patches which are rendered at the next recursion level. For this reason, $\mathbf{p}_n$ is polynomial over each of the subtiles of $[0, 1]^2$ depicted in Fig. 4 where the left, middle, and right part of the figure correspond to $n = 0, 1, 3$, respectively.

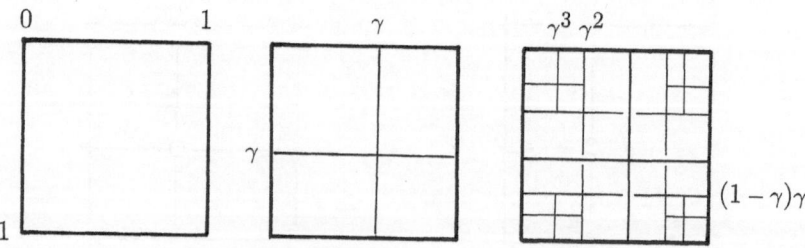

Figure 4. The domains of the polynomial pieces

On summarizing these facts, we can conclude that the limit surface $p(u, v) = \lim_{n \to \infty} p_n(u, v)$ is C^r continuous in $[0, 1]^2 \backslash \{0, 1\}^2$ and interpolates $f(u, v)$ with its cross boundary derivatives up to order r at all $(u, v) \in [0, 1] \times \{0, 1\}^2 \cup \{0, 1\}^2 \times [0, 1]$. It remains to analyze p in $\{0, 1\}^2$.

5. Properties and Assumptions

The analysis of p addresses its smoothness in $\{0, 1\}^2$ and the convergence rate of $(p_n)_{n=0}^{\infty}$. These questions can be settled for each component of p separately. Thus we assume from now on that $a, b, p, \ldots$ are scalar valued functions. Also we consider p over $[0, \gamma]^2$, $[\gamma, 1] \times [0, \gamma]$, $[0, \gamma] \times [\gamma, 1]$ and $[\gamma, 1]^2$ separately. Because of symmetry reasons it suffices to look at p over $[0, \gamma]^2$. Equivalently we can and will analyze $p = \text{BLEND}(a, b, 0, \infty, \mathbf{o}, \mathbf{e})$ over $[0, 1]^2$ where

$$D(a, \cdot) = D(b, \cdot) \qquad \text{at } (0, 1), (1, 0), (1, 1). \tag{5.1}$$

Namely by recursive induction one verifies that p can also be obtained for any $\sigma > 0$ through

$$\text{BLEND}(a(\sigma(u, v)), b(\sigma(u, v)), 0, \infty, \mathbf{o}, \sigma^{-1}\mathbf{e}) = p(\sigma(u, v)). \tag{5.2}$$

The analysis in the next Section rests on a few properties of $\text{BLEND}(\cdot, \cdot) = \text{BLEND}(\cdot, \cdot, m, n, \mathbf{x}, \mathbf{y})$. First there are two linear properties:

$$\text{BLEND}(a + \bar{a}, b + \bar{b}) = \text{BLEND}(a, b) + \text{BLEND}(\bar{a}, \bar{b}) \tag{5.3}$$

$$\text{BLEND}(\lambda a, \lambda b) = \lambda \text{BLEND}(a, b). \tag{5.4}$$

Since $\text{BLEND}(a, a) = a$, we can conclude then

$$\text{BLEND}(a, b) = a + \sum_{i,j=0}^{r} \tau_{ij} \text{BLEND}(0, e_{ij}) \tag{5.5}$$

where

$$b - a = \sum_{i,j=0}^{r} \tau_{ij} e_{ij} \tag{5.6}$$

and e_{ij} are the polynomials defined in (2.2).

Because a and b have the same differentials up to order r at $(0, 0)$, see Section 3, one has

$$\tau_{ij} = 0 \qquad \text{for } i + j \leq r. \tag{5.7}$$

6. Analysis

Because of (5.5) and (5.7) we consider $p_n = \text{BLEND}(a, b, 0, n, \mathbf{o}, \mathbf{e})$ where $a = 0$ and $b = e_{ij}, i + j > r, i \leq r, j \leq r$. Then $d_m := b_1^{(m)} - a_1^{(m)}$ satisfies $D(d_m, \mathbf{o}) = D(e_{ij}, \mathbf{o})$ and $D(d_m, \cdot) = 0$ at $(0, \gamma^m), (\gamma^m, 0)$ and (γ^m, γ^m). Hence

$$d_m(u,v) = \gamma^{m(i+j)} e_{ij}(\gamma^{-m}(u,v)). \tag{6.1}$$

Let

$$q_m(u,v) := \begin{cases} (p_{m+1} - p_m)(u,v) & \text{for } (u,v) \in [0,\gamma^m]^2 \\ 0 & \text{otherwise}. \end{cases}$$

Then $p_n = p_0 + \sum_{m=0}^{n-1} q_m$ where q_m is generated by

$$q_m = \text{BLEND}(a_1^{(m)}, b_1^{(m)}, 0, 1, \mathbf{o}, \gamma^m \mathbf{e}) - \text{BLEND}(a_1^{(m)}, b_1^{(m)}, 0, 0, \mathbf{o}, \gamma^m \mathbf{e})$$

and because of (5.3)

$$q_m = \text{BLEND}(0, d_m, 0, 1, \mathbf{o}, \gamma^m \mathbf{e}) - \text{BLEND}(0, d_m, 0, 0, \mathbf{o}, \gamma^m \mathbf{e})$$

where $a_1^{(0)} = a$, $b_1^{(0)} = b$ and $d_0 = b - a$. Employing (6.1), (5.4), and (5.2) we arrive at

$$q_m(u,v) = \gamma^{m(i+j)} q_0(\gamma^{-m}(u,v))$$

and

$$p(u,v) = p_0(u,v) + \sum_{m=0}^{\infty} \gamma^{m(i+j)} q_0(\gamma^{-m}(u,v)). \tag{6.2}$$

Since $\gamma < 1$ and $i + j > r$, see before (6.1), this infinite sum converges and also its mixed partial derivatives up to order r.

More generally, if $b - a = \sum \tau_{ij} e_{ij}$ as in (5.6) and (5.7), one can define for each component

$$q_{ij}(u,v) = \text{BLEND}(0, e_{ij}, 0, 1, \mathbf{o}, \mathbf{e}) - \text{BLEND}(0, e_{ij}, 0, 0, \mathbf{o}, \mathbf{e})$$

and derive from (6.2) and (5.3)

$$p(u,v) = \frac{1}{2}(a + b)(u,v) + \sum_{i,j} \tau_{ij} \sum_{m=0}^{\infty} \gamma^{m(i+j)} q_{ij}(\gamma^{-m}(u,v)). \tag{6.3}$$

7. Convergence Rate

Applying the operator

$$\mathscr{D} = \frac{\partial^{\rho+\sigma}}{\partial u^\rho \partial v^\sigma}, \qquad \rho \geq 0, \quad \sigma \geq 0, \quad \rho + \sigma \leq r, \tag{7.1}$$

to (6.3) gives

$$\mathscr{D}p = \frac{1}{2}\mathscr{D}(a + b) + \sum_{ij} \tau_{ij} \sum_{m=0}^{\infty} \gamma^{m(i+j-\rho-\sigma)} [\mathscr{D}q_{ij}](\gamma^{-m}(\cdot, \cdot)).$$

Thus the deviation between the limit surface and the surface obtained after n recursions can be estimated as

$$|\mathscr{D}(p - p_n)| \leq \sum_{i,j} \tau_{ij} \sum_{m=n}^{\infty} \gamma^{m(i+j-\rho-\sigma)} \max_{[0,1]^2} |\mathscr{D}q_{ij}(u,v)|. \tag{7.2}$$

The lesser $\rho + \sigma$ the quicker is the convergence. This means that higher derivatives are more expensive to approximate, in general.

The choice of γ governs the convergence rate for (p_n) and also affects the constants $\max |\mathscr{D}q_{ij}|$. In general, $\max |\mathscr{D}q_{ij}|$ becomes bigger for smaller γ. However, for $r = 1$ the influence of γ is bounded. More details are explained in the sequel.

First we observe that $q = q_{ij}$ is a polynomial over each F_i, $i = 1, 2, 3, 4$, where the F_i are as in Fig. 1 for $\mathbf{x} = \mathbf{o}$ and $\mathbf{y} = \mathbf{e}$. Further, $D(q, \cdot) = 0$ at all corners of the F_i except that

$$D(q, (0, \gamma)) = D(e, (0, \gamma))$$

and

$$D(q, (\gamma, 0)) = D(-e, (\gamma, 0))$$

where

$$e = \tfrac{1}{2}(e_{ij} - 0).$$

Over F_1 we have

$$q = \sum_{k,l=0}^{2r+1} \varepsilon_{kl}\gamma^{k'+l'}e_{kl}(\gamma^{-1}(u, v)), \qquad (u, v) \in [0, \gamma]^2$$

where

$$[\varepsilon_{kl}] = \begin{bmatrix} 0 & D(e, (0, \gamma)) \\ D(-e, (\gamma, 0)) & 0 \end{bmatrix}$$

and

$$k' = k \bmod(r + 1), \qquad l' = l \bmod(r + 1).$$

Since

$$\frac{\partial^{\mu+\nu}}{\partial u^\mu \partial v^\nu} e(0, 0) = 0 \qquad \text{for } \mu + \nu \leq r,$$

there are constants C_{kl} such that

$$|\varepsilon_{kl}| \leq \gamma^{(r+1-k'-l')_+} C_{kl}, \qquad \gamma \in [0, 1],$$

where

$$(s)_+ = \begin{cases} s & \text{if } s > 0 \\ 0 & \text{otherwise.} \end{cases}$$

Hence

$$\max_{[0, \gamma]^2} |\mathscr{D}q(u, v)| \leq \sum_{k,l=0}^{2r+1} C_{kl}\gamma^{k'+l'-\rho-\sigma+(r+1-k'-l')_+} \max_{[0, 1]^2} |\mathscr{D}e_{kl}(u, v)|. \qquad (7.3)$$

Recall from (7.1) that $\rho + \sigma \leq r$. Therefore $k' + l' - \rho - \sigma + (r + 1 - k' - l')_+ \geq 1$, i.e.,

$$\max_{[0,\gamma]^2} |\mathscr{D}q(u,v)| \leq \sum_{k,\bar{l}=0}^{2r+1} C_{kl} \max_{[0,1]^2} |\mathscr{D}e_{kl}(u,v)|.$$

Over F_2 things are slightly different. There one has

$$q = \sum_{k,\bar{l}=0}^{2r+1} \varepsilon_{kl} \gamma^{k'} (1-\gamma)^{l'} e_{kl}\left(\frac{u}{\gamma}, \frac{v-\gamma}{1-\gamma}\right), \qquad (u,v) \in [0,\gamma] \times [\gamma,1]$$

where

$$[\varepsilon_{kl}] = \begin{bmatrix} D(e,(0,\gamma)) & 0 \\ 0 & 0 \end{bmatrix}.$$

Again there are constants C_{kl} such that

$$[\varepsilon_{kl}] \leq \gamma^{(r+\frac{1}{2}-k'-l')_+} C_{kl}$$

But different from (7.3) one gets

$$\max_{[0,\gamma]\times[\gamma,1]} |D(q,(u,v))| \leq \sum_{k,\bar{l}=0}^{2r+1} C_{kl} \gamma^{k'-\rho+(r+1-k'-l')_+} (1-\gamma)^{l'-\sigma} \max_{[0,1]^2} |\mathscr{D}e_{kl}(u,v)|.$$

$$(7.4)$$

One obtains an analogous result for q over F_3. q over F_4 is identically 0 and therefore not of interest here. Because of symmetry reasons it suffices to consider $\gamma \in [0,1/2]$. For $r = 1$ one can show that in (7.4) $\alpha = \gamma^{k'-\rho+(r+1-k'-l')_+}(1-\gamma)^{l'-\sigma}$ is bounded in $[0,1/2]$ for $\rho + \sigma \leq 1$. Note that α goes to infinity as γ tends to zero for $(k,l,\rho,\sigma) = (k,1,2,0)$. This means that p has high curvature. Although p solves the problem of Section 3, its shape may be unexpected.

For $r \geq 2$ and $\rho + \sigma \leq r$ there are always unbounded α's. For $(k,l,\rho,\sigma) = (1,r,2,\sigma)$ one has $\alpha \to \infty$ as $\gamma \to 0$.

References

[1] Boehm, W., Farin, G., Kahmann, J.: A survey of curve and surface methods in CAGD. Comp. Aided Geometric Design *1984*, 1–60.
[2] Gregory, J. A.: C^1 rectangular and non-rectangular surface patches. In: Barnhill, Boehm (eds.) Surfaces in CAGD, pp. 25–33. North-Holland 1983.
[3] Prautzsch, H.: Algorithmic blending. Journal of Approx. Theory (1992) (to appear).
[4] Prautzsch, H.: Approximate C^1-blending with triangular cubic patches, to appear in the proceedings of the Eurographics workshop, Santa Margharita Oct. 1991. Berlin, Heidelberg, New York, Tokyo: Springer 1992.
[5] Takai, K., Wang, K. K.: Curvature–continuous Gregory patch: A modification of Gregory patch for continuity of curvature, Proc. Japan—U.S.A., Symposium on Flexible Automation, Kyoto, Japan (1990) 1205–1211.

H. Prautzsch
Fakultät für Informatik der Universität
D-W-7500 Karlsruhe 1
Federal Republic of Germany

Computing Suppl. 8, 259–266 (1993)

Shape Information in Industry Specific Product Data Model

D. Roller, Stuttgart

Abstract. In his paper first an architectural approach to the development of a CIM solution portfolio for a specific type of discrete manufacturing industry is presented. This approach is based on enterprise-wide information modelling. In particular, a product data model for automotive supplier companies is described that has been developed as one of a set of means to get to a consistent and comprehensive CIM software solution. Within the presented product data model the aspect of the shape information is laid out in detail.

Key words: Shape information, product data model, enterprise-wide data model, computer integrated manufacturing.

1. Development of an Industry Specific CIM Solution Portfolio

One of the most significant pressures on the manufacturing industry is the reduction of the time span from product development to the first shipment of a new product series. Frequently, the manufacturing costs need to be minimized at the same time, in order to keep abreast with the increasingly competitive market. An approach that generally is expected to support these goals is the comprehensive use of computer-based solutions. However, these solutions need to be integrated in order to get a maximum leverage and fast throughput.

Of particular relevance are integrated solutions in research and development, production planning and control and the operational part of manufacturing. The use of such integrated computer systems within an industrial company is commonly referred to as computer integrated manufacturing (CIM) [9]. Comprehensive CIM projects have been initiated by large automotive companies as well as in government funded international research projects like CIM-OSA [1]. In this paper an approach is presented that can serve as a means to develop a comprehensive and consistent software portfolio for a manufacturing enterprise.

The needs of industrial companies in terms of a CIM solution typically differ substantially. Even within the class of discrete manufacturers the requirements can vary a lot, depending on the company's structure and the type of products that are manufactured. The development of a CIM solution for one particular company would be very expensive and a solution for a wide class of companies is not possible because of the different requirements. However, the goal of the project, from which a particular aspect is described in this paper, is to develop a solution portfolio for

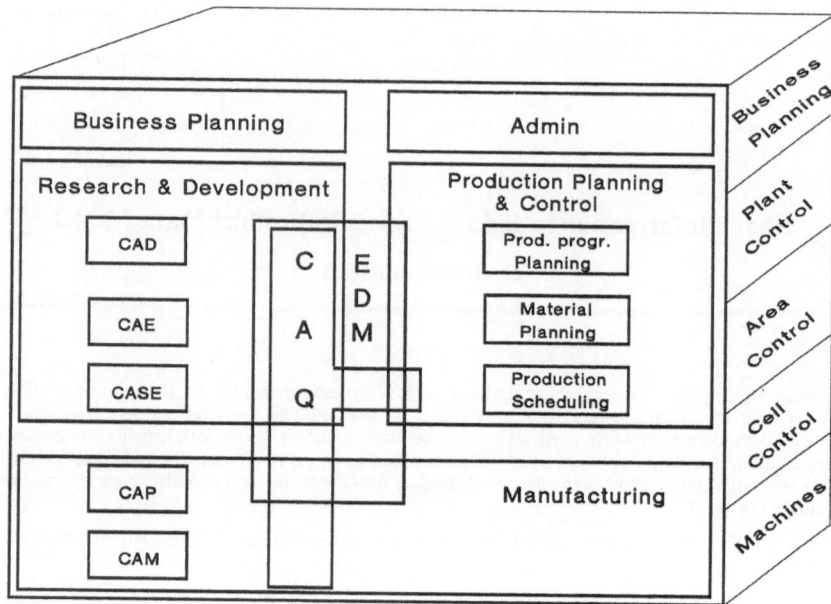

Figure 1. CIM enterprise application areas

a particular type of discrete manufacturing enterprise [8]. For any particular company of this type, only the relevant parts of the portfolio need to be specified in order to get to a particular CIM solution.

In this work the type of enterprises that are covered are automotive supplier companies that manufacture electro-mechanical assemblies. Figure 1 shows the relevant application areas of such type of a company. Obviously, CAD covers mechanical as well as electrical applications. Since many assemblies include electronic control units with firmware, also computer aided software engineering (CASE) needs to be considered. Sometimes the quality of the firmware is directly related to safety aspects, like in anti blocking brake systems.

Note, that some of the areas, particularly computer aided quality management (CAQ) and engineering data management (EDM), are overlapping with various other application areas. It is evident that requirements of all application areas need to be considered in order to optimally integrate an information technology solution.

2. Industry Specific Enterprise-Wide Information Models

In practice it is not viable—if not impossible within given constraints—to design and implement a comprehensive CIM solution from scratch. For the bulk of needed software this is also not necessary. Rather, tools for the selection of a suitable set of solutions are required. It is mandatory, however, that the various software

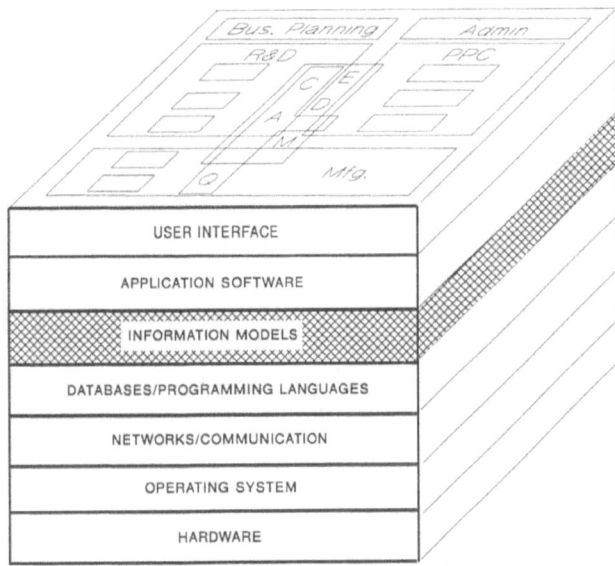

Figure 2. CIM information technology layers

modules fit in the sense, that they make up an integrated solution. This means that data and functions that are relevant in several application areas need to be consistent [9, 6]. From the above outlined situation the importance of an architectural frame work for the development of an effective set of CIM software products becomes clear.

As a basis for an application software architecture, certain industry specific enterprise-wide information models are introduced. In order to be able to cover requirements that are typical within the selected industry, a sufficient level of detail of these models is needed. This obviously leads to industry specific information models. Figure 2 shows how these information models fit into the set of information technology layers, which a CIM solution is typically based on.

More specifically, the models that are used are an *enterprise-wide data model* (EDM) and an *enterprise-wide function model* (EFM). The EDM explicitly captures the generic relations of all data that describe the target industrial enterprise. This information is fundamental for the understanding of how software solutions need to fit from a logical data point of view. However, supporting the right data relations is not sufficient in a CIM application software concept.

A further important requirement is the coverage of the functionality needed in the industry. The EFM provides this information and can serve as a tool to check the completeness of a CIM solution. Figure 3 shows a schematic of the used industry-specific enterprise-wide information models. Note, that in the schematic only extremely small sections of the overall models are shown. If fact, the models developed within this work contain over thousand elements each.

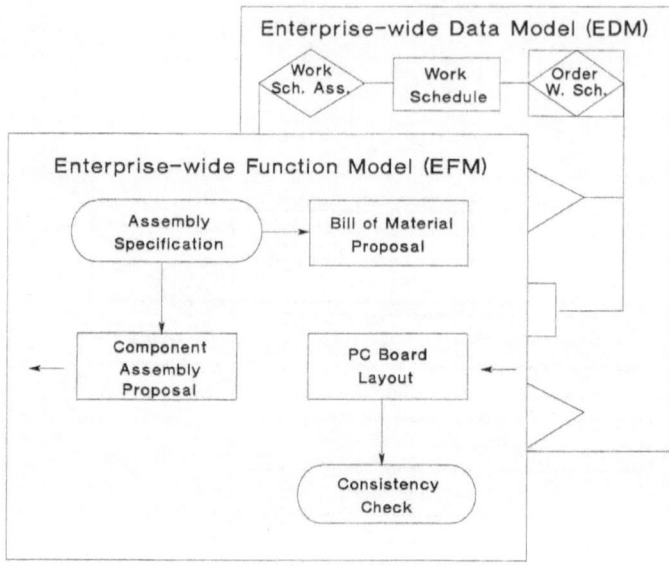

Figure 3. CIM information models

In the next chapter the product related part of the EDM is presented in more detail.

3. Product Data Model

A technique that is widely used for data modelling is the entity relation ship model, first introduced by Chen [2]. Several modifications and enhanced versions have been proposed since then [10, 5]. The work described in this paper is based on the extended entity relationship model from Scheer [10], which has already served as enterprise-wide data modelling technique in other CIM projects.

One aspect that is particularly important for the integration of computer aided design, planning and manufacturing systems, is the product related data. These data and their relations make up a partial model of the EDM, the *product data model* (PDM). Figure 4 shows the different aspects that potentially are part of a PDM. Note, that this is just a simple schematic to illustrate the range of data, it is not intended to represent a true set diagram.

For the purpose of exchanging product data between different software and hardware systems in a standardized format, product data models already have been developed in internationally supported standardization projects, like PDES and STEP [3, 4]. However, these product data models are neither industry specific nor cover they the full spectrum of design data in sufficient detail for the purpose of development of a CIM software platform. Whereas in their current state they are fairly sophisticated in the description of geometry they are not much advanced in

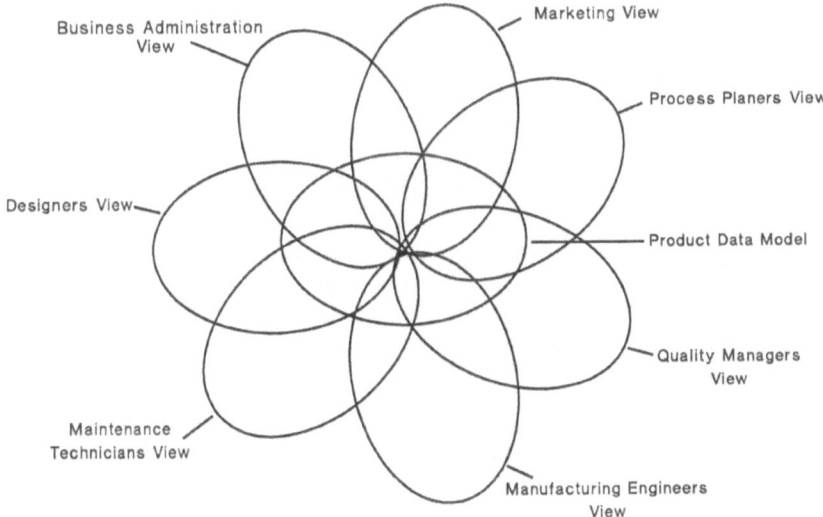

Figure 4. Product data model as a combination of views of a product

other areas, like CAQ. However, particularly the CAQ aspect is of significant importance for the target industry of this work.

The scope of the product data model developed as part of the industry-specific EDM covers all information that is established during the product life cycle, focussing on

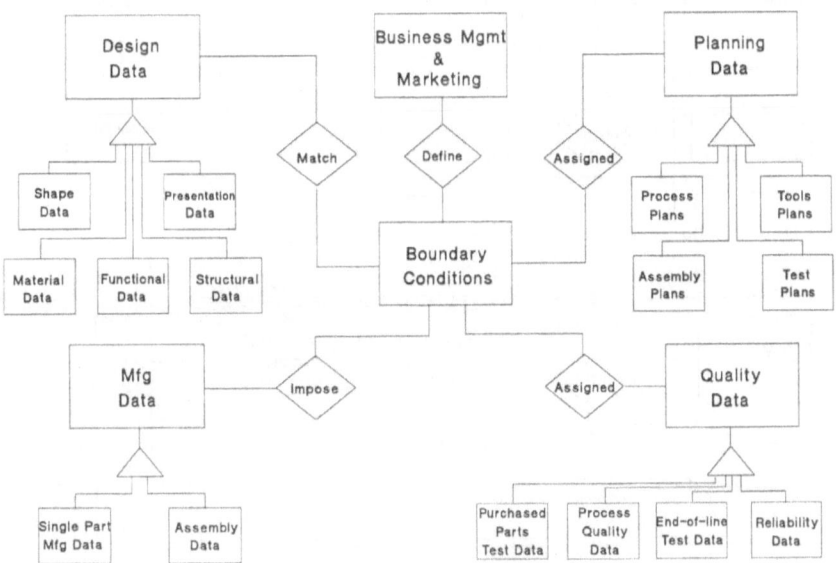

Figure 5. Overview structure of the product data model

design data,
planning data,
manufacturing data,
and quality assurance data.

Figure 5 shows the principal structure of the PDM. In this notion, the rectangles denote data elements, rhombs represent relations between data elements and the triangle mark generalization operators.

On a more detailed level the PDM explicitly captures the relations between data of different functional areas. Thus, it provides a basis for the planning of an integrated set of CA … solutions (CAD, CAP, CAQ, CAM) for the specific industry under consideration. Figure 6 shows a part of the PDM that includes the design data and their external relations.

Figure 7 eventually focuses on the shape data. It shows on the lowest level of detail within the developed PDM, what in Fig. 6 still was just one element. Note, that Fig. 7 for clarity reasons does not include the relations to data elements outside the design data.

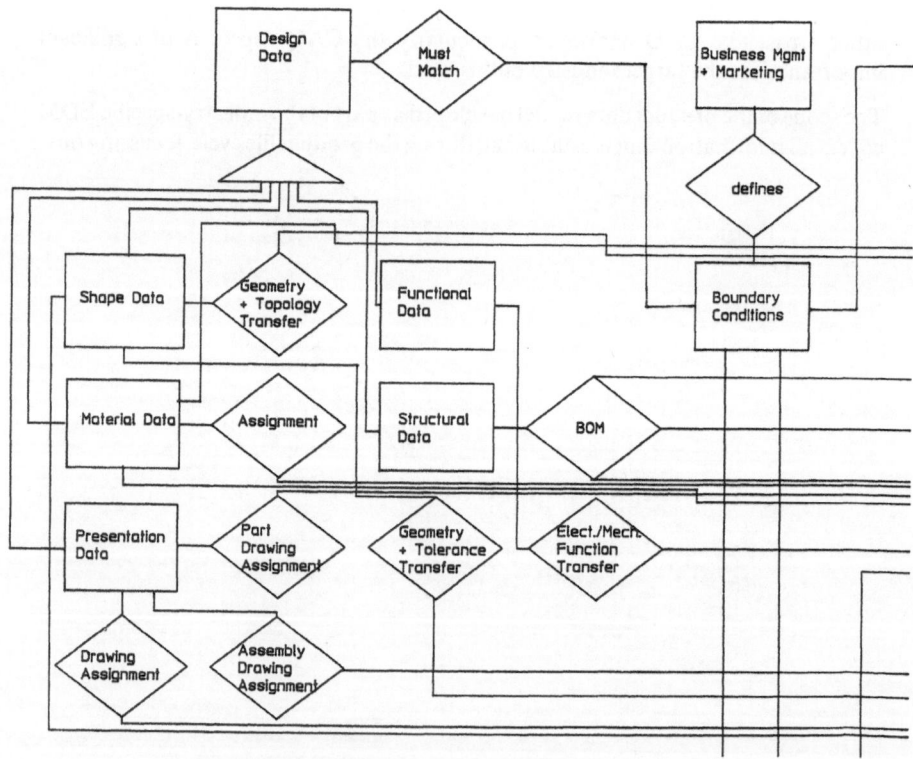

Figure 6. Detail section of design data including external relations

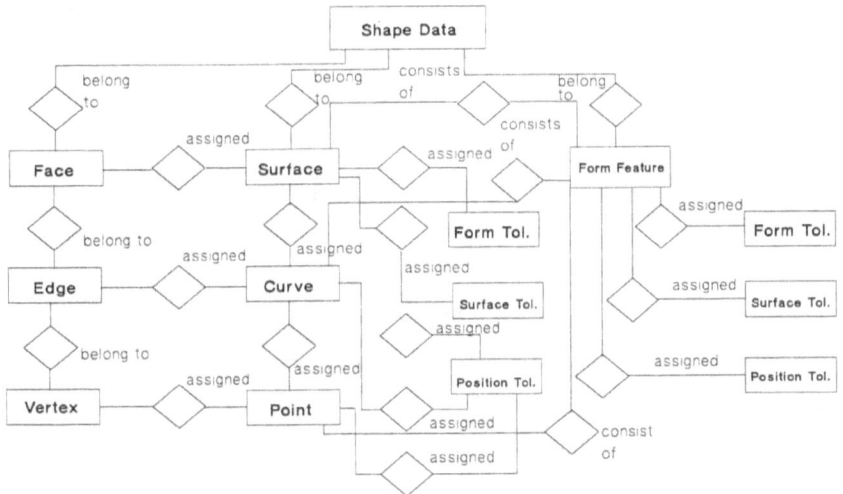

Figure 7. Partial model for shape data

The shape data model essentially includes geometric data elements (surfaces, curves, points), topological data elements (faces, edges, vertexes), tolerance data elements (form tolerances, surface tolerances, position tolerances) and the concept of form features [7]. Tolerances may be assigned to geometric elements as well as to form features. The form features themselves are made up of geometric elements. The requirements for the representation of this type of data stem from model analyses (e.g. topology), manufacturing planning (e.g. form features), quality control (e.g. tolerances) and manufacturing (e.g. geometry).

It should be mentioned that this model for shape data does not prescribe a particular CAD system nor a particular model input technique. However, it specifies which shape describing data should be accessible in principle for applications in the target industry and how these data relate to each other. In fact, contemporary CAD systems typically do not even support the full range of this structure yet.

4. Conclusion

An approach to the development of an industry-specific CIM solution portfolio has been presented. This approach is based on two types of enterprise-wide information models, an enterprise-wide data model and an enterprise-wide function model. In order to reach a sufficient level of detail for the coverage of industry-specific technologies, industry-specific models have been developed. Within the enterprise-wide data model the focus has been set on the product related data. Specifically the shape information was covered in more detail.

The presented types of information models explicitly capture cross-functional relations. This way they support the development of a consistent set of application software in an enterprise. Additionally the developed models turned out to be a good tool to document specific know-how about information technology aspects within a particular industry.

An industry-specific detailed product data model can be a solid starting point for CIM software selection and software development. However, this is not a sufficient criterion. Further aspects like usability, performance, reliability also need to be considered.

References

[1] Beeckman, D.: CIM-OSA: computer integrated manufacturing—open systems architecture. International Journal on Computer Integrated Manufacturing 2 (2), 94–105 (1989).
[2] Chen, P. P.-S.: The entity-relationship model—Toward a unified view of data. ACM Transactions on Database Systems 1 (1), 9–36 (1996).
[3] Integrated Product Information Model (IPIM), ISO Document N 284, (STEP-Registerd Draft Proposal, ISO (1989).
[4] Marczinski, G., Prengemann, U., Holland, M., Mittmann, B.: Anwendungsorientierte Analyse des zukünftigen Schnittstellen-Standards STEP, pp. 456–461. ZwF 84. München: Carl Hanser 1989.
[5] Müller-Ettrich, G. (ed): Effektives Datendesign. Köln: Rudolf Müller 1989.
[6] Roller, D., Ruess, H.: An approach to an open CAD system architecture. In: F.-L. Krause, H. Jansen (eds.) Geometric modelling for engineering application pp. 365–378. Amsterdam: North Holland 1990.
[7] Roller, D.: Design by features: An approach to high level shape manipulations. Computers in Industry, North-Holland 12 (3), 185–191 (1989).
[8] Roller, D.: Automotive industry specific enterprise-wide information modelling: A tool for development of a CIM solution platform. Proceedings of Mechatronics, 24th International Symposium on Automotive Technology & Automation, Florence, May 20–24, pp. 561–568, 1991.
[9] Scheer, A.-W.: CIM Computer steered industry. Berlin, Heidelberg: Springer 1988.
[10] Scheer, A.-W.: Enterprise-wide data modelling—information systems in industry. Berlin, Heidelberg: Springer 1989.

D. Roller
Institut für Informatik
Universität Stuttgart
Breitwiesenstrasse 20–22
D-W-7000 Stuttgart 80

Computing Suppl. 8, 267–281 (1993)

Free Form Deformation with Scattered Data Interpolation Methods

D. Ruprecht and **H. Müller***, Freiburg

Abstract. Free form deformation has applications in computer animation, in data acquisition, and in computer aided geometric design. One approach is to deform the space in order to induce a deformation of an embedded object. A new method of this class is proposed which is based on scattered data interpolation. The usefulness of two selected classes of scattered data interpolation methods in this context is analyzed, with emphasis on the two-dimensional case. It turns out that suitable scattered data methods do exist, and that the new approach should complete well with more complex, computationally expensive physically based models.

Key words: Geometric modelling, computer animation, deformation, scattered data interpolation.

1. Free Form Deformation

Suppose we are given an image of two-dimensional geometric objects. The task is to distort the image irregularly.

In the physical world, this goal can be reached by drawing the image on an elastic skin. The elastic skin is deformed by moving some of its points into a new position. This causes a deformation of the image, too.

In a computer, the elastic skin, the image drawn on it, and the input of the points to be moved can be easily modeled by an interactive program. The user interface of such a program is shown in Fig. 1. Besides some menus it consists of two neighboring rectangular regions. The left region shows the image before transformation, in the example a digitized cross section of a blossom. If a second image is known which is to be used as a target for the deformation, it can be displayed in the right region. The example in Fig. 1a shows another cross section of the blossom at a later stage of evolution. By consecutively choosing a point in the left and a point in the right region, pairs of an original point and its image under the deformation to be carried out are defined. This displacement of control points is displayed by dashed lines. The calculation of the deformation is initiated by selecting a corresponding menu item. The resulting deformed picture is then displayed in the right region, overlaid over the second image if one is present as shown in Fig. 1b.

---

* Institut für Informatik, Universität Freiburg, Rheinstr. 10–12, W-7800 Freiburg i.Br. E-mail: [ruprecht|mueller]@informatik.uni-freiburg.de

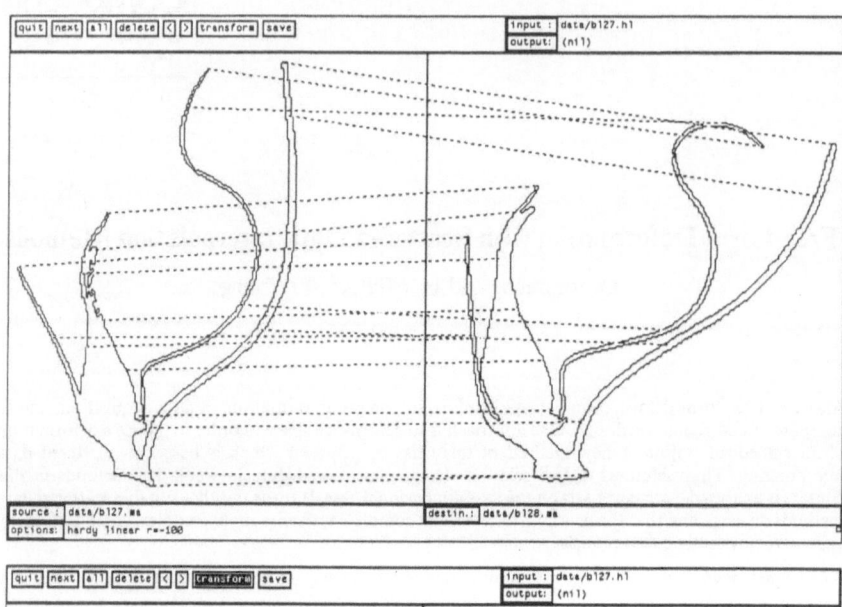

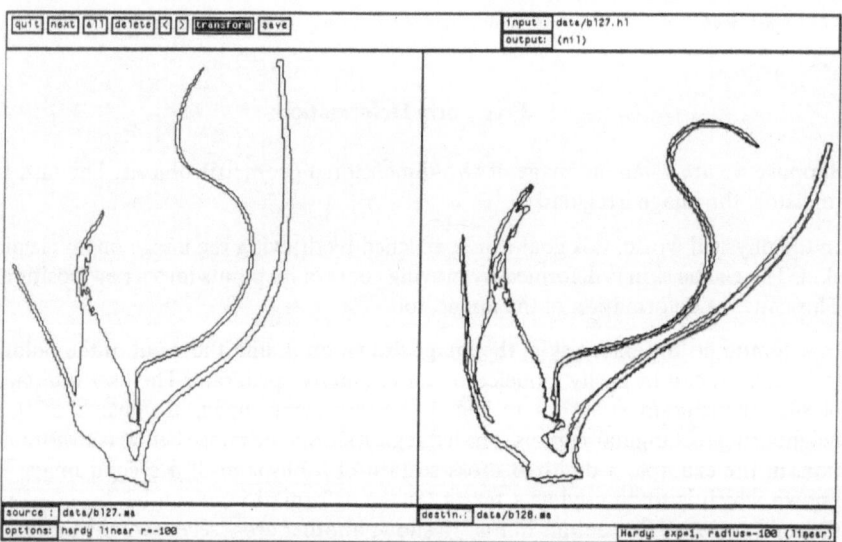

Figure 1. Deformation of a cross section of a blossom. The upper picture shows two given states of evolution of the blossom and an assignment for deformation. In the right region of the picture below the original second state and the deformation of the first state calculated from the assignment are displayed simultaneously

The images deformed in this paper consist of polygonal chains. The deformed picture is obtained by calculating the images under deformation of points on the polygonal chains and then connecting them by line segments in the same way as in the original picture. It is not too difficult to deform raster images too. A possibility is to use a look-up table with positions of pixels of the initial picture as it was done in [Wolberg, Boult, 1989].

An immediate application of image deformation is in two-dimensional computer animation. A given drawing can be animated by continuous transformation. Usually, the animation will not be designed by extrapolating an initial image, but by interpolating between two given keyframes. In applying the technique of deformation, interpolation can be performed approximately by deforming the first keyframe so that it coincides with the second as well as possible, as it was done in Fig. 1. The intermediate frames required for animation may then be obtained by linear interpolation between the first keyframe and its deformation. Let $\mathbf{p}$ be a point in the first keyframe, $\mathbf{f}: \mathbb{R}^2 \to \mathbb{R}^2$ the function describing the deformation. Then the location of $\mathbf{p}$ at time t, $0 < t < 1$, is $(1 - t) \cdot \mathbf{p} + t \cdot \mathbf{f}(\mathbf{p})$.

The initial motivation for the work reported on here was a problem of geometric design. A widely used technique of solid modeling is the reconstruction of the surface of a solid from cross sections. Reconstruction from cross sections finds its application in particular in the acquisition of existing solid structures by techniques of tomography. Tomography is used in medicine, but also in microscopy or geology. One of the most powerful approaches of reconstruction is that of [Boissonnat, 1988]. This approach is based on nearest-neighbor heuristics, and is thus able to interconnect even multiple contours including branches in a usually quite likely way. However, troubles arise if the cross sections are highly dissimilar. The blossoms of Fig. 1 are an example if the time axis is taken as the third spatial axis. Free form deformation allows a distortion of the cross sections with only a few assignments so that Boissonnat's algorithm may find a reconstruction between the deformed cross sections. The actual reconstruction is obtained by back-transformation which can simply be achieved by replacing the deformed coordinates by the original ones. In this manner, a good reconstruction from the original cross sections can often be achieved.

Another problem with techniques of tomography is that their physical properties often cause distortions of the cross sections during catching [Buche, Camillerapp, 1991]. For their correction, free form deformation can be used as well. A last example of an application is the design of free form solids in computer aided geometric design by deforming initially simple solids into complex shapes [Sederberg, Parry, 1986].

The crucial question not addressed yet is the calculation of the deformation. One possibility is the correct simulation of a physical model, for example an elastic skin of rubber. Such models have found application in e.g. face modelling [Guenther, 1989]. These approaches seem to be expensive. In the following a more simple method is proposed. The problem of deformation is interpreted as a problem of interpolation. It turns out that deformation is closely related to the problem of scattered data interpolation. This opens the possibility to refer to well-known techniques in this area [Franke, 1982, Alfeld, 1989]. From the high number of scattered data methods, we have chosen two types for our investigations, the classic Shepard approach with some modifications and interpolation with radial basis functions, in particular Hardy's multiquadrics. The experimental investigations show that with this approach good results can be achieved as long as the dissimilarity between the original image and that endeavored by deformation is not too high. A first impression can be got from Fig. 1, in which interpolation with multiquadrics is used. Despite the low number of assignments, the similarity of both pictures is convincing. Bigger deformations can be performed by decomposition into a sequence of smaller deformations.

The remainder of this contribution is organized as follows. In sect. 2 the relation between the problem of deformation and the problem of scattered data interpolation is established. The scattered data methods and their behavior for the problem of deformation are the subject of sect. 3 and 4. Piecewise deformation is discussed in sect. 5. A comparison between the two approaches of scattered data interpolation, as well as with other methods of deformation known in literature is performed in sect. 6.

2. Deformation and Scattered Data Interpolation

The problem of deformation can be formulated as follows.

Deformation Problem.
Input: n pairs $(\mathbf{p}_i, \mathbf{q}_i)$ of control points, $\mathbf{p}_i, \mathbf{q}_i \in \mathbb{R}^d$, $i = 1, \ldots, n$.
Output: An at least continuous function $\mathbf{f}: \mathbb{R}^d \to \mathbb{R}^d$ with $\mathbf{f}(\mathbf{p}_i) = \mathbf{q}_i$.

For the image deformation problem we have $d = 2$. Besides continuity, other properties of the function $\mathbf{f}$ are desirable which concern the visual pleasantness and which are hard to describe in an exact manner. These properties will become recognizable in the following discussion of concrete propositions.

The problem of scattered data interpolation is to find a real valued multivariate function interpolating a finite set of irregularly located data points.

Scattered Data Interpolation Problem.
Input: n data points $(\mathbf{x}_i, y_i)$, $\mathbf{x}_i \in \mathbb{R}^d$, $y_i \in \mathbb{R}$, $i = 1, \ldots, n$.
Output: An at least continuous function $f: \mathbb{R}^d \to \mathbb{R}$ interpolating the given data points, i.e. $f(\mathbf{x}_i) = y_i$, $i = 1, \ldots, n$.

Scattered data methods can now be simply used by interpolating each component $f_j: \mathbb{R}^d \to \mathbb{R}$, $j = 1, \ldots, d$, of $\mathbf{f}$ separately. The data points to be used are $(\mathbf{p}_i, q_{i,j})$, $q_{i,j}$ the j-th component of $\mathbf{q}_i$.

3. Shepard Interpolation

The Shepard approach of scattered data interpolation [Shepard, 1968] uses a weighted average of the data values at the data points, with weights dependent on the distance of the observed point from the given data points,

$$f(\mathbf{x}) = \sum_{i=1}^{n} w_i(\mathbf{x}) y_i. \tag{1}$$

$w_i: \mathbb{R}^d \to \mathbb{R}$ is the weight function and $y_i \in \mathbb{R}$ the data value at the data point $\mathbf{x}_i \in \mathbb{R}^d$.
The weight functions satisfy the conditions

$$w_i(\mathbf{x}) \geq 0, \quad i = 1, \ldots, n, \quad \sum_{i=1}^{n} w_i(\mathbf{x}) = 1, \quad \text{and} \tag{2}$$

$$w_i(\mathbf{x}_j) = \delta_{ij}, \quad \text{i.e.} \quad w_i(\mathbf{x}_i) = 1 \quad \text{and} \quad w_i(\mathbf{x}_j) = 0, \quad j \neq i, \quad i,j = 1, \ldots, n. \tag{3}$$

These conditions guarantee the property of interpolation.

In [Shepard, 1968] the following weight function was proposed,

$$w_i(\mathbf{x}) = \frac{\sigma_i(\mathbf{x})}{\sum_{j=1}^{n} \sigma_j(\mathbf{x})}, \quad \text{with} \tag{4}$$

$$\sigma_i(\mathbf{x}) = \frac{1}{(d_i(\mathbf{x}))^\mu}, \quad d_i(\mathbf{x}) \text{ the distance between } \mathbf{x} \text{ and } \mathbf{x}_i. \tag{5}$$

The smoothness is determined by the exponent μ. $\mu > 1$ assures the continuity of the first derivative, its value disappears at the data points.

The Shepard interpolation in this form shows an unfavorable behavior when applied to the problem of deformation. Using the weighted average of the given control points causes all other points to be shifted towards the geometric center of the displaced control points. Figure 2a shows the example of a square grid whose four corners are fixed. No further assignments were given apart from these four pairs of identical points. The original grid is shown with dashed lines, the deformation with solid lines. The created image is considerably distorted, which means that a simple Shepard approach will usually not be well suited for the application to deformation.

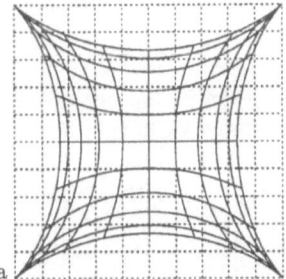
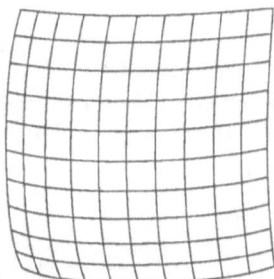

Figure 2. Deformation of a square grid with fixed corners: **a** simple Shepard approach; **b** radial basis functions with Hardy's multiquadrics without polynomial

3.1 Local Interpolants

Several authors, among them [Gordon, Wixom, 1978, Franke, 1980, Lancaster, Salkauskas 1981, Farwig, 1986], have proposed modified approaches which replace the value y_i in Shepard's formula by a local interpolant $f_i(\mathbf{x}): \mathbb{R}^n \to \mathbb{R}$,

$$f(\mathbf{x}) = \sum_{i=1}^{n} w_i(\mathbf{x})f_i(\mathbf{x}), \qquad f_i(\mathbf{x}_i) = y_i, \qquad i = 1, \ldots, n. \tag{6}$$

[Franke, 1980] proposed to compute the local interpolant $f_i(\mathbf{x})$ by minimizing the squared error of the mapping of all other control points $\mathbf{x}_j$ with f_i, weighted with a coefficient w_{ij} to attenuate the influence of distant points. The corresponding error function $E_i(f)$ is

$$E_i(f) = \sum_{j=1, j \neq i}^{n} w_{ij} \cdot \|f(\mathbf{x}_j) - y_j\|^2. \tag{7}$$

The optimum local interpolant is obtained by minimizing $E_i(f)$. The coefficients w_{ij} should depend on the distance between $\mathbf{x}_i$ and $\mathbf{x}_j$ in a manner similar to $w_i(\mathbf{x})$, but it is not possible to use w_i because of (3). A simple possibility is to use $w_{ij} = \sigma_i(\mathbf{x}_j)$. This approach has been used in our examples on local interpolants.

The application of (6) to the problem of deformation leads to

$$\mathbf{f}(\mathbf{p}) = \sum_{i=1}^{n} w_i(\mathbf{p})\mathbf{f}_i(\mathbf{p}), \qquad \mathbf{f}_i(\mathbf{p}_i) = \mathbf{q}_i, \qquad i = 1, \ldots, n. \tag{8}$$

$\mathbf{f}_i: \mathbb{R}^d \to \mathbb{R}^d$ are the local interpolants. They can be rewritten as

$$\mathbf{f}_i(\mathbf{p}) = \mathbf{q}_i + \mathbf{D}_i(\mathbf{p} - \mathbf{p}_i), \qquad \text{with} \qquad \mathbf{D}_i: \mathbb{R}^d \to \mathbb{R}^d, \qquad \mathbf{D}_i(0) = 0. \tag{9}$$

A simple possibility is to choose the $\mathbf{D}_i$ as linear transformations. Then $\mathbf{D}_i$ can be represented by a $d \times d$-matrix $(d_{i,kl})$, $k, l = 1, \ldots, d$. In the two-dimensional case the error function corresponding to a linear transformation $\mathbf{D}$ is

$$E_i(\mathbf{D}) = \sum_{j=1, j \neq i}^{n} w_{ij} \cdot \left\| \mathbf{q}_i + \begin{pmatrix} d_{11} & d_{12} \\ d_{21} & d_{22} \end{pmatrix}(\mathbf{p}_j - \mathbf{p}_i) - \mathbf{q}_j \right\|^2 \tag{10}$$

$$= \sum_{j=1, j \neq i}^{N} w_{ij}[(d_{11}(p_{j,1} - p_{i,1}) + d_{12}(p_{j,2} - p_{i,2}) + q_{i,1} - q_{j,1})^2$$

$$+ (d_{21}(p_{j,1} - p_{i,1}) + d_{22}(p_{j,1} - p_{i,1}) + q_{i,2} - q_{j,2})^2]. \tag{11}$$

The minimum of the error function is obtained with the partial derivatives with respect to the d_{kl}, $k, l = 1, 2$. For $\partial E_i(\mathbf{D})/\partial d_{11}$, for example, we have

$$\frac{\partial E_i(\mathbf{D})}{\partial d_{11}} = \sum_{j=1, j \neq i}^{n} w_{ij}[2d_{11}(p_{j,1} - p_{i,1})^2 + 2d_{12}(p_{j,1} - p_{i,1})(p_{j,2} - p_{i,2})$$

$$+ 2(q_{i,1} - q_{j,1})(p_{j,1} - p_{i,1})]. \tag{12}$$

Hence the derivatives are linear functions of the $d_{i,kl}$. Their second derivatives show that the zeros of the first derivative are indeed a minimum. Thus a linear system of four equations with four unknowns is obtained. The system has an unique solution if its determinant is not equal to zero. The solutions are

$$det_i = \left(\sum_{j=1, j \neq i}^{n} w_{ij}(p_{j,1} - p_{i,1})^2 \right)\left(\sum_{j=1, j \neq i}^{n} w_{ij}(p_{j,2} - p_{i,2})^2 \right)$$

$$- \left[\sum_{j=1, j \neq i}^{n} w_{ij}(p_{j,1} - p_{i,1})(p_{j,2} - p_{i,2}) \right]^2 \tag{13}$$

$$d_{i,11} = \frac{1}{det_i}[(\sum w_{ij}(q_{j,1} - q_{i,1})(p_{j,1} - p_{i,1}))(\sum w_{ij}(p_{j,2} - p_{i,2})^2)$$

$$- (\sum w_{ij}(q_{j,1} - q_{i,1})(p_{j,2} - p_{i,2}))(\sum w_{ij}(p_{j,1} - p_{i,1})(p_{j,2} - p_{i,2}))] \qquad (14)$$

$$d_{i,12} = \frac{1}{det_i}[(\sum w_{ij}(q_{j,1} - q_{i,1})(p_{j,2} - p_{i,2}))(\sum w_{ij}(p_{j,1} - p_{i,1})^2)$$

$$- (\sum w_{ij}(q_{j,1} - q_{i,1})(p_{j,1} - p_{i,1}))(\sum w_{ij}(p_{j,1} - p_{i,1})(p_{j,2} - p_{i,2}))] \qquad (15)$$

$$d_{i,21} = \frac{1}{det_i}[(\sum w_{ij}(q_{j,2} - q_{i,2})(p_{j,1} - p_{i,1}))(\sum w_{ij}(p_{j,2} - p_{i,2})^2)$$

$$- (\sum w_{ij}(q_{j,2} - q_{i,2})(p_{j,2} - p_{i,2}))(\sum w_{ij}(p_{j,1} - p_{i,1})(p_{j,2} - p_{i,2}))] \qquad (16)$$

$$d_{i,22} = \frac{1}{det_i}[(\sum w_{ij}(q_{j,2} - q_{i,2})(p_{j,2} - p_{i,2}))(\sum w_{ij}(p_{j,1} - p_{i,1})^2)$$

$$- (\sum w_{ij}(q_{j,2} - q_{i,2})(p_{j,1} - p_{i,1}))(\sum w_{ij}(p_{j,1} - p_{i,1})(p_{j,2} - p_{i,2}))]. \qquad (17)$$

If $det_i = 0$, a simpler local interpolant with $d_{i,11} = d_{i,22}$ and $d_{i,12} = -d_{i,21}$ is used which allows uniform scaling and rotation, but no shearing. The $d_{i,kl}$ are again calculated by minimizing $E_i(\mathbf{D})$ with the result

$$d_i = \sum w_{ij}((p_{j,1} - p_{i,1})^2 + (p_{j,2} - p_{i,2})^2) \qquad (18)$$

$$d_{i,11} = d_{i,22} = \frac{1}{d_i} \sum w_{ij}((p_{j,1} - p_{i,1})(q_{j,1} - q_{i,1}) + (p_{j,2} - p_{i,2})(q_{j,2} - q_{i,2})) \qquad (19)$$

$$d_{i,12} = -d_{i,21} = \frac{1}{d_i} \sum w_{ij}((p_{j,2} - p_{i,2})(q_{j,1} - q_{i,1}) - (p_{j,1} - p_{i,1})(q_{j,2} - q_{i,2})) \qquad (20)$$

If even this cannot be computed, $\mathbf{D}_i$ is set to an identical transformation.

With a linear local interpolant, the interpolation has linear precision, i.e. if the requested displacements can be satisfied by a linear transformation, this is what will be obtained. For example, a deformation as in Fig. 2, where all control points retain their initial position, will exactly reproduce the initial image.

The examples in Figs. 3–5 demonstrate the effect of the deformation. To make comparison easier, various examples of the deformation of a 10 × 10 square grid are shown. Examples of modelling applications are also given. The same set of displacement assignments for the control points is used throughout each figure. For clarity, the original image and the displacements are only shown in the first example in each figure.

Figure 3a shows a square grid with fixed corners within which a 2 × 2 square has been turned clockwise by 45°. The original grid is displayed with dashed lines and the displacement assignments with solid lines. Figure 3b shows the result of Shepard deformation with $\mu = 2$. The waviness of the edges of the inner square comes from the influence of the fixed corner points of the grid. Increasing μ reduces this waviness, as the influence of the corner points is reduced. Similarly, the grid edges are pulled

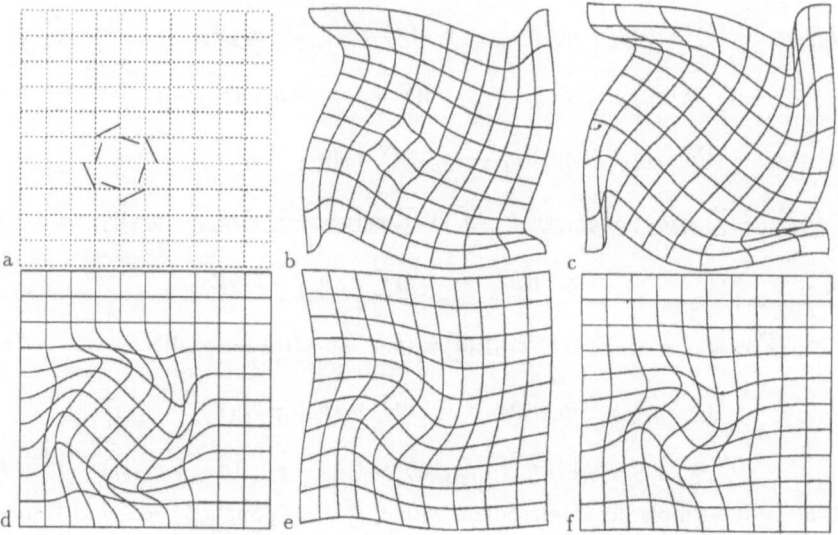

Figure 3. Rotating an embedded 2 × 2-square by 45° within a square grid with fixed corners: **a** Original grid with displacements; **b** Shepard with local interpolation ($\mu = 2$); **c** Shepard with local interpolation ($\mu = 3$); **d** Shepard with weight functions with limited influence ($r = 3$, $\sigma_0 = 1$); **e** Hardy's multiquadrics ($\mu = 1$); **f** Hardy's multiquadrics with limited influence ($r = 3$, $\mu = -8$)

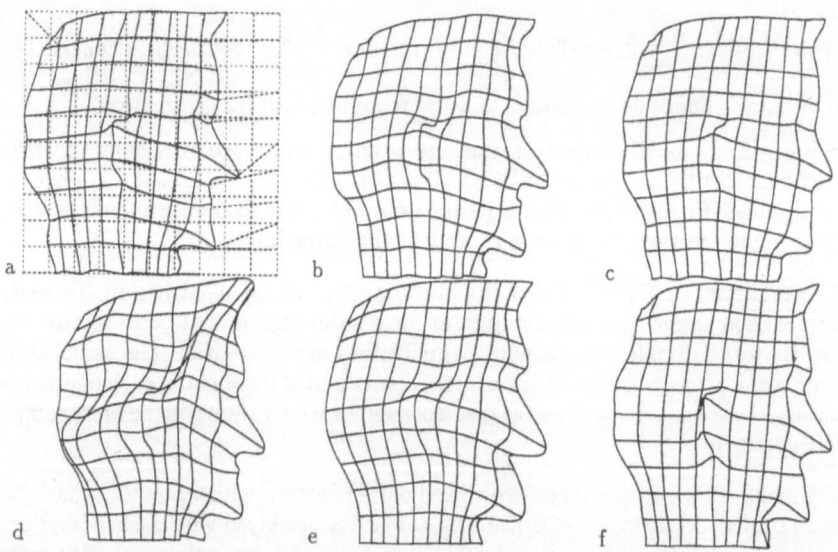

Figure 4. Deformation of a square grid into a head: **a** Shepard with local interpolation ($\mu = 2$); **b** Shepard with local interpolation ($\mu = 3$); **c** Shepard with damping ($\mu = 2$, $c = 0.1$); **d** Shepard with limited influence ($r = 5$, $\sigma_0 = 0.1$); **e** Hardy's Multiquadrics ($\mu = 1$); **f** Hardy's Multiquadrics with limited influence ($\mu = -5$)

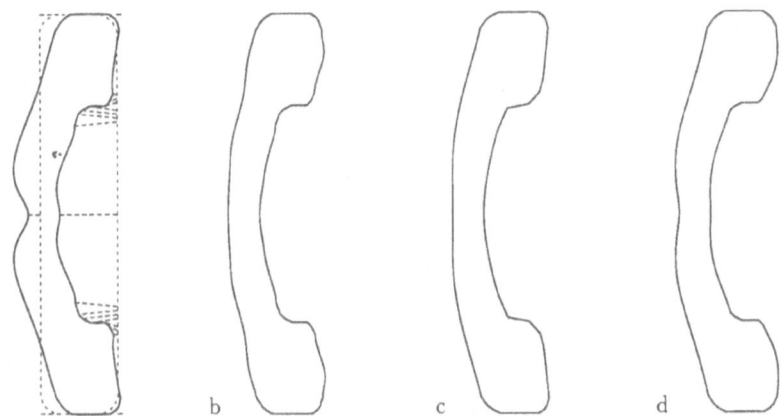

Figure 5. Deformation of a straight bar to a telephone handset: **a** Shepard with local interpolation; **b** Shepard with limited influence ($r = 3$, $\sigma_0 = 1$); **c** Hardy's Multiquadrics; **d** Radial basis function $d(\mathbf{x})^2 \log d(\mathbf{x})$

along by the rotation of the inner square. As some points along the grid edges are actually closer to the inner square than to the corner points, increasing μ does not help here. Figure 3c shows the deformation with $\mu = 3$.

Figures 4a and 4b show the same square grid deformed into a side-view of a head. Figure 4a uses $\mu = 2$, Fig. 4b uses $\mu = 3$. The method reacts to the input of displacements in an intuitive and straightforward manner. Note the differences caused by the different range of influence of the control points.

Figure 5a shows a straight bar which is deformed into a telephone handset with $\mu = 2$. The produced image is not very satisfying. The observed unevenness near the data points is typical for the Shepard approach. Therefore it is better suited for applications where some jaggedness is desirable—as in the head shown in Fig. 4—than for applications needing smooth deformations as is common in Computer Aided Geometric Design.

Higher order polynomials can be used as well. [Franke, Nielson, 1980] recommend the use of second order polynomials. This does not appear to be necessary in the case of deformation. Also, higher order polynomials tend to introduce oscillations which may be acceptable in data interpolation, but undesirable for deformation.

3.2 Relaxing the Interpolation Condition

The smoothness of the deformation can be improved by adding a damping term $c > 0$ to the weight function (5), i.e.

$$\sigma_i(\mathbf{x}) = \frac{1}{(d_i(\mathbf{x}) + c)^\mu} . \tag{21}$$

With this weight function, exact interpolation is not provided, but with c sufficiently small the approximation follows the intention specified in the control point displacements quite well. Figure 4c demonstrates this for a value of $c = 0.1$. This image is noticeable smoother than the images in Fig. 4a and b.

An individual c_i could be specified for each data point, so that the user can specify the desired precision for each control point. This, however, would make the input of displacements more cumbersome.

With $c < 0$, strict adherence to the local interpolant can be enforced for $d_i(\mathbf{x}) \leq c$. This might be useful if special conditions, e.g. conservation of area or angles, have to be met in the vicinity of the control points.

3.3 Locally Bounded Weight Functions

If only part of an image is to be deformed, it is desirable to limit the sphere of influence of the control points, so that the deformation of a point is only influenced by control points at a distance $d_i(x) < r_i$. [Franke, Nielson, 1980] proposed the following weight function:

$$\sigma_i(\mathbf{x}) = \begin{cases} \left(\dfrac{r_i - d_i(\mathbf{x})}{r_i d_i(\mathbf{x})} \right)^{\mu} & \text{for} & d_i(\mathbf{x}) < r_i, \\ 0 & & d_i(\mathbf{x}) \geq r_i. \end{cases} \tag{22}$$

In order to ensure that there is at least one non-zero weight for all points in the source image, a global function $\mathbf{f}_0(\mathbf{x})$ and a corresponding weight $\sigma_0(\mathbf{x})$ should be added to (4) and (8) so that

$$\mathbf{f}(\mathbf{p}) = \sum_{i=0}^{n} w_i(\mathbf{p}) \mathbf{f}_i(\mathbf{p}), \quad \text{with} \quad \mathbf{f}_i(\mathbf{p}_i) = \mathbf{q}_i, \quad i = 1, \ldots, n \quad \text{and} \tag{23}$$

$$w_i(\mathbf{x}) = \frac{\sigma_i(\mathbf{x})}{\sum_{j=0}^{n} \sigma_j(\mathbf{x})}. \tag{24}$$

In the case of localized deformations, identical transformations $\mathbf{f}_0(\mathbf{x}) = \mathbf{x}$ with a constant user-selectable σ_0 perform quite well.

Figure 3d shows the result for the rotated 2×2 square with $r = 3$ and $\sigma_0 = 1$. Areas further than 3 grid units away from the rotated square are completely unaffected by the deformation.

Figure 4d shows the grid deformed into a head with $r = 5$ and $\sigma_0 = 0.1$. The results obtained here are unsatisfying, which is probably due to the sometimes sudden change of spheres of influence from one control point to another.

Figure 5b shows the telephone handset produced from a straight bar with $r = 3$ and $\sigma_0 = 1$. Here the limitation of the influence of the control point has quite a positive effect on the deformation, although the result is still not perfect.

[Franke, Nielson, 1980] alternatively also propose weight functions bounded by a triangulation of the control points. This shows that there is an almost unlimited number of variations on weight functions to suit special purposes.

4. Radial Basis Functions

A further approach to scattered data interpolation is to construct the interpolant as a linear combination of basis functions and then to determine the coefficients of the basis functions,

$$f(\mathbf{x}) = \sum_{i=1}^{n} \alpha_i R(d_i(\mathbf{x})) + p_m(\mathbf{x}). \tag{25}$$

The values of the basis function R usually depend only on the distance from the data point, and are thus called radial. $p_m(\mathbf{x})$ is a polynomial of degree m. It assures a certain degree m of polynomial precision. The coefficients α_i are calculated by putting the data points into (25) and solving the resulting linear system of equations.

Well-known radial basis functions are *Hardy's multiquadrics* [Hardy, 1971],

$$R(d) = (d^2 + r^2)^{\mu/2} \quad \text{with} \quad r > 0 \quad \text{and} \quad \mu \neq 0. \tag{26}$$

Hardy suggests to set p_m identical to 0. This implies a polynomial precision of 0. For the exponent μ, Hardy proposes $\mu = 1$. $\mu = -1$ has also been used, but is generally less stable.

$r > 0$ can be chosen arbitrarily. It determines the smoothness of interpolation at the given data points. [Eck, 1991] proposes to use individual values r_i for each data point $\mathbf{p}_i$, computed from the distance to the nearest neighbor, i.e.

$$r_i = \min_{i \neq j} d_i(\mathbf{x}_j). \tag{27}$$

This causes the deformation to be softer when data points are widely spaced and stronger when they are closer together. This approach has been used in our examples unless otherwise noted.

The application of radial basis functions to the problem of deformation gives

$$\mathbf{f}(\mathbf{p}) = \sum_{i=1}^{n} \alpha_i R(d_i(\mathbf{p})) + \mathbf{p}_m(\mathbf{p}). \tag{28}$$

Now $\alpha_i \in \mathbb{R}^d$ are coefficient vectors and $\mathbf{p}_m \colon \mathbb{R}^d \to \mathbb{R}^d$ is a function with polynomial components of degree m.

It turns out that Hardy's multiquadrics approach without a polynomial term $\mathbf{p}_m$ does not yield satisfactory results as it produces distortions even with the simple transformation of a square grid with fixed corners, as shown in Fig. 2b. This behavior is caused by the non-linearity of the basis functions. Introducing a linear $\mathbf{p}_m$, i.e. $m = 1$, as a global approximation with coefficients computed by the least squares method overcomes the problem. With this, the example yields an identical transformation as desired.

Figure 3e shows a rotated square inside a grid with $\mu = 1$. The deformation is much smoother than that achieved with Shepard methods. Figure 4e shows a head created from a square grid with multiquadrics. The smoothness of the results obtained by this method is obviously not very well suited for this sort of application. Figure 5c, however, shows that the telephone handset created from a straight bar looks very good with multiquadrics.

With a $\mathbf{p}_m$ as introduced above, it is possible to use locally bounded basis functions which disappear for $d_i(\mathbf{x}) > r_i$. This way, radial basis functions can be used for local deformations. For these applications, the identical transformation $\mathbf{p}_m(\mathbf{p}) := \mathbf{p}$ is an adequate polynomial approximation. The effect of multiquadrics with a negative exponent $\mu < -2$ is quite localized even though they are not strictly bounded basis functions. Figure 3f shows the rotated square for $r = 3$ and $\mu = -8$. The locality of the deformation is well demonstrated. Figure 4f shows multiquadrics with $\mu = -5$ applied to the deformation of a grid into a head. The result is better than with $\mu = 1$, although for this type of "artistic" modeling, the results of Shepard's methods are still superior.

Other radial basis functions have been proposed [Duchon, 1977] which allow a physical interpretation as a minimal thin plate functional:

$$R(d) = d^3, \qquad \text{and} \tag{29}$$

$$R(d) = d^2 \cdot \log d. \tag{30}$$

In our experiments, deformations obtained from these basis functions were consistently less satisfactory than those from multiquadrics. Figure 5d shows an example with the telephone handset for the radial basis function (30).

5. Iterated Deformation

The experiments show that in the case of small deformations good results can be achieved with many approaches. For bigger deformations, however, undesired foldover can occur in the deformed picture (Fig. 6). These are caused by the fact

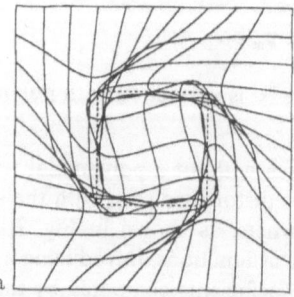 

Figure 6. Rotation of a 4×4 square embedded in a square grid by 90° **a** without iteration, **b** with four steps of iteration

that, contrary to the scattered data problem, the image space is mapped onto itself. The influence of the control points on a transformed point should therefore be dependent not only on the initial relative position, but also on the trajectory along which a control point is moved to its new location.

Moving the control points can be simulated by decomposing the deformation into several smaller deformations which are executed sequentially. When choosing a linear trajectory, the deformation with pairs of control points $(\mathbf{p}_i, \mathbf{q}_i)$ is linearly subdivided into m deformations $(\mathbf{p}_{i,k}, \mathbf{q}_{i,k})$, $\mathbf{q}_{i,k} = \mathbf{p}_{i,k+1}$, $k = 0, \ldots, m-1$, with

$$\mathbf{p}_{i,k} = \mathbf{x}_i + \frac{k}{m} \cdot (\mathbf{q}_i - \mathbf{p}_i), \qquad k = 0, \ldots, m. \tag{31}$$

Figure 6a and b show a square grid with an embedded 4×4 square rotated by $90°$ with a non-iterated transformation and with four iterations, respectively. For the transformations, multiquadrics are used. The foldover in the non-iterated transformation is clearly visible. Note that iteration is not used in any of the other examples. The deformation in these examples is not strong enough to make this necessary.

Splitting a deformation into a series of iterated transformations is also helpful when it is necessary to reverse the deformation. This can be done by deforming the result with the reversed control points $(\mathbf{q}_i, \mathbf{p}_i)$. To reproduce the original image well, the weights w_i, or the α_i for radial basis functions, should not change much between successive iteration steps, i.e. $w_{i,k}(\mathbf{p}_{j,k}) \approx w_{i,k+1}(\mathbf{p}_{j,k+1})$. This can always be enforced by using a sufficiently large m.

For keyframe animation, it appears to be useful to use a non-linear division of the deformation to avoid discontinuities of the movement at the keyframes.

6. Discussion

An approach for the deformation of geometric objects has been presented. The deformation is controlled by pairs of points which can be chosen arbitrarily. From these points, a function of deformation is calculated by methods derived from scattered data interpolation. The function is defined on the whole surrounding space and thus is in particular applied to any objects embedded in this space. The given points are mapped exactly while the other points follow in a natural manner dependent on the method applied.

Two classes of approaches for scattered data interpolation were investigated, the Shepard methods and the method of radial basis functions. Shepard methods have the advantage of more intuitive parameters, with the consequence that the behavior of the deformation can be easily tailored to meet specific needs. They tend to produce a somewhat jagged deformation which is often useful in artistic applications but is undesirable if smooth deformation is sought. Radial basis function methods on the other hand give very smooth and visually pleasant results as usually desired in CAGD applications and are more robust with respect to parameters.

Modifying them to suit special purposes is somewhat less intuitive, however. Both approaches are easily modified to give local deformations. Both behave gracefully outside the convex hull of the control points.

In [Sederbergy, Parry, 1986] an approach was described which also calculates deformations controlled by a finite number of movable points. The points control the deformation by Bernstein polynomials. In contrast to our approach, the control points have to be arranged on a regular grid. Furthermore, their influence is less immediate so that the result of the deformation is harder to estimate. Global mathematical mappings like e.g. those used in [Buche, Camillerap, 1991] can only be used for small deformations or for deformations with a simple mathematical description.

In summary, the approach presented in this contribution has the following advantages:

- The control points can be chosen arbitrarily.
- The deformation is performed exactly at the control points.
- It is easy to control the global influence of local deformations.
- The methods are easily tailored to meet special requirements.

There are other classes of methods of scattered data interpolation not investigated here. One example are the FEM approaches based on triangulation of the given data points. FEM interpolation methods were not taken into consideration since they can be carried out well in $\mathbb{R}^2$ but become computationally expensive and hard to implement in higher dimensions, whereas the methods presented here seem easily extensible. The latter has still to be done, in particular in 3D space.

Each method of interpolation behaves like a deformable material with very own properties. These properties are more or less natural to the user and the given problem. The usefulness depends on the experience of the user who will typically be well acquainted with concrete physical materials, say like an elastic skin of rubber. If a method of interpolation behaves quite different the user may have troubles and may need some time to gain experience with the new material. Then, however, this non-conventional method might be even better suited for solving the given problem of deformation than any real material may be. Nevertheless, a task of further work is to design and implement methods of deformation derived from physical laws.

References

[1] Alfeld, P.: Scattered data interpolation in three or more variables. In: Lyche, T., Schumaker, L. L. (ed.) Mathematical methods in CAGD, pp. 1–33. San Diego: Academic Press 1989.
[2] Boissonat, J.-D.: Shape reconstruction from planar cross sections. Computer Vision, Graphics, and Image Processing 44, 1–19 (1988).
[3] Buche, P., Camillerapp, J.: Serial cuttings matching: An application to muscle fiber characterization. Proc. EUROGRAPHICS '91 1991, 329–340.
[4] Duchon, J.: Splines minimizing rotation invariant semi-norms in Sobolev spaces. In: Schempp, W., Zeller, K. (ed.) Constructive theory of functions of several variables, pp. 85–100. Berlin, New York: Springer, 1977 (Lecture Notes in Mathematics Vol. 571).

[5] Eck, M.: Interpolationsmethoden zur Rekonstruktion von 3D-Oberflächen aus ebenen Schnitt-folgen. CAD und Computergraphik *13*(5), 109–120 (1991).

[6] Farwig, R.: Rate of convergence of Shepard's global interpolation formula. Mathematics of Computation *46*(174), 577–590 (1986).

[7] Franke, R.: Scattered data interpolation: Tests of some methods. Mathematics of Computation *38*(157), 181–200 (1982).

[8] Franke, R., Nielsen, G.: Smooth interpolation of large sets of scattered data. Int. Journal for Numerical Methods in Engineering *15*, 1691–1704 (1980).

[9] Gordon, W. J., Wixom, J. A.: Shepard's method of "metric interpolation" to bivariate and multivariate interpolation. Mathematics of Computation, *32*(141), 253–264 (1978).

[10] Guenther, B.: A system for simulating human facial expression. In: State-of-the-art in Computer Animation, pp. 191–202. Tokyo: Springer 1989.

[11] Hardy, R. L.: Multiquadric equations of topography and other irregular surfaces. J. Geophys. Res. *76*, 1905–1915 (1971).

[12] Lancaster, P., Salkauskas, K.: Surfaces generated by moving least squares methods. Mathematics of Computation *37* (155), 141—158 (1981).

[13] Sederberg, T. W., Parry, S. R.: Free-form deformation of solid geometric models. Computer Graphics *20*(4), 151–160 (1986).

[14] Shepard, D.: A two-dimensional interpolation function for irregularly spaced data. Proc. 23 Nat. Conf. ACM *1968*, 517–524.

[15] Wolberg, G., Boult, T. E.: Separable image warping with spatial lookup tables. Computer Graphics *23*(3), 369–378 (1989).

D. Ruprecht
H. Müller
Institut für Informatik
Universität Freiburg
Rheinstrasse 10–12
D-W-7800 Freiburg i.Br.
Federal Republic of Germany

Computing Suppl. 8, 283–289 (1993)

Computing
© Springer-Verlag 1993

C^1-Smoothing of Multipatch BEZIER Surfaces

W. Schwarz, Mainz

Abstract. Handling and modelling multipatch tensorproduct BEZIER surfaces often causes continuity problems. A method is described how to approximate a given surface in such a way that C^1-continuity at a crossing point of four surface patches can be obtained. The approximated surface uses still a BEZIER representation with the same orders as before. The idea is to choose the four BEZIER control points associated with the mixed derivatives to build a fourside on a tangent hyperbolic paraboloid. This fourside and along with it the hyperbolic paraboloid can easily be obtained by approximation, minimizing the sum of the four distance squares.

Key words: BEZIER surface, continuity.

1. Introduction

Continuity problems often arise during creation or modification of multipatch surfaces. Considerations about the continuity along patch boundaries are necessary for any method of surface modelling which uses patchwise approximation.

An important example occurs in the practical exchange of surfaces between different CAD-systems, if the receiving system cannot handle the high polynomial degree in the surface representation of the sending system. Hoschek, Schneider and Wassum [3] developped a method to approximate the given surface by a representation with reduced polynomial degree. Because the approximation is performed in each patch separately a lack in continuity may occur along patch boundaries.

Another example is the combination of two surfaces, when the continuity along the common boundary has to be newly evaluated. Also, for a fillet surface the continuity to the adjacent surfaces defines constraints for the fillet.

For any C^0-continuous multipatch BEZIER surface C^1-continuity along patch boundaries can be easily be achieved just by putting certain three BEZIER points on one line. However, dependencies occur for the C^1-continuity across patch boundaries at a crossing point of four surface patches.

An approximation method is presented which solves the problem with respect to the dependencies. At first the problem is seen from a geometric point of view introducing the tangent hyperbolic paraboloid. Certain four BEZIER points are chosen to build a fourside on the hyperbolic paraboloid. The error functional is presented as the sum of distance squares of the four desired BEZIER points to the original ones. The approximation problem is solved. Error estimations are presented in the last section.

2. Praepositions

Figure 1 shows the corners of four adjacent patches in a multipatch tensorproduct BEZIER surface. The following considerations only need the four BEZIER points in the corner of each patch. They are denoted with B_{ij} $(i, j \in \{-1, 0, 1\})$.

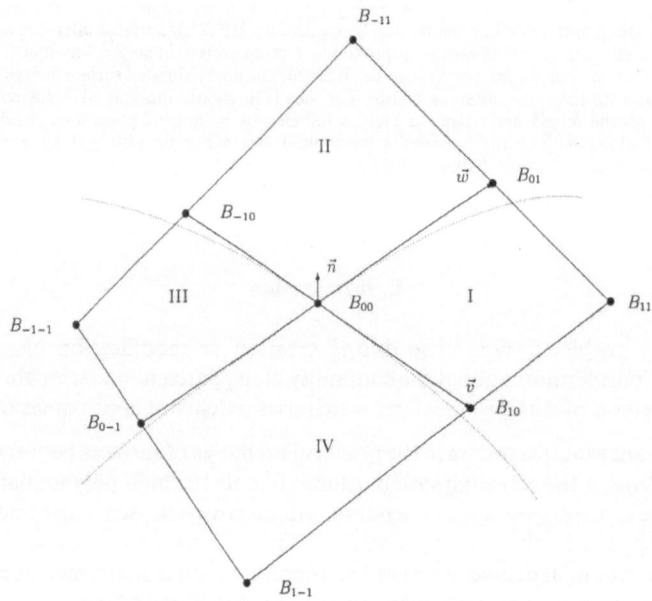

Figure 1. Four surface patches at a crossing point

The following assumptions are mostly made for technical reasons.

1. The surface patches are assumed to be in BEZIER representation.
2. All patches should have the same order in s-direction and the same order in t-direction.
3. C^0-continuity is assumed across the patch boundaries.
4. C^1-continuity in s-direction is assumed at the crossing point. This implies that B_{-10}, B_{00}, B_{10} are collinear.
5. C^1-continuity in t-direction is assumed at the crossing point. This implies that B_{0-1}, B_{00}, B_{01} are collinear.

All coordinates are described in a local, affine coordinate system at the crossing point B_{00} which is set up with the tangent vectors $\vec{v}$, $\vec{w}$ and their crossproduct $\vec{n}$ as axis.

3. Tangent Hyperbolic Paraboloid

The four BEZIER points in the corner of the first patch take the local coordinates:

$$B_{00} = (0, 0, 0)$$
$$B_{10} = (1, 0, 0)$$
$$B_{01} = (0, 1, 0)$$
$$B_{11} = ((1 + \alpha_{11}), (1 + \beta_{11}), \gamma_{11})$$

These four points are the BEZIER points of a hyperbolic paraboloid which describes the surface patch at the crossing point up to continuity order 1. If all four hyperbolic paraboloids are the same then the four surface patches are GC^1-continuous. Along with the assumed C^1-continuity along patch boundaries the four surface patches even are C^1-continuous. In local coordinates this common hyperbolic paraboloid has the parameter representation

$$\tilde{P}(s, t) = \begin{bmatrix} s + \tilde{\alpha}st \\ t + \tilde{\beta}st \\ \tilde{\gamma}st \end{bmatrix}$$

where $\tilde{\alpha}$, $\tilde{\beta}$, $\tilde{\gamma}$ are related to the coordinates of the unknown BEZIER point $\tilde{B}_{11}$.

The C^1-continuity across the patch boundaries in the crossing point is granted if

$$\tilde{B}_{1-1} \quad \tilde{B}_{10} \quad \tilde{B}_{11} \qquad \text{are collinear,}$$
$$\tilde{B}_{11} \quad \tilde{B}_{01} \quad \tilde{B}_{-11} \qquad \text{are collinear,}$$
$$\tilde{B}_{-1-1} \quad \tilde{B}_{-10} \quad \tilde{B}_{-11} \qquad \text{are collinear and}$$
$$\tilde{B}_{1-1} \quad \tilde{B}_{0-1} \quad \tilde{B}_{-1-1} \qquad \text{are collinear,}$$

This implies that $\tilde{B}_{11}$, $\tilde{B}_{-11}$, $\tilde{B}_{-1-1}$, $\tilde{B}_{1-1}$ build a fourside on the common hyperbolic paraboloid.

The coordinates of B_{-10} and B_{0-1} can be derived from the assumed C^1-continuity along patch boundaries.

$$B_{-10} = (-\lambda, 0, 0)$$
$$B_{0-1} = (0, -\mu, 0)$$

λ and μ compensate different lengths of global parameter intervals.

The coordinates of the desired BEZIER points $\tilde{B}_{ij}$ $(i, j \in \{-1, 1\})$ are then determined.

$$\tilde{B}_{11} = \tilde{P}(1, 1) = \begin{pmatrix} (1 + \tilde{\alpha}) \\ (1 + \tilde{\beta}) \\ \tilde{\gamma} \end{pmatrix}$$

$$\tilde{B}_{-11} = \tilde{P}(-\lambda, 1) = \begin{pmatrix} -\lambda(1 + \tilde{\alpha}) \\ (1 - \lambda\tilde{\beta}) \\ -\lambda\tilde{\gamma} \end{pmatrix}$$

$$\tilde{B}_{1-1} = \tilde{P}(1, -\mu) = \begin{pmatrix} (1 - \mu\tilde{\alpha}) \\ -\mu(1 + \tilde{\beta}) \\ -\mu\tilde{\gamma} \end{pmatrix}$$

$$\tilde{B}_{-1-1} = \tilde{P}(-\lambda, -\mu) = \begin{pmatrix} -\lambda(1 - \mu\tilde{\alpha}) \\ -\mu(1 - \lambda\tilde{\beta}) \\ \lambda\mu\tilde{\gamma} \end{pmatrix}$$

4. Error Functional and Solution

The desired $\tilde{B}_{ij}$ are compared to the original B_{ij} $(i, j \in -1, 1)$. Their respective distance squares sum up to the error functional:

$$\begin{aligned}
F(\tilde{\alpha}, \tilde{\beta}, \tilde{\gamma}) = \langle \vec{v}, \vec{v} \rangle &((\tilde{\alpha} - \alpha_{11})^2 + \lambda^2(\tilde{\alpha} - \alpha_{-11})^2 \\
&+ \mu^2(\tilde{\alpha} - \alpha_{1-1})^2 + \mu^2\lambda^2(\tilde{\alpha} - \alpha_{-1-1})^2) \\
&+ 2\langle \vec{v}, \vec{w} \rangle((\tilde{\alpha} - \alpha_{11})(\tilde{\beta} - \beta_{11}) \\
&+ \lambda^2(\tilde{\alpha} - \alpha_{-11})(\tilde{\beta} - \beta_{-11}) \\
&+ \mu^2(\tilde{\alpha} - \alpha_{1-1})(\tilde{\beta} - \beta_{1-1}) \\
&+ \mu^2\lambda^2(\tilde{\alpha} - \alpha_{-1-1})(\tilde{\beta} - \beta_{-1-1})) \\
&+ \langle \vec{w}, \vec{w} \rangle((\tilde{\beta} - \beta_{11})^2 + \lambda^2(\tilde{\beta} - \beta_{-11})^2 \\
&+ \mu^2(\tilde{\beta} - \beta_{1-1})^2 + \mu^2\lambda^2(\tilde{\beta} - \beta_{-1-1})^2) \\
&+ ((\tilde{\gamma} - \gamma_{11})^2 + \lambda^2(\tilde{\gamma} - \gamma_{-11})^2 \\
&+ \mu^2(\tilde{\gamma} - \gamma_{1-1})^2 + \mu^2\lambda^2(\tilde{\gamma} - \gamma_{-1-1})^2)
\end{aligned}$$

where α_{ij}, β_{ij}, γ_{ij} $(i, j \in \{-1, 1\})$ are used to denote the coordinates of the original BEZIER points.

All four original BEZIER points B_{ij} $(i, j \in \{-1, 1\})$ have the same weight in the error functional. It is interesting to remark that there are three degrees of freedom. If one of the desired BEZIER points $\tilde{B}_{ij}$ is chosen the other three are determined by the presumptions and the C^1-continuity condition.

The desired solution is a minimum of the error functional. The solution can be evaluated by differentiations of the error functional and determination of zeroes of the derivations. Up to a constant factor the derivations are:

$$\frac{\partial F}{\partial \tilde{\alpha}} \;(=)\; \tilde{\alpha}\langle \vec{v}, \vec{v}\rangle + \tilde{\beta}\langle \vec{v}, \vec{w}\rangle - \bar{\alpha}\langle \vec{v}, \vec{v}\rangle - \bar{\beta}\langle \vec{v}, \vec{w}\rangle$$

$$\frac{\partial F}{\tilde{\beta}} \;(=)\; \tilde{\alpha}\langle \vec{v}, \vec{w}\rangle + \tilde{\beta}\langle \vec{w}, \vec{w}\rangle - \bar{\alpha}\langle \vec{v}, \vec{w}\rangle - \bar{\beta}\langle \vec{w}, \vec{w}\rangle$$

$$\frac{\partial F}{\partial \tilde{\gamma}} \;(=)\; \tilde{\gamma} - \bar{\gamma}$$

with abbreviations:

$$\bar{\alpha} = \frac{\alpha_{11} + \lambda^2 \alpha_{-11} + \mu^2 \alpha_{1-1} + \lambda^2 \mu^2 \alpha_{-1-1}}{1 + \lambda^2 + \mu^2 + \lambda^2 \mu^2}$$

$$\bar{\beta} = \frac{\beta_{11} + \lambda^2 \beta_{-11} + \mu^2 \beta_{1-1} + \lambda^2 \mu^2 \beta_{-1-1}}{1 + \lambda^2 + \mu^2 + \lambda^2 \mu^2}$$

$$\bar{\gamma} = \frac{\gamma_{11} + \lambda^2 \gamma_{-11} + \mu^2 \gamma_{1-1} + \lambda^2 \mu^2 \gamma_{-1-1}}{1 + \lambda^2 + \mu^2 + \lambda^2 \mu^2}$$

which are weighted averages of the original $\alpha_{ij}, \beta_{ij}, \gamma_{ij}$ $(i,j \in \{-1, 1\})$.

The equation system can easily be solved. $\bar{\alpha}, \bar{\beta}$ and $\bar{\gamma}$ are the solutions.

5. Error Estimations

To achieve C^1-continuity in the assumed situation only one BEZIER point associated with the mixed derivatives must be changed in each patch.

If two BEZIER surface patches S and $\tilde{S}$ only differ in their BEZIER points B_{11} and $\tilde{B}_{11}$,

$$\delta = |B_{11} - \tilde{B}_{11}|_2,$$

their maximum deviation will be

$$\varepsilon = \max_{s,t \in [0,1]} \{|S(s,t) - \tilde{S}(s,t)|_2\}$$

$$= \delta \max_{s \in [0,1]} \{|b_1^n(s)|_2\} \max_{t \in [0,1]} \{|c_1^n(t)|_2\}$$

$$= \delta \left(1 - \frac{1}{n}\right)^{2n-2}$$

where b_1^n and c_1^n are the BERNSTEIN polynomials of degree n.

For BEZIER surfaces of order 4 ($n = 3$) the deviation will be

$$\varepsilon \leq 0.2\delta.$$

The factor will decrease for higher orders but will never be less than

$$\varepsilon \leq \frac{1}{e^2}\delta.$$

6. Applications

As mentioned above, an important application of the described method occurs in the exchange of surfaces between different CAD systems, when the surface representations have to be adjusted. Often a surface representation with a high polynomial degree has to be approximated by a new representation with a lower degree to meet the capabilities of the receiving system. The existing methods (see [2], [3]) approximate a given polynomial multipatch surface by another polynomial surface with a different degree and a different number of patches. For accuracy reasons a degree reduction goes with an increased number of patches. Whereas in general the number of patches can only be decreased if the polynomial degree is enlarged. Because the new representation is evaluated on each new patch separately, a lack of C^1-continuity along patch boundaries may occur.

The method described above corrects this and ensures C^1-continuity all over the new surface representation which is necessary for the generation of cutter paths and other application relevant curves on the surface. The method can be implemented in the VDAFS and IGES processors of the CAD systems. Also, a separate tool is possible which acts on the exchange files directly.

Another example where C^1-discontinuities may occur is the family of fillet surfaces between adjacent surfaces especially with restricted polynomial degrees. Of course, fillet surfaces should always be connected smoothly to their neighbour surfaces but this may put too many constraints on their representation. Again this method can establish C^1-continuity without any change of the polynomial degree.

7. Conclusions

The presented approximation is an easy way to increase the continuity order of a multipatch tensorproduct BEZIER surface from 0 to 1. Moreover, the minimum $F(\bar{\alpha}, \bar{\beta}, \bar{\gamma})$ can be considered as a measure how far a given C^0-continuous surface is away from C^1-continuity.

References

[1] Boehm, W., Farin, G., Kahmann, J.: A survey of Curve and surface methods in CAGD. Computer Aided Geometric Design 1, 1–60 (1984).

[2] Dannenberg, L., Nowacki, H.: Approximate conversion of surface representations with polynomial basis. Computer Aided Geometric Design 2, 123–132 (1985).

[3] Hoschek, J.: Approximate conversion of spline curves. Computer Aided Geometric Design 4, 59–66 (1987).

[4] Schwarz, W.: Konvertierung polynomialer Kurven und Flächendarstellungen. In: Clauer, A., Purgathofer, W. (Hrsg.) AUSTROGRAPHICS '88, 201–214. Wien: Springer 1988.

W. Schwarz
Institut für CAD-Datenaustausch
Kastanienweg 2
D-W-6500 Mainz
Federal Republic of Germany

Computing Suppl. 8, 291–316 (1993)

Geometric Continuity Between Adjacent Rational Bézier Surface Patches

P. Wassum, Darmstadt

Abstract. In this paper, a recurrence formula is presented providing the system of conditions for the geometric continuous joint of arbitrary order h ($h = 1, 2 \ldots$) of two adjacent surface patches along a common boundary curve. This recurrence formula originates from combining the concepts of rescaling and reparametrization. Based on the use of homogeneous coordinates, this procedure has significant advantages for the application to rational Bézier surface patches.

In the case of geometric continuity of order one between two rational tensor-product Bézier patches necessary and sufficient conditions are deduced [10], [21]. Five appropriately choosable form parameters are included in the system of constraints for the involved control points resulting from a special sufficient condition. Thus, this system incorporates one free form parameter more than the constraints previously known. By solving this system for one of the patches, a construction scheme is obtained, which allows a geometric interpretation [25]. In considering the case of an arbitrary value of the continuity order h, necessary and sufficient conditions are obtained by applying an analogous method to that used for $h = 1$. Based on special sufficient conditions, an explicit construction scheme is introduced ensuring the h-th order geometric continuous joint of neighbouring rational tensor-product Bézier patches and allowing a geometric interpretation.

Key words: Geometric continuity, concepts of rescaling and reparametrization, rational Bézier surface patches, space of contact, explicit construction schemes for Bézier control points.

1. Introduction

For describing curves and surfaces in Computer Aided Geometric Design integral as well as rational parametric representation schemes are widely utilized, e.g., Bézier curves and surface patches [14]. The rational representations are advantageous, e.g., they permit the exact description of conic sections, which traditionally support an important part in technically based applications.

The modelling of complex surfaces through combining a collection of surface patches emphasises the importance of considering continuity conditions for adjacent patches. It is commonly agreed that the appropriate approach for the linking together of two neighbouring surface patches along their common boundary curve is the concept of geometric continuity of order h ($h = 1, 2, \ldots$), i.e., in case of $h = 1$ they share the same tangent plane (GC^1-joint), and in case of $h = 2$ they additionally share the same Dupin indicatrix (GC^2-joint). This approach is characterized by its independence of parametrization and supplies additional scalar parameters for controlling the shape.

Different concepts of geometric continuity have been described [17], [8], [23]. These concepts are basically equivalent to the differential geometric concept of order of contact, introduced by Cauchy [7], [24]. For the construction of the geometric continuous joint between adjacent surface patches, conditions for geometric continuity need to be considered.

Numerous papers contain approaches to the construction of integral Bézier surface patches that share the tangent plane along their common boundary, like [11], in which Farin introduced an elegant construction for a GC^1-joint between neighbouring patches, or [22]. These approaches utilize conditions that are in general sufficient for tangent plane continuity. Generalizing these constructions by avoiding unnecessary assumptions, necessary and sufficient conditions for geometric continuity between adjacent patches were developed in [20] (GC^1-case) and [26] (GC^2-case). DeRose proved in [10], based on an example, the general independence of the constraints derived from these conditions (GC^1-case). Degen deduced explicit conditions for first and second order geometric continuity, i.e., solved them for both of the patches [9].

The system of explicit (in general) sufficient constraints for tangent plane continuity between two rational Bézier surface patches in [25], solved for one of the patches, incorporates four appropriately choosable form parameters and yields a construction method (counterpart to Farin's construction in the integral case), which is linked to a geometric interpretation. In [21] necessary and sufficient GC^1-conditions were addressed. The general independence of the constraints obtained from the necessary and sufficient conditions for the GC^1-case is shown in [10].

Based on the concepts of rescaling and reparametrization, a recurrence formula is developed in this paper providing the system of conditions which characterizes the geometric continuous joint of arbitrary order h ($h = 1, 2, \dots$) between two adjacent surface patches, along a common boundary curve. The involved surface patches are given by regular parametric representations of their homogeneous coordinates. The derivation of continuity conditions using homogeneous coordinates provides significant advantages for rational Bézier surface patches. The investigations deal with the combination variant of two tensor-product Bézier patches. By including also triangular Bézier patches into the considerations, three further combination variants of two adjacent rational Bézier patches, i.e., combination variants of the types 'tensor-product—triangle', 'triangle—tensor-product' and 'triangle—triangle', can be specified. As indicated in [27], with minor modifications these combination variants can be treated in an analogous manner. The procedure employed for deducing necessary and sufficient conditions for geometric continuity of order one between two rational Bézier patches is developed. It will be shown that through the use of an analogous method necessary and sufficient conditions for geometric continuity can be derived for an arbitrary value of h. The system of constraints resulting from a special sufficient condition for tangent plane continuity is solved for one of the neighbouring patches and generally provides five appropriately choosable form parameters, thus including one free form parameter more than the constraints known up to now. The construction method presented in [25] is aptly generalized to a construction scheme which is associated with an arbitrary value of order h and allows a straightforward geometric interpretation.

2. Conditions for Geometric Continuity According to Surface Representations in Homogeneous Coordinates

Geometric continuity of order h has the feature of being invariant with respect to parametric transformations. At present, the geometric continuity conditions of arbitrary order h will be derived using homogeneous coordinates. This concept provides advantages for the application to rational Bézier surface patches considered in the successive sections.

The relationship between the homogeneous and the cartesian coordinates of a point (of the three-dimensional object space) is characterized by a normalization, which is a perspective map H of $\mathbb{R}^4 \backslash \{0\}$ on to the hyperplane $h = 1$, carried out from the origin of $\mathbb{R}^4$:

$$H\{(x, y, z, h)\} \\ = H\{\mathbf{X} = (X, h)\} := \begin{cases} \left(\dfrac{x}{h}, \dfrac{y}{h}, \dfrac{z}{h}\right)^T & \text{if } h \neq 0 \\ \text{direction vector from the origin} & \text{if } h = 0. \\ \text{through the point } (x, y, z)^T \end{cases}$$

The parametric representation of a surface patch (of the three-dimensional object space) in homogeneous coordinates can be rescaled [6]. An arbitrary scalar function $q(u, v)$ (denoted as rescaling function) with no zeros over the considered parametric domain $S := [\alpha_1, \alpha_2] \times [\beta_1, \beta_2]$ with $\alpha_1 < \alpha_2, \beta_1 < \beta_2$ is introduced. Applying the procedure of rescaling to the representation $\overline{\mathbf{Y}}(u, v)$ of a surface patch in homogeneous coordinates, the correspondence

$$\mathbf{Y}(u, v) = q(u, v) \quad \overline{\mathbf{Y}}(u, v), \quad \mathbf{Y}(u, v), \overline{\mathbf{Y}}(u, v) \in \mathbb{R}^4, \quad (u, v) \in S, \\ q(u, v) \neq 0 \ \forall (u, v) \in S \tag{2.1}$$

is obtained, involving a surface patch $\mathbf{Y}(u, v)$ which is linked to the same surface patch (of the three dimensional object space) as $\overline{\mathbf{Y}}(u, v)$, whereas in $\mathbb{R}^4$ $\mathbf{Y}(u, v)$ and $\overline{\mathbf{Y}}(u, v)$ represent different surface patches. Surface points $\mathbf{Y}(\bar{u}, \bar{v})$ and $\overline{\mathbf{Y}}(\bar{u}, \bar{v})$, linked to the same parameter values $(u, v) = (\bar{u}, \bar{v}) \in S$, lie as pre-image points on the same straight line passing through the origin $\mathbf{0}$ of $\mathbb{R}^4$.

We will investigate now, how the h-th order geometric continuity of two neighbouring surface patches is transferred to their homogeneous coordinates.

Two surface patches, given by regular parametric representations of their homogeneous coordinates

$$\mathbf{X}: S_x = [\alpha_1, \alpha_2] \times [\beta_1, \beta_2] \to \mathbb{R}^4,$$

$$(r, s) \to \mathbf{X}(r, s) \qquad \alpha_1, \alpha_2, \beta_1, \beta_2, \gamma_1, \gamma_2 \in \mathbb{R}$$

and

$$\mathbf{Y}: S_y = [\gamma_1, \gamma_2] \times [\beta_1, \beta_2] \to \mathbb{R}^4, \qquad \alpha_1 < \alpha_2, \beta_1 < \beta_2, \gamma_1 < \gamma_2$$

$$(u, v) \to \mathbf{Y}(u, v),$$

$$\tag{2.2}$$

are considered (with $\alpha_1 = 0, \gamma_2 = 1$). Assume, the two surface patches are identified with the parametric representations $\mathscr{X} = \mathscr{X}(r, s)$ and $\mathscr{Y} = \mathscr{Y}(u, v)$ of their coordinates.

The assumption of regularity necessitates the requirement, that the absolute values of the alternating products (being described by vector-valued polynomials) are unequal to zero:

$$\left| \mathscr{X}(r,s) \wedge \frac{\partial}{\partial r}\mathscr{X}(r,s) \wedge \frac{\partial}{\partial s}\mathscr{X}(r,s) \right| \neq 0 \qquad \text{for each } (r,s) \in S_x,$$

$$\left| \mathscr{Y}(u,v) \wedge \frac{\partial}{\partial u}\mathscr{Y}(u,v) \wedge \frac{\partial}{\partial v}\mathscr{Y}(u,v) \right| \neq 0 \qquad \text{for each } (u,v) \in S_y.$$

According to the two surface representations the common boundary curve is assumed to be given the same description and parametrization:

$$\mathbf{B}(v) = \mathscr{Y}(u,v)|_{\mathscr{B}} = (q(u,v)\mathscr{X}(r(u,v),s(u,v)))|_{\mathscr{B}} \qquad \text{with } s(u,v)|_{\mathscr{B}} = v \qquad (2.3)$$

(geometric C^0-joint of the two surface patches along $\mathbf{B}(v)$).

The joint of the adjacent surface patches along their common boundary curve should be geometric continuous in the following differential geometric sense of order of contact.

Definition: Two surface patches, given by regular parametric representations $\mathscr{X}(r, s)$ and $\mathscr{Y}(u, v)$ of their homogeneous coordinates, join h-th order geometric continuously along a common boundary curve $\mathscr{Y}(u,v)|_{\mathscr{B}} = (q(u,v)\mathscr{X}(r,s))|_{\mathscr{B}}$ where $s(u,v)|_{\mathscr{B}} = v$ – with the rescaling function $q(u,v) \neq 0$ for all $(u,v) \in S_y$,

corresponding to the rescaling procedure $\mathscr{Y}(u,v) = q(u,v)\bar{\mathscr{Y}}(u,v)$ –

iff a regular orientation preserving parametric transformation $(r,s) \to (r(u,v),s(u,v))$ for the parametric representation $\mathscr{X}(r,s)$ exists such that

$$\frac{\partial^{i+j}}{\partial u^i \partial v^j}\mathscr{Y}(u,v)|_{\mathscr{B}} = \frac{\partial^{i+j}}{\partial u^i \partial v^j}(q(u,v)\mathscr{X}(r(u,v),s(u,v)))|_{\mathscr{B}},$$

$$s(u,v)|_{\mathscr{B}} = v \qquad (i+j \leq f, f = 1(1)h) \tag{2.4}$$

holds, i.e., the two surface patches join h-th order continuously with respect to the (global) parameters u and v along the common boundary curve (combination of the concepts of rescaling and reparametrization).

As shown in [27], the geometric C^h-joint between two adjacent surface patches along their common boundary curve is characterized by the following system of h conditions:

$$\frac{\partial^f}{\partial u^f}\mathscr{Y}(u,v)|_{\mathscr{B}} = \frac{\partial^f}{\partial u^f}(q(u,v)\mathscr{X}(r(u,v),s(u,v)))|_{\mathscr{B}} \qquad (f = 1(1)h), s(u,v)|_{\mathscr{B}} = v \quad (2.5)$$

Hence, geometric continuity of order h between the two surface patches along their common boundary does not presume geometric continuity between their homogeneous coordinates, interpreted as surface patches in $\mathbb{R}^4$ (see Fig. 2.1).

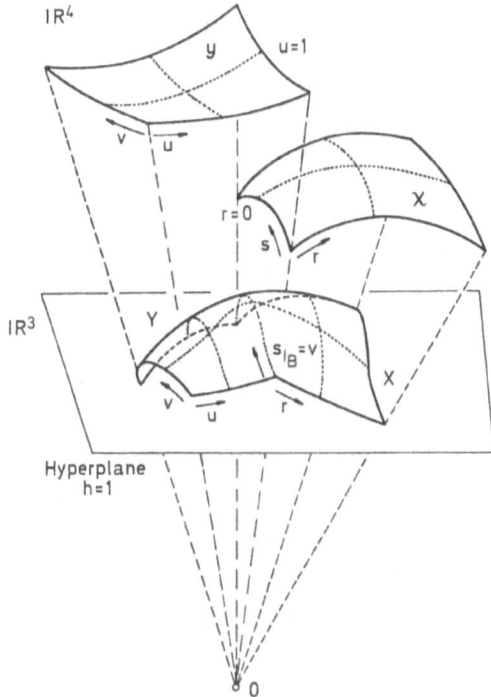

Figure 2.1. Geometric C^h-joint between two surface patches along a common boundary curve—Perspective map of the homogeneous coordinates on to the cartesian coordinates

Assuming, the boundary curve of patch $\mathscr{Y}$ linked to $u = \gamma_2 = 1$, $v \in [\beta_1, \beta_2]$ coincides with the boundary curve of patch $\mathscr{X}$ associated with $r = \alpha_1 = 0$, $s = v \in [\beta_1, \beta_2]$, and using the notations

$$(u_B = 1, v) := (u, v)|_{\mathscr{B}}, \qquad (r_B = 0, v) := (r(u, v), s(u, v))|_{\mathscr{B}}, \qquad q_{0,B}(v) := q(u, v)|_{\mathscr{B}},$$

the geometric C^0-joint condition between the two surface patches $\mathscr{X}$ and $\mathscr{Y}$ is given by

$$\mathscr{Y}(1, v) = q_{0,B}(v)\mathscr{X}(0, v). \tag{2.6}$$

After carrying out the differentiations in (2.5), one obtains a system of conditions which involves the rescaling function (denoted by $q_{0,B}(v)$) as well as the derivatives of the rescaling function $q(u, v)$ and the parameter functions $r(u, v)$, $s(u, v)$ of order j ($j = 1(1)h$) with respect to u, considered along the common boundary:

$$\frac{\partial^j}{\partial u^j}q(u, v)|_{\mathscr{B}} = q_{j,B}(v), \quad \frac{\partial^j}{\partial u^j}r(u, v)|_{\mathscr{B}} = r_{j,B}(v), \quad \frac{\partial^j}{\partial u^j}s(u, v)|_{\mathscr{B}} = s_{j,B}(v), \quad (j = 1(1)h). \tag{2.7}$$

The functions $q_{0,B}(v)$, $q_{j,B}(v)$, $r_{j,B}(v)$ and $s_{j,B}(v)$ ($j = 1(1)h$) can be regarded as the starting coefficients of appropriate Taylor expansions of the functions $q(u,v)$, $r(u,v)$ and $s(u,v)$.

The coefficient function $r_{1,B}(v)$ is strictly positive for each value of $v \in [\beta_1, \beta_2]$, because of the required regularity and the orientation preserving properties of the parametric transformation $(r,s) \to (r(u,v), s(u,v))$.

For the purpose of abbreviation, the simplifying notations

$$\mathcal{Y}^{(f,0)}(1,v) := \frac{\partial^f}{\partial u^f} \mathcal{Y}(u,v)|_{\mathscr{B}}, \qquad \mathcal{X}^{(i,k)}(0,v) := \frac{\partial^{i+k}}{\partial r^i \partial s^k} \mathcal{X}(r,s)|_{\mathscr{B}} \qquad (2.8)$$

are introduced.

The system of geometric C^f-joint conditions ($f = 1(1)h$), e.g., $h = 2$:

$f = 1$: $\mathcal{Y}^{(1,0)}(1,v)$

$$= q_{1,B}(v)\mathcal{X}(0,v) + q_{0,B}(v) \cdot (r_{1,B}(v)\mathcal{X}^{(1,0)}(0,v) + s_{1,B}(v)\mathcal{X}^{(0,1)}(0,v)), \qquad (2.9a)$$

$f = 2$: $\mathcal{Y}^{(2,0)}(1,v) = q_{2,B}(v)\mathcal{X}(0,v) + (2q_{1,B}(v)r_{1,B}(v) + q_{0,B}(v)r_{2,B}(v))\mathcal{X}^{(1,0)}(0,v)$

$$+ (2q_{1,B}(v)s_{1,B}(v) + q_{0,B}(v)s_{2,B}(v))\mathcal{X}^{(0,1)}(0,v)$$

$$+ q_{0,B}(v)(r_{1,B}(v))^2\mathcal{X}^{(2,0)}(0,v) + 2q_{0,B}(v)r_{1,B}(v)s_{1,B}(v)\mathcal{X}^{(1,1)}(0,v)$$

$$+ q_{0,B}(v)(s_{1,B}(v))^2\mathcal{X}^{(0,2)}(0,v), \qquad (2.9b)$$

includes the scalar functions $q_{0,B}(v)$, $q_{f,B}(v)$, $r_{f,B}(v)$, $s_{f,B}(v)$ ($f = 1(1)2$) and characterizes the second order geometric continuous joint of the surface patches along the common boundary curve.

The system of conditions for the geometric C^h-joint (according to an arbitrary value of h) between two surface patches $\mathcal{X}$ and $\mathcal{Y}$ along their common boundary can be derived from a recurrence formula.

Theorem:
Assume, that with $\mathcal{X}(r,s)$ and $\mathcal{Y}(u,v)$ regular parametric representations of two surface patches, and with $\mathscr{B}(v)$ a regular parametric representation of their common boundary curve, are given in homogeneous coordinates. The system of conditions characterizing the geometric continuous joint of order h between the two surface patches $\mathcal{X}$ and $\mathcal{Y}$ along their common boundary

$$\mathscr{B}(v) = \mathcal{Y}(1,v) = q_{0,B}(v)\mathcal{X}(0,v)$$

(geometric C^0-joint between the two surface patches along $\mathscr{B}(v)$) results from successive evaluation in accordance with

$$\mathcal{Y}^{(f,0)}(1,v) = \sum_{j=0}^{f} \left(\binom{f}{j} q^{(f-j)}(1,v) \cdot \sum_{i=0}^{f-i}\sum_{k=0}^{f-i} [\Psi_{jik}|_{\mathscr{B}} \, \mathcal{X}^{(i,k)}(0,v)] \right) \qquad (f = 1(1)h) \quad (2.10)$$

with $\Psi_{110}|_{\mathscr{B}} = r_{1,B}(v)$, $\qquad \Psi_{101}|_{\mathscr{B}} = s_{1,B}(v)$

$$\Psi_{j00}|_{\mathscr{B}} = 0 \quad (j = 1(1)f), \qquad \Psi_{fik}|_{\mathscr{B}} = 0 \ (f < i + k)$$

and the recurrence for the coefficients

$$\Psi_{fik}|_{\mathscr{B}} = (r_{1,B}(v)\Psi_{f-1,i-1,k} + s_{1,B}(v)\Psi_{f-1,i,k-1} + E\Psi_{f-1,i,k})|_{\mathscr{B}}.$$

The shift-operator E is defined by the property

$$E([r_{k,B}(v)]^i[s_{j,B}(v)]^q) := i[r_{k,B}(v)]^{i-1} r_{k+1,B}(v)[s_{j,B}(v)]^q$$
$$+ q[r_{k,B}(v)]^i [s_{j,B}(v)]^{q-1} s_{j+1,B}(v).$$

The proof is based on the method of induction [27].

The geometric C^h-joint can be interpreted geometrically under the required regularity conditions: Two surface patches in $\mathbb{R}^3$ joining first order geometric continuously along a common boundary curve, have a common tangent plane in each point of their common boundary, the geometric C^2-joint along their common boundary corresponds to their property of sharing the same tangent plane and also the same Dupin indicatrix in each point of their common boundary. For the cases with $h > 2$ the following geometric interpretation is possible (valid in each point $\mathscr{P}$ located along the common boundary): Corresponding to each curve segment on the surface patch $\mathscr{X}$ with the starting point $\mathscr{P}$ there exists one curve segment on the surface patch $\mathscr{Y}$, such that these two curve segments join h-th order geometric continuously in $\mathscr{P}$ [24].

The function $q_{0,B}(v)$ in the geometric C^0-joint condition $\mathscr{Y}(1,v) = q_{0,B}(v)\mathscr{X}(0,v)$ is determined as the quotient of the corresponding homogenizing coordinate functions. An examination of the set of conditions (2.10) indicates that the function $q_{0,B}(v)$ appears as a factor in the terms present in the geometric C^f-joint conditions $(f = 1(1)h)$.

To simplify calculations, let us assume throughout the following, that the homogeneous coordinates of $\mathscr{X}$ and $\mathscr{Y}$ interpreted as surface patches in $\mathbb{R}^4$, join zeroth order geometric continuously, i.e., they have a common boundary curve. Then the function $q_{0,B}(v)$ is identical to one: $q_{0,B}(v) \equiv 1$.

Remark: If this assumption is not in force, an analogous deduction to the one employed in the successive sections can be applied to the corresponding considerations.

3. Combination Variant of Tensor-Product Rational Bézier Surface Patches

By taking rational Bézier surface patches of both tensor-product and triangular type into consideration the four combination variants 'tensor-product–tensor-product', 'triangular–tensor-product', 'tensor-product–triangular' and 'triangular–triangular' of two adjacent patches sharing a common boundary can be specified.

The investigations in the subsequent sections are confined to the combination variant of neighbouring tensor-product Bézier patches which are given by parametric representations of their homogeneous coordinates.

The methods undertaken in the following sections with appropriate modifications can be applied analogously to the other combination variants.

By acknowledging the fact that the polynomial degrees of the adjacent patches along their common boundary as well as the degree of this boundary can be selected differently, further joint variants for the type 'tensor-product–tensor-product' result.

3.1 Rational Tensor-Product Bézier Surface Patches

Assume, the parametric representations of the homogeneous coordinates

$$\mathscr{X}(r,s) = \sum_{i=0}^{n} \sum_{j=0}^{m} \mathbf{E}_{ij} B_i^n(r) B_j^m(s) \qquad \mathbf{E}_{ij} \in \mathbb{R}^4, \quad (r,s) \in [0,1] \times [0,1]$$

and   (3.1)

$$\mathscr{Y}(u,v) = \sum_{j=0}^{n^*} \sum_{j=0}^{m^*} \mathbf{F}_{ij} B_i^{n^*}(u) B_j^{m^*}(v) \qquad \mathbf{F}_{ij} \in \mathbb{R}^4, \quad (u,v) \in [0,1] \times [0,1]$$

of the tensor-product Bézier surfaces $\mathscr{X}$ and $\mathscr{Y}$ of degree (n,m) and (n^*,m^*) are given. The points $\{\mathscr{E}_{ij}\}$ and $\{\mathscr{F}_{ij}\}$ are denoted as homogeneous control points, the $B_j^n(t)$ are the well known Bernstein polynomials of degree n.

The image of the homogeneous coordinates $\mathscr{X}(r,s)$ by the perspective map H is given by the rational cartesian coordinates

$$X(r,s) := H\{\mathscr{X}(r,s)\} = \frac{\displaystyle\sum_{j=0}^{m} \sum_{k=0}^{m} e_{jk} E_{jk} B_j^n(r) B_k^m(s)}{\displaystyle\sum_{j=0}^{n} \sum_{k=0}^{m} e_{jk} B_j^n(r) B_k^m(s)},$$

$$e_{jk} \in \mathbb{R}, \quad E_{jk} \in \mathbb{R}^3 \ (j = 0(1)n, k = 0(1)m), \qquad (r,s) \in [0,1] \times [0,1].$$

In this representation the control points are now described inhomogeneously: $\mathscr{E}_{jk} = (e_{jk} E_{jk}, e_{jk})^T$. The homogenizing factors e_{jk}, denoted as weights, are shape parameters. The coefficients $E_{jk} = H\{\mathscr{E}_{jk}\}$ characterize the Bézier net in $\mathbb{R}^3$ (analogously for patch $\mathscr{Y}$, see Fig. 3.1).

The common boundary curve, attached to patch $\mathscr{X}$, is described by

$$r_{\mathscr{A}} = 0, s \in [0,1]: \mathscr{X}(r,s)|_{\mathscr{A}} = \mathscr{X}(0,s) \text{ and to patch } \mathscr{Y} \text{ by}$$ (3.2)

$$u_{\mathscr{A}} = 1, v \in [0,1]: \mathscr{Y}(u,v)|_{\mathscr{A}} = \mathscr{Y}(1,v).$$

Along the common boundary curve the parameters s and v coincide:

$$s|_{\mathscr{A}} = v \qquad \text{where } v \in [0,1].$$

3.2 Rational Bézier Curve—Common Boundary

Assume, the parametric representation of the homogeneous coordinates

$$\mathscr{B}(v) = \sum_{i=0}^{\bar{p}} \mathbf{E}_i B_i^{\bar{p}}(v) \qquad \mathbf{E}_i \in \mathbb{R}^4, v \in [0,1]$$ (3.3)

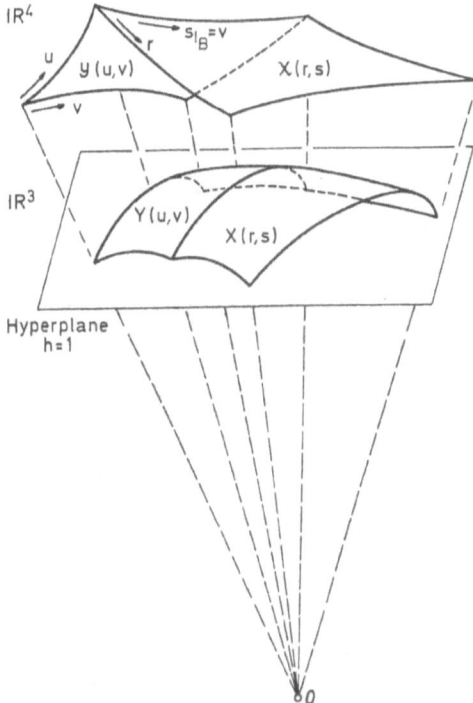

Figure 3.1. Assumption of the geometric C^0-joint (common boundary curve) of the homogeneous coordinates, interpreted as Bézier surface patches in $\mathbb{R}^4$—combination variant of the type 'tensor-product–tensor-product'

is allocated to the common boundary $\mathscr{B} = \mathscr{B}(v)$ of degree $\bar{p}$ with control points $\{\mathscr{E}_i\}$. The image of the homogeneous coordinates $\mathscr{B}(v)$ by H is given by the rational cartesian coordinates

$$B(v) := H\{\mathscr{B}(v)\} = \frac{\sum\limits_{i=0}^{\bar{p}} e_i E_i B_i^{\bar{p}}(v)}{\sum\limits_{i=0}^{\bar{p}} e_i B_i^{\bar{p}}(v)}, \qquad e_i \in \mathbb{R},\ E_i \in \mathbb{R}^3,\ v \in [0,1].$$

In this representation the control points are now described inhomogeneously: $\mathscr{E}_i = (e_i E_i, e_i)^T$. The weights e_i are shape parameters and the coefficients $E_i = H\{\mathscr{E}_i\}$ fix the Bézier polygon of the common boundary in $\mathbb{R}^3$.

The system of geometric C^f-joint conditions (2.10) ($f = 1(1)h$) involves derivatives of $\mathscr{B}(v)$ with respect to the parameter v as well as cross and mixed derivatives of the patches $\mathscr{X}$ and $\mathscr{Y}$ along their common boundary. In the following, appropriate notations for the partial derivatives of $\mathscr{X}$ and $\mathscr{Y}$ along their common boundary are introduced. Utilizing the difference vectors $\mathbf{a}_j^{(k)} := \mathbf{E}_{k,j} - \mathbf{E}_{k-1,j}$, $\mathbf{c}_j^{(k)} := \mathbf{F}_{n^*-k+1,j} -$

$\mathscr{F}_{n^*-k,j}$ between Bézier points of adjacent rows of corresponding control nets, the cross and mixed derivatives along the common boundary

$$D_u^{(f)}\mathscr{Y}(v) := \frac{\partial^f}{\partial u^f}\mathscr{Y}(u,v)|_{\mathscr{B}}, \qquad D_r^{(f)}\mathscr{X}(v) := \frac{\partial^f}{\partial r^f}\mathscr{X}(r,s)|_{\mathscr{B}},$$

$$D_{rv}^{(i,t)}\mathscr{X}(v) := \frac{\partial^{i+t}}{\partial r^i \partial s^t}\mathscr{X}(r,s)|_{\mathscr{B}} \qquad (k = 1(1)f, f = 1(1)h)$$

are described by

$$D_u^{(f)}\mathscr{Y}(v) := \frac{n^*!}{(n^*-f)!}\sum_{j=0}^{m^*}\sum_{k=0}^{f-1}(-1)^k(-1)^{f-1}\binom{f-1}{k}\mathbf{c}_j^{(f-k)}B_j^{m^*}(v),$$

$$D_r^{(f)}\mathscr{X}(v) := \frac{n!}{(n-f)!}\sum_{j=0}^{m}\sum_{k=0}^{f-1}(-1)^k\binom{f-1}{k}\mathbf{a}_j^{(f-k)}B_j^m(v), \tag{3.4}$$

$$D_{rv}^{(i,t)}\mathscr{X}(v) := \frac{n!}{(n-i)!}\frac{m!}{(m-t)!}\sum_{j=0}^{m-t}\sum_{k=0}^{i-1}\sum_{s=0}^{t}(-1)^k(-1)^s\binom{i-1}{k}\binom{t}{s}\mathbf{a}_{j+t-s}^{(i-k)}B_j^{m-t}(v).$$

For convenience later on, the following abbreviations are introduced:

$$D_u\mathscr{Y}(v) := \frac{\partial}{\partial u}\mathscr{Y}(u,v)|_{\mathscr{B}} = n^*\sum_{j=0}^{m^*}\mathbf{c}_j^{(1)}B_j^{m^*}(v) =: \sum_{j=0}^{m^*}\bar{\mathbf{c}}_j B_j^{m^*}(v),$$

$$D_r\mathscr{X}(v) := \frac{\partial}{\partial r}\mathscr{X}(r,s)|_{\mathscr{B}} = n\sum_{j=0}^{m}\mathbf{a}_j^{(1)}B_j^m(v) =: \sum_{j=0}^{m}\bar{\mathbf{a}}_j B_j^m(v). \tag{3.5}$$

The difference vectors $\partial_j := \mathbf{E}_{j+1} - \mathbf{E}_j$ between adjacent Bézier points of the common boundary are incorporated in the representation of the derivatives of $\mathscr{B}(v)$ with respect to v:

$$D_v^{(f)}\mathscr{X}(v) := \frac{d^f}{dv^f}\mathscr{B}(v) = \frac{\bar{p}!}{(\bar{p}-f)!}\sum_{j=0}^{\bar{p}-f}\sum_{k=0}^{f-1}(-1)^k\binom{f-1}{k}\partial_{j+f-1-k}B_j^{\bar{p}-f}(v). \tag{3.6}$$

The first order derivative of $\mathscr{B}(v)$ is denoted by

$$D_v\mathscr{X}(v) := \frac{d}{dv}\mathscr{B}(v) = \bar{p}\sum_{j=0}^{\bar{p}-1}\partial_j B_j^{\bar{p}-1}(v) =: \sum_{j=0}^{\bar{p}-1}\bar{\partial}_j B_j^{\bar{p}-1}(v) \tag{3.7}$$

(see Fig. 3.2. The control points in $\mathbb{R}^3$ and associated weights are shown).

The weights attached to the control points of the Bézier curves and Bézier surface patches are presumed to be positive real numbers. Hence, the denominator terms are strictly positive over the considered parametric domain and infinite points are omitted from further discussion.

It should be noted that the system of geometric C^f-joint conditions ($f = 1(1)h$), characterizing the geometric C^h-joint of two Bézier patches $\mathscr{X}$ and $\mathscr{Y}$, involves the row of control points associated with the common boundary curve $\mathscr{B}$ as well as the first h rows of control points of $\mathscr{X}$ and $\mathscr{Y}$ on either side of $\mathscr{B}$.

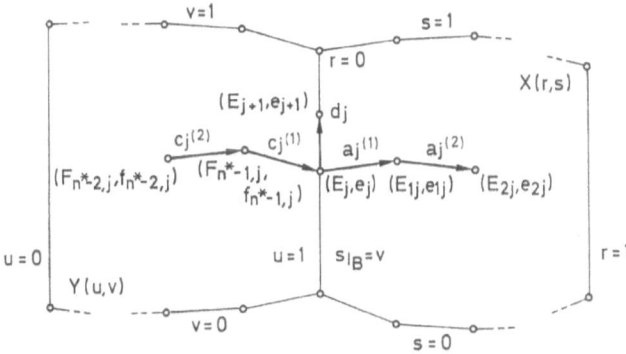

Figure 3.2. Notations for the difference vectors between Bézier points involved in the representation of derivative vectors—combination variant of the type 'tensor-product–tensor-product'

4. Geometric C^1-Joint Between Rational Tensor-Product Bézier Patches

4.1 Rational Coefficient Functions and Determinant Approach

Initially, let us agree upon a notation: The three dimensional subspace of $\mathbb{R}^4$, spanned by the linear independent vectors $\dfrac{\partial}{\partial r}\mathscr{X}(r,s)|_{\mathscr{P}}$, $\dfrac{\partial}{\partial s}\mathscr{X}(r,s)|_{\mathscr{P}}$, $\mathscr{X}(r,s)|_{\mathscr{P}}$ in the point $\mathscr{X}(r,s)|_{\mathscr{P}}$ of a surface patch in $\mathbb{R}^4$ (as the homogeneous coordinates in the regular parametric representation $\mathscr{X}(r,s)$ are interpreted) will be denoted as 'space of contact' of this surface patch in the point $\mathscr{X}(r,s)|_{\mathscr{P}}$.

Hence, the elements of the basis systems

$$\{D_r\mathscr{X}(v), D_v\mathscr{X}(v), \mathscr{B}(v)\} \qquad \text{and} \qquad \{D_u\mathscr{Y}(v), D_v\mathscr{X}(v), \mathscr{B}(v)\}$$

span corresponding spaces of contact of the involved surface patches in $\mathbb{R}^4$, respectively, in each point of the common boundary curve $\mathscr{B}$.

The geometric C^1-joint condition

$$D_u\mathscr{Y}(v) = q_{1,B}(v)\mathscr{B}(v) + r_{1,B}(v)D_r\mathscr{X}(v) + s_{1,B}(v)D_v\mathscr{X}(v) \tag{4.1}$$

is linked to the geometric interpretation, that the adjacent surface patches in $\mathbb{R}^4$ have common spaces of contact in each point $\mathscr{B}(\bar{v})$, with $v = \bar{v} \in [0,1]$.

The determinant condition

$$\Phi_1(v) := \det\{D_u\mathscr{Y}(v), \mathscr{B}(v), D_r\mathscr{X}(v), D_v\mathscr{X}(v)\} = 0 \tag{4.2}$$

depending on the parameter v, indicates a necessary and sufficient condition characterizing the continuity of the spaces of contact of the neighbouring surface patches in $\mathbb{R}^4$ (i.e., tangent plane continuity of the corresponding rational surface patches in $\mathbb{R}^3$). To analyze the type of the coefficient functions $q_{1,B}(v)$, $r_{1,B}(v)$ and $s_{1,B}(v)$ in (4.1), the dual basis system $\{\mathscr{Q}_1(v), \mathscr{Q}_2(v), \mathscr{Q}_3(v)\}$ to the system $\{\mathscr{B}(v), D_r\mathscr{X}(v), D_v\mathscr{X}(v)\}$ is introduced.

The coefficient functions can be described as the following scalar products:

$$q_{1,B}(v) = \langle \mathcal{Q}_1(v), D_u \mathcal{Y}(v) \rangle, \qquad r_{1,B}(v) = \langle \mathcal{Q}_2(v), D_u \mathcal{Y}(v) \rangle,$$

$$s_{1,B}(v) = \langle \mathcal{Q}_3(v), D_u \mathcal{Y}(v) \rangle,$$

and by simple algebraic manipulations one obtains

$$q_{1,B}(v) = \frac{\langle D_u \mathcal{Y}(v) \wedge D_r \mathcal{X}(v) \wedge D_v \mathcal{X}(v), \mathcal{B}(v) \wedge D_r \mathcal{X}(v) \wedge D_v \mathcal{X}(v) \rangle}{(\mathcal{B}(v) \wedge D_r \mathcal{X}(v) \wedge D_v \mathcal{X}(v))^2},$$

$$r_{1,B}(v) = \frac{\langle \mathcal{B}(v) \wedge D_u \mathcal{Y}(v) \wedge D_v \mathcal{X}(v), \mathcal{B}(v) \wedge D_r \mathcal{X}(v) \wedge D_v \mathcal{X}(v) \rangle}{(\mathcal{B}(v) \wedge D_r \mathcal{X}(v) \wedge D_v \mathcal{X}(v))^2}, \qquad (4.3)$$

$$s_{1,B}(v) = \frac{\langle \mathcal{B}(v) \wedge D_r \mathcal{X}(v) \wedge D_u \mathcal{Y}(v), \mathcal{B}(v) \wedge D_r \mathcal{X}(v) \wedge D_v \mathcal{X}(v) \rangle}{(\mathcal{B}(v) \wedge D_r \mathcal{X}(v) \wedge D_v \mathcal{X}(v))^2}.$$

The expression $p = a \wedge b \wedge c$ denotes the alternating product of the three vectors $a, b, c \in \mathbb{R}^4$. The coordinates of $p \in \mathbb{R}^4$ are evaluated as

$$p_i = (-1)^{i-1} \det \begin{bmatrix} a_j & b_j & c_j \\ a_k & b_k & c_k \\ a_l & b_l & c_l \end{bmatrix}$$

with $(i, j, k, l) \in \{(1,2,3,4), (2,3,4,1), (3,4,1,2), (4,1,2,3)\}$.

Therefore, the coefficient functions $q_{1,B}(v), r_{1,B}(v)$ and $s_{1,B}(v)$ prove out to be rational functions of the parameter v with a common denominator polynomial (in general, before one starts to cancel possible common linear factors).

4.2 Systems of Necessary and Sufficient Continuity Constraints

From the determinant condition (4.2), a system of necessary and sufficient scalar constraints for the geometric C^1-joint of two adjacent patches, which is formulated directly in terms of involved control points, can be derived. The approach used to provide this system of constraints is to require the vanishing of all Bézier ordinates of the determinant function $\Phi_1(v)$.

Inserting the expressions (3.3), (3.5), (3.7) into condition (4.2) and using multilinearity of determinants, $\Phi_1(v)$ can be rewritten as

$$\Phi_1(v) = \sum_{i_1, i_2, i_3, i_4} \det(\bar{c}_{i_1}, \mathscr{E}_{i_2}, \bar{a}_{i_3}, \bar{\partial}_{i_4}) B_{i_1}^{m*}(v) B_{i_2}^{\bar{p}}(v) B_{i_3}^m(v) B_{i_4}^{\bar{p}-1}(v). \qquad (4.4)$$

Employing a new summation index ξ, where $\xi := i_1 + i_2 + i_3 + i_4$, and utilizing the identity

$$B_f^n(t) B_k^m(t) = \frac{\binom{n}{f}\binom{m}{k}}{\binom{n+m}{f+k}} B_{f+k}^{n+m}(t), \qquad (4.5)$$

the terms in the summation can be recombined by grouping together all the terms refering to a common value of ξ:

$$\Phi_1(v) = \sum_{\xi=0}^{m^*+m+2\bar{p}-1} \left[\sum_{i_1+i_2+i_3+i_4=\xi}^{m^*+m+2\bar{p}-1} \frac{1}{\dbinom{m^*+m+2\bar{p}-1}{\xi}} \right.$$
$$\left. \cdot \left\{ \binom{m^*}{i_1}\binom{\bar{p}}{i_2}\binom{m}{i_3}\binom{\bar{p}-1}{i_4} \det(\bar{c}_{i_1}, \mathscr{E}_{i_2}, \bar{a}_{i_3}, \bar{\partial}_{i_4}) \right\} \right] B_\xi^{m^*+m+2\bar{p}-1}(v). \quad (4.6)$$

The term in square brackets represents the ξ-th Bézier coefficient φ_ξ of $\Phi_1(v)$.

The determinant condition (4.2) is satisfied, iff each of the Bézier coefficients φ_ξ of $\Phi_1(v)$ with $\xi = 0(1)m^* + m + 2\bar{p} - 1$ vanishes.

Theorem:
Two rational tensor-product Bézier patches $\mathscr{X}$ and $\mathscr{Y}$, given by parametric representations of their homogeneous coordinates and sharing a common boundary curve, join first order geometric continuously, iff the following system of scalar constraints holds for $\xi = 0(1)m^* + m + 2\bar{p} - 1 =: g_\Phi + 1$:

$$\varphi_\xi = 0 \quad \text{respectively}$$

$$\sum_{i_1+i_2+i_3+i_4=\xi} \det\left\{ \binom{m^*}{i_1}\bar{c}_{i_1}, \binom{\bar{p}}{i_2}\mathscr{E}_{i_2}, \binom{m}{i_3}\bar{a}_{i_3}, \binom{\bar{p}-1}{i_4}\bar{\partial}_{i_4} \right\} = 0. \quad (4.7)$$

The determinant function $\Phi_1(v)$ is described by a scalar polynomial with the highest possible degree g_Φ, which is a direct consequence of the fact that the determinant in (4.2) involves with the vector polynomials $\mathscr{B}(v)$ and $D_v\mathscr{X}(v)$ a function and its derivative [27].

Hence, the Bézier representation of the determinant function $\Phi_1(v)$ in (4.6) results from a Bézier representation of $\Phi_1(v)$, associated with the polynomial degree g_Φ by degree elevation.

Assume now, the determinant function $\Phi_1(v)$ is considered in a form with the polynomial degree g_Φ, from which, due to the independence of the Bernstein polynomials, a system of scalar constraints, denoted as S_1^*, is obtained.

DeRose's considerations [10] are based on an equivalent form of the determinant condition (4.2) for the geometric C^1-joint of two rational Bézier patches, resulting in a system of necessary and sufficient constraints, where the system S_1^* represents an equivalent system of constraints.

The idea employed in [10] to prove the minimal property of the system (i.e., the contained constraints are in general independent) is applied in [27] to show the minimal property of the system S_1^*: It is proved that the dimension of the space of polynomials of the form given in (4.2) is equal to $g_\Phi + 1$. Hence, $\Phi_1(v)$ can be a fully general polynomial of highest degree g_Φ, and the constraints in the system S_1^*, which correspond to the Bézier coefficients of the (degree reduced) determinant function $\Phi_1(v)$, are in general independent.

In the case, e.g., $m^* = m = \bar{p} = p$, a system consisting of $4p - 1$ constraints has in general to be studied to control the tangent plane continuous joint of the adjacent tensor-product Bézier patches.

To develop a system of necessary and sufficient constraints for tangent plane continuity between two adjacent Bézier patches $\mathscr{X}$ and $\mathscr{Y}$ along their common boundary curve from the geometric C^1-joint condition (4.1), the lowest upper bounds for the degrees of the numerator polynomials and the common denominator polynomial of the coefficient functions $q_{1,B}(v), r_{1,B}(v), s_{1,B}(v)$ have to be determined.

First of all, let us consider four alternating products, involving in each instance three out of the four vector polynomials (i.e., vector-valued polynomials) $D_u\mathscr{Y}(v)$, $\mathscr{B}(v)$, $D_r\mathscr{X}(v)$, $D_v\mathscr{X}(v)$ and evaluated by

$$\mathbf{N}_{t,1}(v) := \mathscr{B}(v) \wedge D_r\mathscr{X}(v) \wedge D_v\mathscr{X}(v), \qquad \mathbf{N}_{q,1}(v) := D_u\mathscr{Y}(v) \wedge D_r\mathscr{X}(v) \wedge D_v\mathscr{X}(v),$$

$$\mathbf{N}_{r,1}(v) := \mathscr{B}(v) \wedge D_u\mathscr{Y}(v) \wedge D_v\mathscr{X}(v), \qquad \mathbf{N}_{s,1}(v) := \mathscr{B}(v) \wedge D_r\mathscr{X}(v) \wedge D_u\mathscr{Y}(v).$$

$$(4.8)$$

These vector polynomials $\mathbf{N}_{t,1}(v)$, $\mathbf{N}_{q,1}(v)$, $\mathbf{N}_{r,1}(v)$ and $\mathbf{N}_{s,1}(v)$ can respectively be described in the form of a product

$$\mathbf{N}_{t,1}(v) = t_1(v)\mathbf{P}(v), \qquad \mathbf{N}_{q,1}(v) = q_1(v)\mathbf{P}(v),$$
$$\mathbf{N}_{r,1}(v) = r_1(v)\mathbf{P}(v), \qquad \mathbf{N}_{s,1}(v) = s_1(v)\mathbf{P}(v),$$

$$(4.9)$$

of a corresponding scalar polynomial $t_1(v)$, $q_1(v)$, $r_1(v)$ and $s_1(v)$ with the common irreducible vector polynomial $\mathbf{P}(v)$. Hence, the linear factors common to each four coordinate polynomials are combined into the polynomials $t_1(v)$, $q_1(v)$, $r_1(v)$ and $s_1(v)$.

The irreducible vector polynomial $\mathbf{P}(v)$ (with $\mathbf{P}(v) \neq 0 \ \forall v \in \mathbb{R}$) is orthogonal to the derivative vectors $D_u\mathscr{Y}(v)$, $D_r\mathscr{X}(v)$, $D_v\mathscr{X}(v)$ and to the vector $\mathbf{B}(v)$ in each point $\mathscr{B}(\bar{v})$, with $v = \bar{v} \in [0,1]$ of the common boundary curve. Therefore, $\mathbf{P}(v)$ indicates in each point of the boundary the normal direction to the related space of contact, which is common to the adjacent surface patches in $\mathbb{R}^4$.

The product representations (4.9) for the vector polynomials $\mathbf{N}_{t,1}(v), \mathbf{N}_{q,1}(v), \mathbf{N}_{r,1}(v)$ and $\mathbf{N}_{s,1}(v)$ are inserted into the description (4.3) of the coefficient functions $q_{1,B}(v)$, $r_{1,B}(v)$, $s_{1,B}(v)$. Cancelling the factor $(\mathbf{P}(v))^2$, one obtains each of the coefficient functions

$$q_{1,B}(v) = \frac{q_1(v)}{t_1(v)}, \qquad r_{1,B}(v) = \frac{r_1(v)}{t_1(v)}, \qquad s_{1,B}(v) = \frac{s_1(v)}{t_1(v)} \qquad (4.10)$$

as the quotient of two out of the four scalar polynomials (denoted as form functions) $t_1(v)$, $q_1(v)$, $r_1(v)$ and $s_1(v)$.

To avoid a sharp edge of the adjacent patches along their common boundary curve, it is a necessary requirement that both (with $|\mathbf{N}_{t,1}(v)| \neq 0$ and $|\mathbf{N}_{r,1}(v)| \neq 0 \ \forall v \in [0,1]$ because of the assumption of regularity) collinear vector polynomials

$N_{t,1}(v)$, $\mathscr{N}_{r,1}(v)$ have the same orientation in each point of the shared boundary. Hence, the coefficient function $r_{1,B}(v)$ is strictly positive on the parametric interval $I = [0,1]$. Also the corresponding normal vectors in $\mathbb{R}^3$, being orthogonal to the common tangent plane of the rational Bézier patches, have the same orientation along the common boundary curve.

Due to the relation between the two vector polynomials $\mathscr{B}(v)$ and $D_v\mathscr{X}(v)$, simultaneously appearing as factors of the alternating products $\mathscr{N}_{t,1}(v)$ and $\mathscr{N}_{r,1}(v)$, the highest degree in v of each vector polynomial $\mathscr{N}_{t,1}(v)$ and $\mathscr{N}_{r,1}(v)$ is equal to the sum of the degrees of the vector polynomials involved minus one.

Therefore, the highest degrees of the vector polynomials $\mathscr{N}_{t,1}(v)$, $\mathscr{N}_{q,1}(v)$, $\mathscr{N}_{r,1}(v)$ and $\mathscr{N}_{s,1}(v)$ are equal to $m + 2\bar{p} - 2$, $m^* + m + \bar{p} - 1$, $m^* + 2\bar{p} - 2$ and $m^* + m + \bar{p}$.

Taking the example $m^* = m = \bar{p} = p$ it is assumed that the vector polynomials $\mathscr{N}_{t,1}(v)$, $\mathscr{N}_{q,1}(v)$, $\mathscr{N}_{r,1}(v)$ and $\mathscr{N}_{s,1}(v)$ have exactly the polynomial degrees $3p - 2$, $3p - 1$, $3p - 2$ and $3p - 1$ respectively.

Given the control points of patch $\mathscr{X}$, one obtains from the determinant condition (4.2), due to the independence of the Bernstein polynomials, a system of homogeneous linear constraints for the vector polynomial $D_u\mathscr{Y}(v)$ (i.e., for the first interior row of unknown control points of the patch $\mathscr{Y}$). This system consists of $4p - 1$ constraints with $4p + 4$ unknowns and has therefore at least five linear independent solutions. According to the general situation where $g_{\mathscr{P}}$, denoting the polynomial degree of the irreducible vector polynomial $\mathscr{P}(v)$, has the value $g_{\mathscr{P}} = 3p - 2$, the general solution for the unknown derivative vector $D_u\mathscr{Y}(v)$ takes the form

$$D_u\mathscr{Y}(v) = (\eta_{1,0} + \eta_{1,1}v)\mathscr{B}(v) + \mu_{1,0}D_r\mathscr{X}(v) + (\delta_{1,0} + \delta_{1,1}v + \delta_{1,2}v^2)D_v\mathscr{X}(v),$$

$$(\eta_{1,i}, \mu_{1,0}, \delta_{1,i} \in \mathbb{R}), \qquad \text{where } \eta_{1,1} + p\delta_{1,2} = 0$$

and the condition of regularity $\mu_{1,0} > 0$ also holds.

In [18] a criterion for the number of linear independent solutions of (4.2) is developed in relation to the degree of the irreducible normal vector $\mathscr{P}(v)$.

The difference values between the degrees of the vector polynomials $\mathscr{N}_{t,1}(v)$, $\mathscr{N}_{q,1}(v)$, $\mathscr{N}_{r,1}(v)$, $\mathscr{N}_{s,1}(v)$ and $g_{\mathscr{P}}$:

$$\text{degr}(t_1(v)) = 3p - 2 - g_{\mathscr{P}}, \qquad \text{degr}(q_1(v)) = 3p - 1 - g_{\mathscr{P}},$$

$$\text{degr}(r_1(v)) = 3p - 2 - g_{\mathscr{P}}, \qquad \text{degr}(s_1(v)) = 3p - 1 - g_{\mathscr{P}} \tag{4.11}$$

state the polynomial degrees of the form functions $t_1(v)$, $q_1(v)$, $r_1(v)$ and $s_1(v)$.

The indicated degrees of the form functions have to be viewed as their highest values, if it is not explicitly presumed that the degree of each of the vector polynomials $\mathscr{N}_{t,1}(v)$, $\mathscr{N}_{q,1}(v)$, $\mathscr{N}_{r,1}(v)$ and $\mathscr{N}_{s,1}(v)$ is equal to its maximum value.

When setting up the highest values in (4.11) for the polynomial degrees of the form functions, corresponding to an arbitrarily selected but fixed chosen value of $g_{\mathscr{P}}$ out of the range $0 \leq g_{\mathscr{P}} \leq 3p - 2$, the whole spectrum of possible joint configurations of two, first order geometric continuously joining tensor-product patches with $m^* = m = \bar{p} = p$ is covered.

The totality of all geometric C^1-joint configurations of two tensor-product Bézier patches is taken into consideration by assigning all possible values to the degree $g_{\mathscr{P}}$ of the irreducible normal vector $\mathscr{P}(v)$.

Combinations of adjacent tensor-product Bézier patches with different degrees along their common boundary curve can be addressed by analogous considerations.

Consequently, the highest polynomial degrees of the form functions $t_1(v)$, $q_1(v)$, $r_1(v)$ and $s_1(v)$, corresponding to a system of necessary and sufficient constraints for the geometric C^1-joint between two adjacent Bézier patches, are determined by:

$$\max \operatorname{degr}(t_1(v)) = m + 2\bar{p} - 2, \qquad \max \operatorname{degr}(q_1(v)) = m^* + m + \bar{p} - 1,$$

$$\max \operatorname{degr}(r_1(v)) = m^* + 2\bar{p} - 2, \qquad \max \operatorname{degr}(s_1(v)) = m^* + m + \bar{p},$$

In [21] a formal method was used to determine the lowest upper bounds for the degrees of the polynomial form functions $t_1(v)$, $q_1(v)$, $r_1(v)$ and $s_1(v)$.

The geometric C^1-joint condition

$$t_1(v)\boldsymbol{D}_u\mathscr{Y}(v) = q_1(v)\mathscr{B}(v) + r_1(v)\boldsymbol{D}_r\mathscr{X}(v) + s_1(v)\boldsymbol{D}_v\mathscr{X}(v) \tag{4.12}$$

shows a symmetric form: Each of the four vector polynomials involved is linked to a polynomial form function.

The polynomial form functions in (4.12) can be expressed in terms of Bernstein polynomials

$$t_1(v) = \sum_{i=0}^{m+2\bar{p}-1} \tau_{1,i} B_i^{m+2\bar{p}-1}(v), \qquad q_1(v) = \sum_{i=0}^{m^*+m+\bar{p}-1} \upsilon_{1,i} B_i^{m^*+m+\bar{p}-1}(v),$$

$$r_1(v) = \sum_{i=0}^{m^*+2\bar{p}-1} \rho_{1,i} B_i^{m^*+2\bar{p}-1}(v), \qquad s_1(v) = \sum_{i=0}^{m^*+m+\bar{p}} \sigma_{1,i} B_i^{m^*+m+\bar{p}}(v), \tag{4.13}$$

containing form parameters $\tau_{1,i}, \upsilon_{1,i}, \rho_{1,i}, \sigma_{1,i}$.

Having substituted $t_1(v)$, $q_1(v)$, $r_1(v)$ and $s_1(v)$ in (4.12) by (4.13) as well as $\boldsymbol{D}_v\mathscr{Y}(v)$, $\boldsymbol{D}_r\mathscr{X}(v)$, $\boldsymbol{D}_v\mathscr{X}(v)$, $\mathscr{B}(v)$ by the notations (3.3), (3.5), (3.7) and used identity (4.5), one obtains, due to the independence of the Bernstein polynomials, a system of linear constraints, combining difference vectors between control points of the three involved rows of Bézier points and unknown form parameters.

Theorem:
A system of necessary and sufficient constraints for the geometric C^1-joint of two adjacent tensor-product Bézier patches $\mathscr{X}$ and $\mathscr{Y}$, given by parametric representations of their homogeneous coordinates and joining along a common boundary curve $\mathscr{B}$ is given by

$$\sum_{\substack{i=0 \\ i+j=v}}^{m+2\bar{p}-1} \sum_{j=0}^{m^*} \binom{m + 2\bar{p} - 1}{i} \binom{m^*}{j} \tau_{1,i} \bar{c}_j$$

$$= \sum_{\substack{i=0 \\ i+j=v}}^{m^*+m+\bar{p}-1} \sum_{j=0}^{\bar{p}} \binom{m^* + m + \bar{p} - 1}{i} \binom{\bar{p}}{j} v_{1,i} \bar{\mathscr{e}}_j$$

$$+ \sum_{\substack{i=0 \\ i+j=v}}^{m^*+2\bar{p}-1} \sum_{j=0}^{m} \binom{m^* + 2\bar{p} - 1}{i} \binom{m}{j} \rho_{1,i} \bar{a}_j$$

$$+ \sum_{\substack{i=0 \\ i+j=v}}^{m^*+m+\bar{p}} \sum_{j=0}^{\bar{p}-1} \binom{m^* + m + \bar{p}}{i} \binom{\bar{p} - 1}{j} \sigma_{1,i} \bar{\partial}_j$$

(4.14)

with $v = 0(1)m^* + m + 2\bar{p} - 1$.

Remark: The polynomial functions $t_1(v)$ and $r_1(v)$ in (4.13) have been described in a degree elevated form, and the system (4.14) results from a corresponding degree elevated representation of the geometric C^1-condition (4.12).

4.3 Geometric Construction Scheme Corresponding to a System of Special Sufficient Continuity Constraints

The approach to reduce the obviously large number of form parameters, included in the system of necessary and sufficient geometric C^1-constraints (4.14), is to determine various systems of sufficient geometric C^1-constraints. These systems correspond (after having cancelled common linear factors of the terms in (4.12)) to different degrees of the form functions $T_1(v)$, $Q_1(v)$, $R_1(v)$ and $S_1(v)$ in the geometric C^1-condition

$$T_1(v)n^* \sum_{j=0}^{m^*} c_j^{(1)} B_j^{m^*}(v) = Q_1(v) \sum_{j=0}^{\bar{p}} \mathscr{e}_j B_j^{\bar{p}}(v) + R_1(v)n \sum_{j=0}^{m} a_j^{(1)} B_j^m(v)$$

$$+ S_1(v)\bar{p} \sum_{j=0}^{\bar{p}-1} \partial_j B_j^{\bar{p}-1}(v)$$

(4.15)

and provide different numbers of unknown form parameters.

Further information on the procedure of deriving systems of special sufficient geometric C^1-constraints can be acquired in [27].

The rows of control points $\mathscr{e}_{0j}$, $\mathscr{e}_{1j}$ $(j = 0(1)m)$ linked to patch $\mathscr{X}$ are assumed to be given and the information $m^* = m = \bar{p} = p$ about the chosen degrees along the common boundary curve provides the system of equations

$$\mathscr{F}_{n \cdot j} = \mathscr{e}_j = \mathscr{e}_{0j} \qquad (j = 0(1)p),$$

characterizing the relationship between the Bézier points of the common boundary and the Bézier points of the involved rows of the patches $\mathscr{X}$ and $\mathscr{Y}$.

Inserting the system of selected form functions (compare with section 4.2)

$$T_1(v) = 1, \qquad Q_1(v) = v_{1,0} - \sigma_{1,2}pv, \qquad R_1(v) = \rho_{1,0},$$
$$S_1(v) = \sigma_{1,0}(1 - v) + \sigma_{1,1}v + \sigma_{1,2}v^2, \tag{4.16}$$

into the geometric C^1-condition (4.15), one obtains, due to the independence of the Bernstein polynomials, a system of $p + 1$ constraints describing the dependency of the unknown control points $\mathscr{F}_{n^*-1\,j}$ ($j = 0(1)p$) to the given Bézier points and five appropriately choosable form parameters.

Setting the form parameter $\sigma_{1,2}$ equal to zero, one derives the system of geometric C^1-constraints developed in [25].

Each control point $\mathscr{F}_{n^*-1,j}$ of the first interior row of control points linked to patch $\mathscr{Y}$ is determined as the convex combination

$$\mathscr{F}_{n^*-1,j} = (1 - \zeta)\mathscr{E}^{(1)}_{0j,1} + \zeta\mathscr{E}^{(1)}_{0j,2} \qquad (j = 0(1)p) \tag{4.17}$$

of the two auxiliary points

$$\mathscr{E}^{(1)}_{0j,1} = (\eta_{01} - \gamma_{10})\mathscr{E}_{0j} + \gamma_{10}\mathscr{E}_{1j} + \gamma_{01}\mathscr{E}_{0,j+1},$$
$$\mathscr{E}^{(1)}_{0j,2} = (\varepsilon_{0,-1} - \delta_{10})\mathscr{E}_{0j} + \delta_{10}\mathscr{E}_{1j} + \delta_{0,-1}\mathscr{E}_{0,j-1}, \tag{4.18}$$

evaluated in the first construction step.

The parameters employed in the construction scheme and the form parameters of the form functions (4.16) are related in the following way

$$\zeta = \frac{j}{p}, \quad \gamma_{10} = \delta_{10} = -\rho_{1,0}\frac{n}{n^*}, \quad \gamma_{01} = -\sigma_{1,0}\frac{p}{n^*}, \quad \delta_{0,-1} = \sigma_{1,1}\frac{p}{n^*} + p\sigma_{1,2},$$

$$\eta_{01} = (n^* - v_{1,0} + \sigma_{1,0}p)/n^*, \qquad \varepsilon_{0,-1} = (n^* - v_{1,0} - \sigma_{1,1}p)/n^*.$$

Please note, that the coefficients contained in either of the two equations corresponding to the corner situations $j = 0, j = p$:

$$\mathscr{F}_{n^*-1,0} = (\eta_{01} - \gamma_{01})\mathscr{E}_{00} + \gamma_{10}\mathscr{E}_{10} + \gamma_{01}\mathscr{E}_{01},$$
$$\mathscr{F}_{n^*-1,p} = (\varepsilon_{0,-1} - \delta_{0,-1})\mathscr{E}_{0p} + \delta_{10}\mathscr{E}_{1p} + \delta_{0,-1}\mathscr{E}_{0,p-1},$$

don't sum to one. This means, in general none of the Bézier points $\mathscr{F}_{n^*-1,0}, \mathscr{F}_{n^*-1,p}$ lies in the corresponding plane, determined by the triples of non-collinear control points $\mathscr{E}_{00}, \mathscr{E}_{10}, \mathscr{E}_{01}$ and $\mathscr{E}_{0p}, \mathscr{E}_{1p}, \mathscr{E}_{0,p-1}$ respectively. Analogous properties are shared by the $p - 1$ auxiliary points $\mathscr{E}^{(1)}_{0j,1}, \mathscr{E}^{(1)}_{0j,2}$ with $j = 1(1)p - 1$. In consequence, the homogeneous coordinates of $\mathscr{X}$ and $\mathscr{Y}$ interpreted as surface patches in $\mathbb{R}^4$ have, in general, no common tangent plane along their common boundary curve.

The cross ratio of four collinear points

$$cr(\mathscr{P}_1, \mathscr{P}_2, \mathscr{P}_3, \mathscr{P}_4) := \text{sign}(cr_{1,2,3,4})\frac{\overline{\mathscr{P}_1\mathscr{P}_2}\ \overline{\mathscr{P}_3\mathscr{P}_4}}{\overline{\mathscr{P}_1\mathscr{P}_3}\ \overline{\mathscr{P}_2\mathscr{P}_4}}, \qquad \text{where sign}(cr_{1,2,3,4}) := \pm 1,$$

depending on the relative order of the involved points, is projectively invariant. When choosing $\mathscr{P}_1 = \mathscr{E}^{(1)}_{0j,1}, \mathscr{P}_2 = \mathscr{F}_{n^*-1,j}, \mathscr{P}_3 = \frac{1}{2}(\mathscr{E}^{(1)}_{0j,1} + \mathscr{E}^{(1)}_{0j,2}), \mathscr{P}_4 = \mathscr{E}^{(1)}_{0j,2}$, one ob-

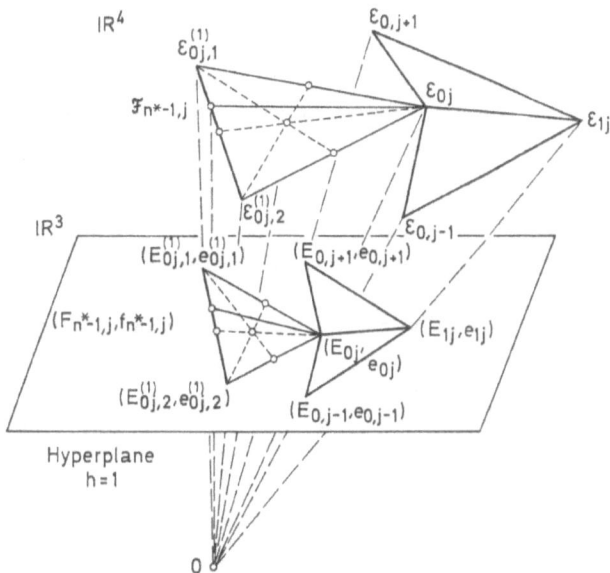

Figure 4.1. Construction of the points $F_{n^{*}-1,j}$ for the geometric C^1-joint of two adjacent tensor-product-Bézier-patches according to the proposed construction scheme—Perspective map of the homogeneous coordinates on to the cartesian coordinates—Farin's auxiliary points

tains $cr(\mathscr{E}_{0j,1}^{(1)}, \mathscr{F}_{n^{*}-1,j}, \frac{1}{2}(\mathscr{E}_{0j,1}^{(1)} + \mathscr{E}_{0j,2}^{(1)}), \mathscr{E}_{0j,2}^{(1)}) = \dfrac{j}{p-j}$. Hence, this value varies with the running index j according to the considered quadruple of points [25] (see Fig. 4.1).

The information for the weights linkèd to corresponding control points, is attached to the control net by means of auxiliary points introduced in [12], denoted in the following as 'Farin points'. Interpreting these weights as physical weights attached to each pair of neighbouring control points, the matching Farin point indicates the position at which an imaginary fulcrum should be placed to balance the edge with the involved control points.

With respect to the control nets of tensor-product Bézier surface patches it is necessary to take into account, that along any closed path in the control net, having fixed all but one of these fulcrum positions, the last one is already uniquely determined.

The lines joining each projected vertex with the position of the fulcrum on the opposite side are the projections of the medians of a corresponding triangle in $\mathbb{R}^4$, and thus intersect at a point, which is the projected centroid. By assigning appropriate physical weights to the triangle corner points, this projected centroid characterizes the triangle's fulcrum position.

5. Geometric C^k-Joint Between Rational Tensor-Product Bézier Patches

Utilizing an analogous procedure, the considerations for the geometric C^1-joint between two adjacent tensor-product Bézier patches along a common boundary curve in the previous section will be extended to the case of higher order geometric continuity.

5.1 Rational Coefficient Functions and Determinant Approach

The successive evaluation of the recurrence formula, developed in section 2, (making use of the notations (2.7), (3.3), (3.4), (3.6))

$$D_u^{(f)}\mathcal{Y}(v) - \sum_{j=1}^{f-1} \left[\binom{f}{j} q_{f-j,B}(v) \cdot \left[\sum_{i=0}^{j} \sum_{\substack{k=0 \\ i+k>0}}^{j-i} [\Psi_{jik}|_\mathcal{B} \, D_{rv}^{(i,k)}\mathcal{X}(v)] \right] \right]$$

$$- \sum_{i=0}^{f} \sum_{\substack{k=0 \\ i+k>1}}^{f-i} [\Psi_{fik}|_\mathcal{B} \, D_{rv}^{(i,k)}\mathcal{X}(v)] \tag{5.1}$$

$$= q_{f,B}(v)\mathcal{B}(v) + r_{f,B}(v)D_r\mathcal{X}(v) + s_{f,B}(v)D_v\mathcal{X}(v) \qquad (f = 1(1)h)$$

with $\Psi_{110}|_\mathcal{B} = r_{1,B}(v), \qquad \Psi_{101}|_\mathcal{B} = s_{1,B}(v), \qquad q_{0,B}(v) \equiv 1,$

and the recurrence for the coefficients

$$\Psi_{fik}|_\mathcal{B} = (r_{1,B}(v)\Psi_{f-1,i-1,k} + s_{1,B}(v)\Psi_{f-1,i,k-1} + E\Psi_{f-1,i,k})|_\mathcal{B},$$

as well as the shift-operator E, defined by

$$E[(r_{i,B}(v))^f (s_{j,B}(v))^k] := f(r_{i,B}(v))^{f-1} r_{i+1,B}(v) (s_{j,B}(v))^k$$
$$+ k(r_{i,B}(v))^f (s_{j,B}(v))^{k-1} s_{j+1,B}(v),$$

yields a system S_H consisting of h conditions and characterizing the h-th order geometric continuous joint between two surface patches $\mathcal{X}$ and $\mathcal{Y}$ along their common boundary curve $\mathcal{B}$.

Let us refer to the condition where $f = k$ as the geometric C^k-joint condition and to the system S_H as the system of the geometric C^f-joint conditions with $f = 1(1)h$.

The geometric C^f-joint condition ($f = 1(1)h$) in (5.1) describes the corresponding combination of derivative vectors on the left hand side in linear dependency to the vector polynomials $\mathcal{B}(v)$, $D_r\mathcal{X}(v)$ and $D_v\mathcal{X}(v)$.

Geometrically, this states that this vector combination lies in each point $\mathcal{B}(\bar{v})$ of the common boundary curve (where $\bar{v} = v \in [0, 1]$) in the attached space of contact.

The geometric C^f-joint condition includes derivative vectors up to order f and coefficient functions $q_{1,B}(v), r_{1,B}(v), s_{1,B}(v)$ as well as $q_{i,B}(v), r_{i,B}(v), s_{i,B}(v)$ ($i = 2(1)f$). Successively, both the coefficient functions $q_{1,B}(v), r_{1,B}(v), s_{1,B}(v)$ are exchanged for rational expressions given by (4.3), and analogously, $q_{i,B}(v), r_{i,B}(v), s_{i,B}(v)$

$(i = 2(1)f - 1)$ for rational expressions, where appropriate vector combinations take the place of $D_u \mathscr{Y}(v)$.

The multiplication of the corresponding determinant condition (stated in rational form) with the denominator polynomial

$$\langle \mathscr{B}(v) \wedge D_r \mathscr{X}(v) \wedge D_v \mathscr{X}(v) \rangle^{2(f-1)} = \langle \mathscr{N}_{t,1}(v), \mathscr{N}_{t,1}(v) \rangle^{(f-1)}$$

yields a polynomial determinant condition

$$\Phi_{f,0}(v) := \det \left\{ \left[D_u^{(f)} \mathscr{Y}(v) - \sum_{j=1}^{f} \left[\binom{f}{j} q_{f-j,B}(v) \cdot \left[\sum_{\substack{i=0 \\ i+g>1}}^{j} \sum_{g=0}^{j-i} [\Psi_{jig}|_{\mathscr{B}} D_{rv}^{(i,g)} \mathscr{X}(v)] \right] \right] \right] \right.$$

$$\left. \cdot \langle \mathscr{N}_{t,1}(v), \mathscr{N}_{t,1}(v) \rangle^{(f-1)}, \mathscr{B}(v), D_r \mathscr{X}(v), D_v \mathscr{X}(v) \right\} = 0. \tag{5.2}$$

Let us agree on the following abbreviation denoting the vector combination

$$D_f \mathscr{Y}(v) := \left[D_u^{(f)} \mathscr{Y}(v) - \sum_{j=1}^{f} \left[\binom{f}{j} q_{f-j,B}(v) \cdot \left[\sum_{\substack{i=0 \\ i+k>1}}^{j} \sum_{k=0}^{j-i} [\Psi_{jik}|_{\mathscr{B}} D_{rv}^{(i,k)} \mathscr{X}(v)] \right] \right] \right]$$

$$\cdot [t_1(v)]^{2(f-1)},$$

where the multiplication of the rational term within the round brackets with the factor $[t_1(v)]^{2(f-1)}$ is necessary to achieve a polynomial representation.

The system of the geometric C^f-joint conditions

$$D_f \mathscr{Y}(v) = \mathscr{q}_{f,B}(v) \mathscr{B}(v) + \imath_{f,B}(v) D_r \mathscr{X}(v) + \mathfrak{d}_{f,B}(v) D_v \mathscr{X}(v) \qquad (f = 2(1)h) \tag{5.3}$$

includes the rational coefficient functions

$$\mathscr{q}_{f,B}(v) = [t_1(v)]^{2(f-1)} q_{f,B}(v),$$

$$\imath_{f,B}(v) = [t_1(v)]^{2(f-1)} \sum_{j=1}^{f} \left[\binom{f}{j} q_{f-j,B}(v) \Psi_{j10}|_{\mathscr{B}} \right], \qquad (f = 2(1)h)$$

$$\mathfrak{d}_{f,B}(v) = [t_1(v)]^{2(f-1)} \sum_{j=1}^{f} \left[\binom{f}{j} q_{f-j,B}(v) \Psi_{j01}|_{\mathscr{B}} \right],$$

which by simple algebraic manipulations can be described in the following way:

$$\mathscr{q}_{f,B}(v) = \frac{\langle D_f \mathscr{Y}(v) \wedge D_r \mathscr{X}(v) \wedge D_v \mathscr{X}(v), \mathscr{B}(v) \wedge D_r \mathscr{X}(v) \wedge D_v \mathscr{X}(v) \rangle}{(\mathscr{B}(v) \wedge D_r \mathscr{X}(v) \wedge D_v \mathscr{X}(v))^2},$$

$$\imath_{f,B}(v) = \frac{\langle \mathscr{B}(v) \wedge D_f \mathscr{Y}(v) \wedge D_v \mathscr{X}(v), \mathscr{B}(v) \wedge D_r \mathscr{X}(v) \wedge D_v \mathscr{X}(v) \rangle}{(\mathscr{B}(v) \wedge D_r \mathscr{X}(v) \wedge D_v \mathscr{X}(v))^2}, \tag{5.4}$$

$$\mathfrak{d}_{f,B}(v) = \frac{\langle \mathscr{B}(v) \wedge D_r \mathscr{X}(v) \wedge D_f \mathscr{Y}(v), \mathscr{B}(v) \wedge D_r \mathscr{X}(v) \wedge D_v \mathscr{X}(v) \rangle}{(\mathscr{B}(v) \wedge D_r \mathscr{X}(v) \wedge D_v \mathscr{X}(v))^2}.$$

$$(f = 2(1)h)$$

Therefore, the coefficient functions $\mathscr{q}_{f,B}(v)$, $\imath_{f,B}(v)$, $\mathfrak{d}_{f,B}(v)$ $(f = 2(1)h)$ prove out to be rational functions of the parameter v with a common denominator polynomial (in general, before one starts to cancel possible common linear factors).

5.2 Necessary and Sufficient Continuity Constraints

According to the set of h determinant conditions $\Phi_{f,0}(v) = 0$ $(f = 1(1)h$, section 5.1), one obtains, due to the independence of the Bernstein polynomials, a total system consisting of h subsystems of constraints expressed directly in terms of involved control points. This total system provides necessary and sufficient conditions for the geometric C^h-joint between two tensor-product Bézier surface patches along their common boundary curve.

Based on the approach to consider divisibility properties of terms contained in the determinant conditions, the number of independent constraints can be determined. In the case of $m^* = m = \bar{p} = p$, $4p - 1$ constraints have in general to be considered for each value of f. Hence, to control the h-th order geometric continuous joint of the tensor-product patches a total system of $h(4p - 1)$ scalar conditions has to be examined.

For further considerations, the vector polynomials

$$\mathcal{N}_{t,f}(v) := \mathcal{B}(v) \wedge D_r\mathcal{X}(v) \wedge D_v\mathcal{X}(v) = \mathcal{N}_{t,1}(v),$$

$$\mathcal{N}_{q,f}(v) := D_f\mathcal{Y}(v) \wedge D_r\mathcal{X}(v) \wedge D_v\mathcal{X}(v), \qquad \mathcal{N}_{r,f}(v) := \mathcal{B}(v) \wedge D_f\mathcal{Y}(v) \wedge D_v\mathcal{X}(v),$$

$$\mathcal{N}_{s,f}(v) := \mathcal{B}(v) \wedge D_r\mathcal{X}(v) \wedge D_f\mathcal{Y}(v), \qquad (f = 2(1)h), \tag{5.5}$$

evaluated as alternating products and involving in each instance three out of the four vector polynomials $D_f\mathcal{Y}(v)$, $\mathcal{B}(v)$, $D_r\mathcal{X}(v)$, $D_v\mathcal{X}(v)$ are introduced and respectively described in the form of a product

$$\mathcal{N}_{t,f}(v) = t_f(v)\mathcal{P}(v) = t_1(v)\mathcal{P}(v), \qquad \mathcal{N}_{q,f}(v) = q_f(v)\mathcal{P}(v),$$
$$\mathcal{N}_{r,f}(v) = r_f(v)\mathcal{P}(v), \qquad \mathcal{N}_{s,f}(v) = s_f(v)\mathcal{P}(v), \qquad (f = 2(1)h) \tag{5.6}$$

of a corresponding scalar polynomial $t_f(v)$, $q_f(v)$, $r_f(v)$ or $s_f(v)$ with the common irreducible vector polynomial $\mathcal{P}(v)$ introduced in section 4.2. Hence, the polynomials $t_f(v)$, $q_f(v)$, $r_f(v)$ and $s_f(v)$ respectively incorporate the linear factors, occuring in the corresponding four coordinate polynomials of the vector polynomials $\mathcal{N}_{t,1}(v)$, $\mathcal{N}_{q,f}(v)$, $\mathcal{N}_{r,f}(v)$ and $\mathcal{N}_{s,f}(v)$.

Resulting from $t_1(v) = t_f(v)$ $(f = 2(1)h)$ (which are obtained from (5.6)), the equations

$$q_{1,B}(v)q_f(v) = \varphi_{f,B}(v)q_1(v), \qquad r_{1,B}(v)r_f(v) = \imath_{f,B}(v)r_1(v),$$
$$s_{1,B}(v)s_f(v) = \jmath_{f,B}(v)s_1(v), \qquad (f = 2(1)h) \tag{5.7}$$

link the rational coefficient functions, involved in the geometric C^1- and C^f-joint conditions and polynomial form functions.

Inserting the product representations (5.6) for the vector polynomials $\mathcal{N}_{t,1}(v)$, $\mathcal{N}_{q,f}(v)$, $\mathcal{N}_{r,f}(v)$, $\mathcal{N}_{s,f}(v)$ into the description (5.4) of the coefficient functions $\varphi_{f,B}(v)$, $\imath_{f,B}(v)$, $\jmath_{f,B}(v)$ and cancelling the common factor $(\mathcal{P}(v))^2$ yields each of the coefficient functions

$$\varphi_{f,B}(v) = \frac{q_f(v)}{t_f(v)}, \qquad \imath_{f,B}(v) = \frac{r_f(v)}{t_f(v)}, \qquad \jmath_{f,B}(v) = \frac{s_f(v)}{t_f(v)} \qquad (f = 2(1)h) \tag{5.8}$$

as the quotient of two out of the four polynomial form functions $t_1(v)$, $q_f(v)$, $r_f(v)$ and $s_f(v)$, the lowest upper bounds of which can be determined by methods analogous to those employed in section 4.2.

Each of the geometric C^f-joint conditions

$$t_f(v)D_f\mathscr{Y}(v) = q_f(v)\mathscr{B}(v) + r_f(v)D_r\mathscr{X}(v) + s_f(v)D_v\mathscr{X}(v) \qquad (f = 2(1)h) \qquad (5.9)$$

shows a symmetric form: Each of the four vector polynomials involved is attached to a polynomial form function.

When setting up the lowest upper bounds for the polynomial degrees of the form functions $t_f(v)$, $q_f(v)$, $r_f(v)$ and $s_f(v)$, one obtains due to the independence of the Bernstein polynomials, a system of necessary and sufficient constraints for the geometric C^f-joint between two adjacent Bézier patches. The single systems of necessary and sufficient constraints corresponding to $f = 1(1)h$ have to be combined into a total system, characterizing the h-th order geometric continuous joint of two tensor-product Bézier patches $\mathscr{X}$ and $\mathscr{Y}$ along a common boundary curve.

Different total systems, consisting of corresponding subsystems of sufficient constraints, concerning the h-th order geometric continuous joint between two Bézier patches along a common boundary curve, can be derived in an analogous way to section 4.3.

5.3 Geometric Construction Scheme Corresponding to Systems of Special Sufficient Continuity Constraints

Assume, the common boundary $\mathscr{B}$ and patch $\mathscr{X}$ are given and in addition $m^* = m = \bar{p} = p$. By adopting the approach to formulate the coefficient functions $q_{f,B}(v)$, $r_{f,B}(v)$ and $s_{f,B}(v)$ as constants and linear polynomials

$$q_{0,B}(v) = v_{0,0} = 1 \qquad (f = 1(1)h)$$

$$q_{f,B}(v) = v_{f,0}, \qquad r_{f,B}(v) = \rho_{f,0}, \qquad s_{f,B}(v) = \sigma_{f,0}(1 - v) + \sigma_{f,1}v, \qquad (5.10)$$

the representations for the geometric C^f-joint conditions result by successively utilizing the calculation procedure

$$D_u^{(f)}\mathscr{Y}(v) - \sum_{j=1}^{f} \left[\binom{f}{j}v_{f-j,0} \cdot \left[\sum_{i=0}^{j}\sum_{\substack{k=0\\i+k>1}}^{j-i} [\Psi_{jik}|_\mathscr{B}D_{rv}^{(i,k)}\mathscr{X}(v)] \right] \right]$$

$$= \bar{v}_{f,0}\mathscr{B}(v) + \bar{\rho}_{f,0}D_r\mathscr{X}(v) + [\bar{\sigma}_{f,0}(1 - v) + \bar{\sigma}_{f,1}v]D_v\mathscr{X}(v) \qquad (f = 1(1)h)$$

with $\Psi_{110}|_\mathscr{B} = \rho_{1,0}$, $\Psi_{101}|_\mathscr{B} = \sigma_{1,0}(1 - v) + \sigma_{1,1}v$ (5.11)

and the recurrence for the coefficients

$$\Psi_{fik}|_\mathscr{B} = (\rho_{1,0}\Psi_{f-1,i-1,k} + [\sigma_{1,0}(1 - v) + \sigma_{1,1}v]\Psi_{f-1,i,k-1} + E\Psi_{f-1,i,k})|_\mathscr{B}$$

as well as the shift operator E defined by

$$E(\rho_{i,0}^f[\sigma_{j,0}(1-v) + \sigma_{j,1}v]^k) := f\rho_{i,0}^{f-1}\rho_{i+1,0}[\sigma_{j,0}(1-v) + \sigma_{j,1}v]^k$$
$$+ k\rho_{i,0}^f[\sigma_{j,0}(1-v) + \sigma_{j,1}v]^{k-1}$$
$$\cdot [\sigma_{j+1,0}(1-v) + \sigma_{j+1,1}v].$$

The following relationships apply to the involved scalar form parameters:

$$\bar{v}_{f,0} = v_{f,0}, \qquad \bar{\rho}_{f,0} = \sum_{j=1}^{f}\left[\binom{f}{j}v_{f-j,0}\Psi_{j10}|_{\mathscr{B}}\right],$$

$$\bar{\sigma}_{f,0}(1-v) + \bar{\sigma}_{f,1}v = \sum_{j=1}^{f}\left[\binom{f}{j}v_{f-j,0}\Psi_{j01}|_{\mathscr{B}}\right]. \qquad (f = 1(1)h)$$

Due to the independence of the Bernstein polynomials, a system consisting of $p + 1$ constraints emerges for each value of f, with $f = 1(1)h$. The h single subsystems have to be combined into a total system, providing a construction concept for determining the unknown Bézier points $\mathscr{F}_{n^*-f,j}$ (with $f = 1(1)h$, $j = 0(1)p$) of the first f rows of control points of patch $\mathscr{Y}$, adjacent to boundary curve $\mathscr{B}$.

For a fixed value $f = k$, the corresponding system of $p + 1$ constraints involves known control points of the boundary row, and the first k interior rows of patch $\mathscr{X}$, as well as the control points of the first k interior rows of patch $\mathscr{Y}$.

The Bézier points of the first $k - 1$ interior rows of patch $\mathscr{Y}$ can be evaluated using the constraints of the subsystems linked to $f = 1(1)k - 1$, hence, the system for $f = k$ yields the constraints for the $p + 1$ unknown control points $\mathscr{F}_{n^*-k,j}$.

The construction scheme to be proposed is composed of $f + 1$ construction steps, shows symmetric properties and represents a direct extension of the construction method in section 4.3, indicating a straightforward geometric interpretation.

During each of the first f construction steps a set of new auxiliary points is determined in an appropriate manner. The construction step following an executed construction step only includes the auxiliary points evaluated in the previous step.

The $(f + 1)$-th construction step yields each of the control points $\mathscr{F}_{n^*-f,j}$ ($j = 0(1)p$) as a convex combination of the auxiliary points $\mathscr{E}_{0j,1}^{(f)}$, $\mathscr{E}_{0j,2}^{(f)}$.

The projectively invariant cross ratio

$$cr(\mathscr{E}_{0j,1}^{(f)}, \mathscr{F}_{n^*-f,j}, \tfrac{1}{2}(\mathscr{E}_{0j,1}^{(f)} + \mathscr{E}_{0j,2}^{(f)}), \mathscr{E}_{0j,2}^{(f)}) = \frac{j}{p-j}$$

of the four collinear points $\mathscr{E}_{0j,1}^{(f)}$, $\mathscr{F}_{n^*-f,j}$, $\tfrac{1}{2}(\mathscr{E}_{0j,1}^{(f)} + \mathscr{E}_{0j,2}^{(f)})$, $\mathscr{E}_{0j,2}^{(f)}$ varies with the running index j.

The construction procedure to be employed during the single construction steps is shown in a schematic way in Fig. 5.1 (with the involved control points in $\mathbb{R}^3$) for the value $f = 3$. The geometric constructions concerning the weights, illustrated in Fig. 4.1 and discussed in section 4.3, can be appropriately integrated in the suggested construction scheme.

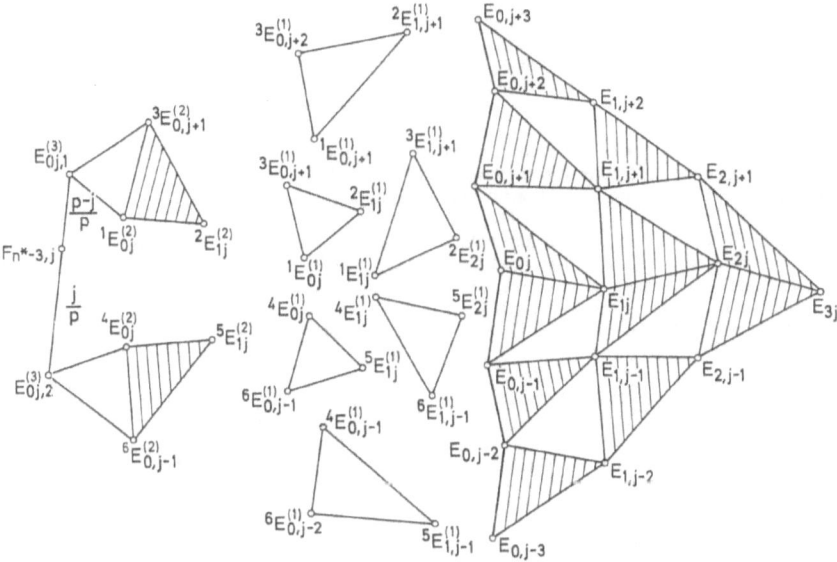

Figure 5.1. Construction of the points $F_{n^*-f,j}$ for the geometric C^f-joint of two adjacent tensor-product Bézier patches; according to the proposed construction scheme (for instance $f = 3$)

References

[1] Boehm, W., Farin, G., Kahmann, J.: A survey of curve and surface methods in CAGD. Computer Aided Geometric Design *1*, 1–60 (1984).
[2] Boehm, W.: Curvature continuous curves and surfaces. Computer Aided Geometric Design *2*, 313–323 (1985).
[3] Boehm, W.: Smooth curves and surfaces. In: Farin, G. (ed.) Geometric modeling: algorithms and new trends. SIAM (1987) 175–184.
[4] Boehm, W.: Visual continuity. Computer-Aided Design *20*, 307–311 (1988).
[5] Boehm, W.: On G^2-continuity of surfaces. Presented at Topics in CAGD. Jerusalem June '88.
[6] Bol, G.: Projektive Differentialgeometrie, vol. 1. Göttingen Vandenhoeck u. Ruprecht: 1967.
[7] Cauchy, A.: Lecons sur les Applications du calcul infinitésimal a la géométrie, De Bure fréres, Librairie du Roi et Bibliotheque du Roi, Paris (1826).
[8] Degen, W.: Some remarks on Bézier curves. Computer Aided Geometric Design *5*, 259–268 (1988).
[9] Degen, W.: Explicit continuity conditions for adjacent Bézier surface patches. Computer Aided Geometric Design *7*, 181–189 (1990).
[10] DeRose, T.: Necessary and sufficient conditions for tangent plane continuity of Bézier surfaces. Computer Aided Geometric Design *7*, 165–179 (1990).
[11] Farin, G.: A construction for visual C^1-continuity of polynomial surface patches. Computer Graphics and Image Processing *20*, 272–282 (1982).
[12] Farin, G.: Algorithms for rational Bezier curves. Computer–Aided Design *15*, 73–77 (1983).
[13] Farin, G.: Triangular Bernstein-Bézier patches. Computer Aided Geometric Design *3*, 83–127 (1986).
[14] Farin, G.: Curves and surfaces for computer aided geometric design. San Diego: Academic Press 1990.
[15] Gregory, J. A.: Geometric Continuity. In: Schumaker, L. L., Lyche, T. (ed.) Mathematical methods in computer aided geometric design, 353–371. Academic Press 1989.
[16] Herron, G.: *Techniques for visual continuity*. In: Farin, G. (ed.) Geometric modeling: algorithms and new trends, 163–174 SIAM 1987.

[17] Hoschek, J., Lasser, D.: Grundlagen der geometrischen Datenverarbeitung. Stuttgart Teubner: 1989.

[18] Jüttler, B., Wassum, P.: Some remarks on geometric continuity of rational surface patches. Computer Aided Geometric Design 9, 143–157 (1992).

[19] Kahmann, J.: Continuity of curvature between adjacent Bézier patches. In: Barnhill, R. E., Böhm, W. (ed.) Surfaces in CAGD, 65–75. Amsterdam: North-Holland 1983.

[20] Liu, D., Hoschek, J.: GC^1-continuity conditions between adjacent rectangular and triangular Bézier surface patches. Computer-Aided Design 21 194–200 (1989).

[21] Liu, D.: GC^1-continuity conditions between two adjacent rational Bézier surface patches. Computer Aided Geometric Design 7, 151–163 (1990).

[22] Piper, B. R.: Visually smooth interpolation with triangular Bézier patches. In: Farin, G. (ed.) Geometric modeling: algorithms and new trends, 221–233. Siam (1987).

[23] Pottmann, H.: Projectively invariant classes of geometric continuity for CAGD. Computer Aided Geometric Design 6, 307–321 (1989).

[24] Scheffers, G.: Anwendung der Differential- und Integralrechnung auf die Geometrie. Berlin und Leipzig: Walter de Gruyter & Co. 1923.

[25] Vinacua, A., Brunet, P.: A construction for VC^1 continuity of rational Bézier patches. In: Schumaker, L. L., Lyche, T. (ed.) Mathematical methods in computer aided geometric design. Academic Press 1989.

[26] Wassum, P.: Conditions and constructions for GC^1-/GC^2-continuity between adjacent integral Bézier surface patches. In: Hagen, H. (ed.) Topics in surface modeling. SIAM, 187–218 (1992).

[27] Wassum, P.: Bedingungen und Konstruktionen zur geometrischen Stetigkeit und Anwendungen auf approximative Basistransformationen. Dissertation, Fachbereich Mathematik, Technische Hochschule Darmstadt (1991).

P. Wassum
Department of Mathematics
Darmstadt University of Science and
    Technology
D-W-6100 Darmstadt,
Federal Republic of Germany

Computing

Archives for Informatics and Numerical Computation
Archiv für Informatik und Numerik

Editorial Board
Herausgeber-Kollegium
R. Albrecht, Innsbruck
W. Hackbusch, Kiel
W. Knödel, Leipzig
W. L. Miranker, Yorktown Heights
H. J. Stetter, Wien

Advisory Board / Fachbeirat
G. Alefeld, Karlsruhe
H. Brunner, St. John's, Nfld.S
R. E. Burkard, Graz
J. Dai, Nanjing
R. Dutter, Wien
H. Engl, Linz
W. H. Enright, Toronto, Ont.
M. Gössel, Potsdam
G. Gottlob, Wien
M. Grötschel, Berlin
H. Müller, Freiburg
A. Neumaier, Freiburg
H. Neunzert, Kaiserslautern
B. Radig, München
A. Reusken, Eindhoven
J. W. Schmidt, Dresden
H. J. Schneider, Erlangen
H. Schwetlick, Halle
G. Tinhofer, München
W. Törnig, Darmstadt
H. F. Wedde, Detroit, MI

Editorial Assistants
Redaktionssekretäre
J. Schneid and E. Weinmüller, Wien

Presenting the latest research results from computer science and numerical computation, **Computing** is an international journal intended for professionals and students in all fields of scientific computing, for computer center staff, and software and hardware manufacturers. Each issue features original papers and short communications from a wide range of areas: discrete algorithms, symbolic computations, performance and complexity evaluation, operating systems, scheduling, software engineering, picture processing, parallel computation, classical numerical analysis, numerical software, numerical statistics, optimization, computer arithmetic, interval analysis, plotting.

ISSN 0010-485X Title No. 607
Subscription Information:
1993. Vols. 50–51 (4 issues each):
DM 816,–, öS 5712,–
plus carriage charges

Springer-Verlag Wien New York

Surveys on Mathematics for Industry

The main goal of this journal is to bridge the gap between university and industry by
- the presentation of mathematical methods relevant for industry
- the exposition of industrial problems which are of interest to mathematicians.

To achieve this goal, the journal publishes (exclusively in English):

- Surveys on new mathematical techniques
- Surveys on established mathematical techniques with a new range of applications
- Surveys on industrial problems for which appropriate mathematical models or methods are not yet available
- Articles comparing mathematical models or methods for particular industrial problems
- Articles describing mathematical modelling techniques
- Broad historical surveys
- Articles of general interest about the use of mathematics in industry
- Occasional book reviews and reports about conferences in the field of Industrial Mathematics.

Papers will either be solicited by a member of the Editorial Board or should be submitted to the Managing Editor who also welcomes suggestions for possible topics by prospective authors. All papers will be refereed.

ISSN 0938-1953 Title No. 724

Subscription Information:

1993. Vols. 3 (4 issues):
for institutional subscribers:
DM 240,–, öS 1680,–, plus carriage charges
for individual subscribers:
DM 144,–, öS 1008,–, plus carriage charges
Special rates for individual members of
DMV, ECMI, GAMM, JSIAM, ÖMG, and SIMAI

Prices are subject to change without notice

Springer-Verlag Wien New York